Physics of Transitional Shear Flows

FLUID MECHANICS AND ITS APPLICATIONS
Volume 98

Aims and Scope of the Series

The purpose of this series is to focus on subjects in which fluid mechanics plays a fundamental role.

As well as the more traditional applications of aeronautics, hydraulics, heat and mass transfer etc., books will be published dealing with topics which are currently in a state of rapid development, such as turbulence, suspensions and multiphase fluids, super and hypersonic flows and numerical modeling techniques.

It is a widely held view that it is the interdisciplinary subjects that will receive intense scientific attention, bringing them to the forefront of technological advancement. Fluids have the ability to transport matter and its properties as well as to transmit force, therefore fluid mechanics is a subject that is particularly open to cross fertilization with other sciences and disciplines of engineering. The subject of fluid mechanics will be highly relevant in domains such as chemical, metallurgical, biological and ecological engineering. This series is particularly open to such new multidisciplinary domains.

The median level of presentation is the first year graduate student. Some texts are monographs defining the current state of a field; others are accessible to final year undergraduates; but essentially the emphasis is on readability and clarity.

For further volumes:
www.springer.com/series/5980

Andrey V. Boiko • Alexander V. Dovgal •
Genrih R. Grek • Victor V. Kozlov

Physics of Transitional Shear Flows

Instability and Laminar–Turbulent Transition in Incompressible Near-Wall Shear Layers

Springer

Prof. Dr. Andrey V. Boiko
Inst. Theoretical & Applied Mechanics
Russian Academy of Sciences
Novosibirsk 630090
Russia
boiko@itam.nsc.ru

Prof. Dr. Genrih R. Grek
Inst. Theoretical & Applied Mechanics
Russian Academy of Sciences
Novosibirsk 630090
Russia
grek@itam.nsc.ru

Prof. Dr. Alexander V. Dovgal
Inst. Theoretical & Applied Mechanics
Russian Academy of Sciences
Novosibirsk 630090
Russia
dovgal@itam.nsc.ru

Prof. Dr. Victor V. Kozlov
Inst. Theoretical & Applied Mechanics
Russian Academy of Sciences
Novosibirsk 630090
Russia
kozlov@itam.nsc.ru

Additional material to this book can be downloaded from http://extras.springer.com

ISSN 0926-5112 Fluid Mechanics and Its Applications
ISBN 978-94-007-2497-6 e-ISBN 978-94-007-2498-3
DOI 10.1007/978-94-007-2498-3
Springer Dordrecht Heidelberg London New York

Library of Congress Control Number: 2011938475

Cover design: VTeX UAB, Lithuania

Printed on acid-free paper

Springer is part of Springer Science+Business Media (www.springer.com)

'... it is far easier to produce numbers from a computer than understand their relevance to real flows.'

P. G. Saffman & G. R. Baker, Ann. Rev. Fluid Mech., *1979.*

Foreword

Hydrodynamic instability and transition to turbulence in boundary layers has been an intensive research interest during the last century and continues to be so in the 21st century. There is no dearth of research results on the topic and many contributions appear annually in journal papers and conference proceedings. As a consequence, the laminar-turbulent transition problem is a manifold branch of fluid mechanics with a variety of theoretical and experimental approaches to basic and engineering issues that are still open. Naturally, there is a need to summarize from time to time the current state in transition studies through review papers and monographs. That is the goal of this book presented by the Novosibirsk group of researchers who are well-experienced in long-term experimental modeling of transition physics in boundary layers. In fact, the Novosibirsk group, centered at the Institute of Theoretical and Applied Mechanics, has been amongst the leaders in foundational contributions to the area for decades and it is therefore appropriate and welcome that they contribute a monograph on the topic.

Starting from fundamentals of classical stability theory, the authors give an overview of the transition phenomena in subsonic, wall-bounded shear flows. At first, the consideration focuses on elementary small-amplitude velocity perturbations of laminar shear layers, i.e. instability waves, in the simplest canonical configurations of a plane channel flow and a flat-plate boundary layer. Then the linear stability problem is expanded to include the effects of pressure gradients, flow curvature, boundary-layer separation, wall compliance, etc. related to applications. Beyond the amplification of instability waves is the non-modal growth of local stationary and non-stationary shear flow perturbations which are discussed as well. Moreover, the authors proceed with the key aspect of the transition process, that is, receptivity of convectively unstable shear layers to external perturbations, summarizing main paths of the excitation of laminar flow disturbances. The remainder of the book addresses the instability phenomena found at late stages of transition. These include secondary instabilities and nonlinear features of boundary-layer perturbations that lead to the final breakdown to turbulence. Thus, the reader is provided with a step-by-step approach that covers the milestones and recent advances in the laminar-turbulent transition.

Special aspects of instability and transition are discussed through the book and are intended for research scientists, while the main target of the book is the student in the fundamentals of fluid mechanics. Computational guides, recommended exercises, and PowerPoint multimedia notes based on results of real scientific experiments supplement the monograph. These are especially helpful for the neophyte to obtain a solid foundation in hydrodynamic stability.

William S. Saric PhD, NAE
University Distinguished Professor
Texas A&M University

Preface

Overview

The origin of turbulence in fluids is a long-standing problem that has been the focus of research for decades due to its great importance in a variety of engineering applications. Moreover, studying of onset of turbulence is part of the fundamental physical problem of turbulence description and the philosophical problem of determinism and chaos.

In the present book we try to give a panoramic view of the transition of turbulence in near-wall shear flows by binding its different aspects for incompressible flows according to our knowledge obtained during decades of researches. To be in line with this general idea, there was a need to pick out the material for inclusion from plenty of available sources. In this way, our priority was a conceptual treatment of various both classical and modern aspects of the problems under consideration.

The scope and the resulting structure of the book were chosen based on experience collected owing to one semester lecture courses for PhD students read by some of us at Novosibirsk State University, Korean Advanced Institute of Science and Technology, and Pusan National University.

Thus, the book is intended first of all for beginners in this field of science. We expect that the result of our efforts will also be of interest to researchers and high-school teachers involved in hydro-aerodynamics, stability, and laminar–turbulent transition problems.

Prerequisites

Readers of this book are expected to be familiar with the concepts of fluid mechanics of viscous incompressible flow and Prandtl boundary-layer theory. It is also desirable, but not required, for the reader to know basic ideas of linear algebra and to have some experience with matrix computations and MATLAB programming.

Organization of the Book

In Chaps. 1 and 2, we begin with basic observations and the fundamentals of the classical stability theory, which describes low-intensity shear-layer perturbations as instability waves.

Chapters 3–7 are devoted to various generic problems of the theory to show how different isolated factors such as curvature, pressure gradients, flow separation, etc. affect the linear stability of shear flows. The palette of these factors certainly is not exhaustive. However, the set is diverse enough to provide a general view of the basic aspects of the stability analysis.

In Chaps. 8–11, some special but very important topics on linear stability are highlighted. Particularly, various aspects of the initial-value problem with respect to development of small-amplitude disturbances are considered in Chaps. 8–9. The problem of excitation of instability waves by external flow disturbances is considered in Chap. 10. When a linear instability disturbance grows beyond a certain amplitude threshold, it enters the region of its essentially nonlinear, but still deterministic development. The flow modulated by such nonlinear structures is frequently subjected to the phenomenon known as the secondary instability whose initial stage can be described by linear equations of motion. The secondary instability is discussed in Chap. 11.

The nonlinear development of disturbances is accompanied by wave interactions, base-flow distortions, and appearance of turbulent spots. The last part of the book (Chaps. 12–13) is devoted to the description of the main physical mechanisms and phenomena responsible for the final onset of turbulence.

In Appendix, some fundamentals of modern approaches for the prediction and control of the transition are discussed.

Supplementary Material

An interested reader can find details of different aspects of the transition of turbulence in near-wall shear flows in original publications, which are cited in *References* and *Further Reading* sections in the end of each chapter.

We decided to publish the manuscript as an e-book having in mind to make available a set of multimedia files of different value containing supplementary material including video clips and animations.

For those who like to feel the nature through numbers, the book is supplemented by a set of 'toy' MATLAB functions, which allow a reader to understand 'how it works.' The core routines cited in the book are accompanied by interface programs having the same generic names ending with 'Test.' To simplify the process and to give some guidelines, many chapters contain *Exercises*, which can be fulfilled with the help of the functions as a starting point.

Acknowledgements

A lot of original material presented in the book was obtained in collaboration with our colleagues from the Khristianovich Institute of Theoretical and Applied Mechanics, Novosibirsk, as well as with Henrik Alfredsson, Barbro Klingmann, Masaharu Matsubara, Alfons Michalke, Oleg Ryzhov, and William Saric, who provided exploration results included into the present book.

We are grateful to Mashahito Asai, Fabio Bertolotti, Hans Bippes, Alessandro Bottaro, Ho Hwan Chun, Michael Gaster, Stefan Hein, Dan Henningson, Thorwald Herbert, Markus Kloker, Yasuaki Kohama, Inwon Lee, Paolo Luchini, Michio Nishioka, Ulrich Rist, Frank Smith, Hyung Jin Sung, Vassilios Theofilis, Israel Wygnanski, and many others for useful contacts and discussions.

Our thanks also go to the Stuttgart University for providing some multimedia material of F. X. Wortmann and M. Strunz. Special thanks are due to the Alexander von Humboldt Foundation for supporting the research of the first two authors.

Finally, we are grateful to our families for their patience and their recognition of our efforts.

Novosibirsk,
December 2010

Andrey V. Boiko

Alexander V. Dovgal

Genrih R. Grek

Victor V. Kozlov

Contents

List of Programs

List of Multimedia Files

Nomenclature

Only major symbols are given in the following lists.

A	Amplitude
a	Characteristic height
$\mathbf{c}$	Complex phase velocity
$\mathbf{c}_\mathrm{g}$	Complex group velocity
d	Characteristic width
E	Kinetic energy
$\mathbf{e}$	Unit basis vectors
$\mathcal{F}$	Dispersion relation
F	Dimensionless frequency parameter
f	Frequency
G	Growth rate of kinetic energy
$\overline{G}$	Growth rate of kinetic energy scaled with the Reynolds number
H	Shape factor
h	Height
k	Curvature
l	Length
P	Mean pressure
p	Disturbance pressure
R	Radius
$\mathbf{r}$	Radius-vector
T	Temperature
t	Time
$\mathscr{U}$	Full velocity vector
$\mathcal{U}$, $\mathcal{V}$, $\mathcal{W}$	Eulerian components of the full velocity vector
$\mathbf{U}$	Basic flow velocity vector
U, V, W	Eulerian components of the basic flow (mean) velocity vector
$\mathbf{u}$	Disturbance velocity vector
u, v, w	Eulerian components of the fluid disturbance velocity vector
$\overline{U}$	Basic velocity projection to the direction of the wavevector

$\overline{u}$	Disturbance velocity projection to the direction of the wavevector
$\overline{W}$	Basic velocity projection to the direction normal to the wavevector
$\overline{w}$	Disturbance velocity projection to the direction normal to the wavevector
$\mathscr{V}$	Volume of integration

Coordinate systems

x, y, z	Cartesian coordinates (model-fitted)
$\tilde{x}$, y, $\tilde{x}$	Local (external-flow streamline-fitted) Cartesian coordinates
$\bar{x}$, y, $\bar{x}$	Wavevector-fitted Cartesian coordinates
r, θ, z	Cylindrical coordinates

Parameters

De	Dean number
Gö	Görtler number
Re	Reynolds number
Ta	Taylor number
Tu	External flow (free-stream) turbulence level

Greek symbols

Γ	Circulation
Λ	Wavelength parameter
Ψ	Stream function
Ω	Angular velocity
α	Complex streamwise wavenumber
β	Complex transverse wavenumber
β_{H}	Hartree pressure gradient parameter
γ	Angle between κ and streamwise axis
δ	Boundary layer scale
δ^*	Displacement thickness
ε	Small parameter or value
η	Displacement in y
$\hat{\eta}$	Complex amplitude function of normal vorticity mode
$\tilde{\eta}$	Complex amplitude function of normal vorticity initial disturbance
ϑ	Momentum thickness
κ	Wavevector

$\varkappa$	Wavevector length
λ	Wavelength
μ	Dynamic fluid viscosity
ν	Kinematic fluid viscosity
ξ	Displacement in x
ρ	Fluid density
σ	Growth rate
ϕ	The angle between the direction of the external-flow streamline and the wedge chord
ζ	Displacement in z
ψ	Phase variations of a complex amplitude
ω	Complex circular frequency

Tensors

S	Deformation tensor of the basic flow
s	Deformation tensor of the flow disturbances
τ_{ij}	Shear stress tensor component

Operators

∇^2	Laplace operator
Δ	Difference operator
$\mathcal{D}$	Derivative operator in wall-normal direction
$\mathscr{L}$, $\mathscr{M}$	Linear operators
$\mathscr{S}$	Bilinear concomitant operator
$\mathscr{P}$	Linear propagator

Accents and superscripts

$\hat{}$, $\tilde{}$, $\bar{}$	Refers to complex amplitude function
$'$	Root-mean-square value
$\dagger$	Adjoint quantity
$*$	Complex conjugate
H	Hermitian conjugate (complex conjugate and transpose)
T	Transpose

Subscripts

0	Refers to initial value
∞	Refers to infinity
2D	Refers to two-dimensional
a	Refers to axial value
ampl	Refers to amplitude
c	Refers to critical layer value
ch	Refers to characteristic value
coat	Refers to coating
E	Refers to critical values of loss of monotonic stability
e	Refers to free-stream (external-flow) value
eff	Refers to effective value
f	Refers to final value
f	Refers to frequency f
i	Refers to imaginary part
L	Refers to critical values of loss of linear stability
G	Refers to critical values of loss of global stability
lam	Refers to laminar regime
m	Refers to mean value
max	Refers to maximum of corresponding value
min	Refers to minimum of corresponding value
NS	Refers to Navier–Stokes equations
OS	Refers to Orr–Sommerfeld equation
p	Refers to particles
r	Refers to real part
s	Refers to inflection point value
Sq	Refers to Squire equation
T	Refers to values of onset of turbulence
x	Refers to the x-coordinate
y	Refers to the y-coordinate
z	Refers to the z-coordinate
r, t, x, y, z, η	Indicate partial derivatives with respect to the corresponding variables
w	Refers to wall
β	Refers to wavenumber β

Acronyms

2D	two-dimensional
3D	three-dimensional
CFI	Crossflow instability
CIFI	Compliance-induced flow instability

DNS	Direct numerical simulation
LNSEs	Locally-nonparallel stability equations
LPSEs	Locally-parallel stability equations
NSEs	Nonparallel stability equations
PIV	Particle image velocimetry
PSEs	Parabolized stability equations
TSI	Tollmien–Schlichting instability

Part I
Fundamentals of the linear stability theory

Chapter 1
Concept of hydrodynamic stability

In the beginning of this chapter, based on elementary physical example, a general idea of stability of motion with respect to disturbances is introduced. Then it is conceptualized for fluid dynamic applications. Stability of fluid flows is categorized in regards of the behavior of kinetic energy of a disturbance in time and space. Finally, the notion of critical parameters at which the motion character changes from stable to unstable is formulated.

1.1 Hydrodynamic stability

A general and indicative definition of stability was given by Betchov and Criminale (1967): 'stability can be defined as the quality of being immune to small disturbances.' This general property as applied to stability of mechanical systems is illustrated by simple examples shown in Fig. 1.1.

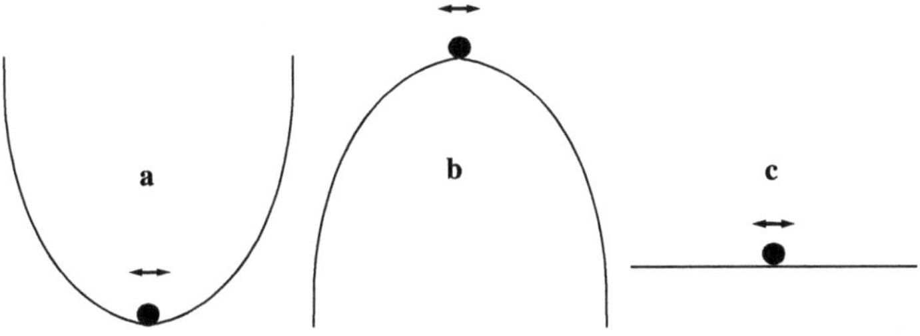

Fig. 1.1 Simple mechanical examples of equilibrium states: stable state (**a**); unstable state (**b**); neutral (indefinite) state (**c**)

Similarly to the examples given above, it is known that, for certain parameters of a hydrodynamic system, the governing equations with precise stationary (laminar) solutions at constant boundary conditions cannot be implemented in practice: the

A.V. Boiko et al., *Physics of Transitional Shear Flows*,
Fluid Mechanics and Its Applications 98, DOI 10.1007/978-94-007-2498-3_1,

motion is unstable. For example, if certain disturbances are inserted into the flow, and the latter returns to the initial state (speaking in the language of the theory of dynamic systems, is 'attracted') in time and/or space, the flow is stable to these disturbances. If the disturbances convert the flow to another (not necessarily turbulent) state, the flow is unstable.

Sometimes the transition to chaotic motion, at least at the initial stage, occurs through a chain of (quasi-)stable states and consequent loss of stability (bifurcation) in each of them. The phenomenon is denoted as the change of stability. An example of this phenomenon are the Bénard cells that appear due to the Rayleigh–Bénard convection sometimes even in an ordinary heated frying pan with a layer of, for instance, some oil.

In considering flow stability, an asymptotic (after a long period) response of a system affected by a disturbance is studied frequently. Situations are possible, however, in which the disturbance at the beginning, during its establishment, experiences a transient growth and only then decays. It is a typical phenomenon that is found in many other branches of physics, for instance, at breaking of an electric circuit, which can lead to a light bulb break. Similarly, if the disturbance becomes in some sense 'dangerously large' during this transient growth, it can trigger the laminar–turbulent transition.

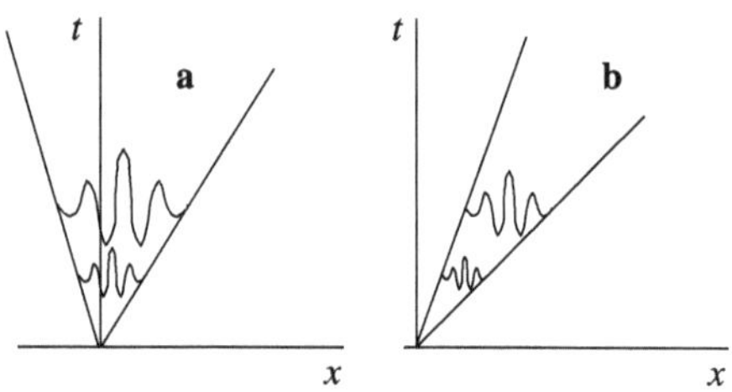

Fig. 1.2 Illustration of propagation of disturbances in systems with different types of instability: absolutely unstable system (**a**); convectively unstable system (**b**)

Hence, due to the variety of the instability phenomena, it is hardly possible and productive to give a unique rational definition of the hydrodynamic stability. Instead, let us find in the beginning some foundations for classification of different ways for the flow to lose its original state.

The equations of motion (e.g., Navier–Stokes equations) can give us some hints in this direction. Namely, two classes of physical problems regarding the propagation of disturbances in hydrodynamic systems can be selected (Fig. 1.2):

1. The problem of initial conditions, or stability in time. If the initial disturbance decays in time at each fixed point of space (or, at least, does not monotonically grow), the system is called stable to these disturbances. Otherwise, if the initial disturbance monotonically grows in time at a fixed point, the system is absolutely unstable.
2. The problem of boundary conditions, or stability in space. If an external signal at the entrance to the system decays whilst propagating in it, it is said that spatial

attenuation (non-transmission of a signal) takes place. Otherwise, there is spatial amplification, and the system is convectively unstable.

It may happen in practice that the disturbance grows in a reference system in both space and time simultaneously, as in certain regions of wakes behind bluff bodies or heated jets. Obviously, the classification depends on the frame of reference chosen. Hence, convective disturbances, can always be transformed to 'developing in time' and vice versa by the Galilei transformations.

Since the velocity distributions and boundary conditions are not invariant with respect to the coordinate transformation, however, these problems are not identical from the mathematical viewpoint. In practice, there is usually a 'natural' laboratory reference system, fitted to wind-tunnel walls, airfoil, disturbance source, etc., in which the classification is conducted. In a rare case when such a reference system is not obvious, it is always possible to formulate a problem of finding a reference system, in which the instability is, e.g., absolute with the maximum increment.

Below in this Chapter, we outline an approach to instability of fluid motion in time, as it appeared that a formal mathematical analysis of spatial amplification is more difficult. Hence, we postpone it to Sect. 2.2.

1.2 Stability of fluid motion in time

First of all, let us specify a measure to describe the disturbance 'growth' or 'decay.' Will it be the growth of the geometrical size of the disturbance, e.g., the disturbed volume or the absolute value of vorticity, or something else? Generally, any of these characteristics can serve as the desired measure. However, probably, the most natural choice in most cases is the disturbance kinetic energy.

Therefore, we give more formal definitions of the stability based on the kinetic energy of disturbance velocity $\mathbf{u}$, integrated over the entire volume $\mathscr{V}$ in which the disturbance is localized $E_{\mathscr{V}} = \int_{\mathscr{V}} (\mathbf{u}^2/2) \mathrm{d}\mathscr{V}$. The following definitions are used in applications:

Asymptotic stability: A flow is asymptotically stable to disturbances if

$$\lim_{t\to\infty} \frac{E_{\mathscr{V}}(t)}{E_{\mathscr{V}}(0)} \to 0\,,$$

where t is time.

Conditional stability: If there is a value $\varepsilon > 0$, such that the solution of equations of motion is asymptotically stable when $E_{\mathscr{V}}(0) < \varepsilon$, the solution is called *conditionally stable.* The value of ε (the attraction radius) determines a set of initial conditions attracting to the undisturbed solution. If the disturbance kinetic energy is $E_{\mathscr{V}}(0) \geq \varepsilon$, the disturbance grows or forms a new stable state (exchange of stabilities).

Global stability: If $\varepsilon \to \infty$, the solution is *globally or unconditionally stable.*

Monotonic stability: If the flow is asymptotically stable and $\mathrm{d}E_{\mathscr{V}}/\mathrm{d}t \leq 0$ for all $t > 0$, the solution is *monotonically stable.*

As seen, each next definition imposes new restrictions on the stability.

Hence, different routes of the laminar–turbulent transition, each of which being associated with corresponding physical phenomena occurring when the basic flow loses asymptotic, conditional, global, or monotonic stability, are possible.

1.2.1 Critical parameters for onset of instability

Quite naturally, the instability is defined as a state when the corresponding stability conditions are broken. In accordance with the principle of similarity first established experimentally by Reynolds (1883) for a flow in a channel, the conditions for instability appearance can depend on certain dimensionless ratios of problem parameters such as viscosity, density, velocity, temperature, etc. For many flows, such a fundamental quantity is the ratio between inertial and friction forces, the Reynolds number $\mathrm{Re} = Ul/\nu$, where U is the characteristic flow velocity, l is the characteristic length, and ν is the kinematic viscosity, which in turn is the ratio of dynamic viscosity of the fluid and density.

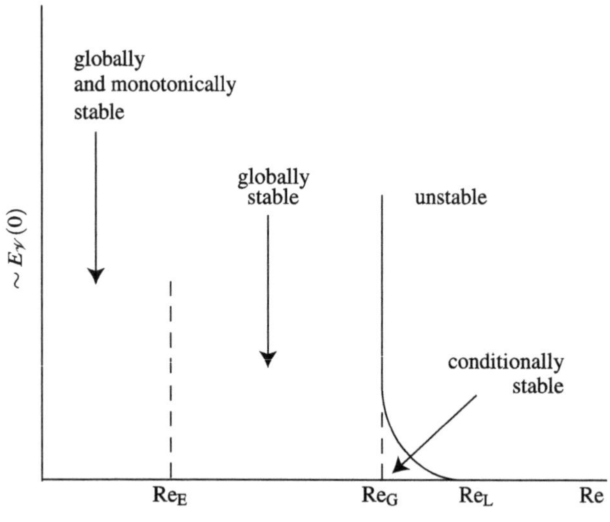

Fig. 1.3 Critical Reynolds numbers and flow stability

Based on the stability definitions given above, the *critical Reynolds numbers* separating the regions of stable and unstable motion are defined below (Fig. 1.3). As we know that the transition to turbulence occurs at high Reynolds numbers, while flows with relatively low Reynolds numbers are always laminar, we conjecture that the less restrictive is the definition of instability from the above list, the higher is the Reynolds number corresponding to it.

Hence, if a flow is capable of becoming turbulent at a finite Reynolds number $\mathrm{Re_T} < \infty$, there should be at least $\mathrm{Re_E} \leq \mathrm{Re_T}$ at which the flow loses monotonic stability; i.e., flow disturbances are possible, whose kinetic energy can experience a transient growth. For example, in a plane channel flow (also called a plane Poiseuille flow) $\mathrm{Re_E} = U_a h/\nu = 49.6$.[1]

At $\mathrm{Re} \geq \mathrm{Re_G}$, the flow loses global stability—it becomes conditionally stable. Usually this Reynolds number is also less than or equal to the transition Reynolds number $\mathrm{Re_T}$. The sense of conditional stability is illustrated by the simple mechanical model in Fig. 1.4: the system is stable to infinitely small (infinitesimal) disturbances, but unstable to finite disturbances exceeding a certain threshold value in amplitude (the 'condition'). This is in contrast to the 'global' stability case in Fig. 1.1**a**.

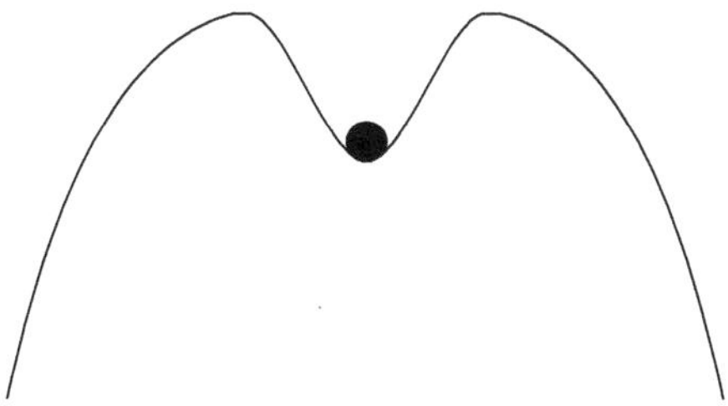

Fig. 1.4 Example of a mechanical system that is stable to 'small' and unstable to 'large' disturbances

In other words, at $\mathrm{Re} \geq \mathrm{Re_G}$, there may be initial disturbances that can, as a minimum, not decay in time and, as a maximum, cause a transition to turbulence. For such flows, where the principle of exchange of stabilities holds, the transition Reynolds number $\mathrm{Re_T}$ obviously differs from $\mathrm{Re_G}$, i.e., stable-not-turbulent equilibrium solutions have to be taken into account.

At $\mathrm{Re} \geq \mathrm{Re_L}$, the flow is linearly unstable, i.e., there is an infinitesimal disturbance that does not decrease in time. The Plane Poiseuille flow and the Blasius boundary layer become linearly unstable at certain finite Reynolds numbers. However, there are flows, such as the flow in a round pipe, that are linearly stable at any Re. In such flows the transition to turbulence obviously is triggered by loosing either monotonic or global stability.

For more complex flows, there appear some other parameters, e.g., the Görtler number Gö, which is a dimensionless measure of wall curvature. For such flows, the definitions of the critical numbers given above should serve as a template, that is, e.g., $\mathrm{Gö_E}$, $\mathrm{Gö_T}$, etc. instead of $\mathrm{Re_E}$, $\mathrm{Re_T}$, etc. should be defined.

[1] In what follows we use Re with lower-case letter subscripts to denote the characteristic length and with upper-case letter subscripts to denote some characteristic points in the Reynolds number range.

1.2.2 Conditional stability

For fluid mechanics applications, the role of conditional instability is seen from Landau's example (Landau and Lifshitz, 1959). In particular, for almost neutral disturbances, i.e., at $|\mathrm{Re}-\mathrm{Re_L}| \ll 1$, a multiple-scale analysis of the equations of motion is allowed. After certain mathematical simplifications (see, e.g., Drazin and Reid, 1981, for details), the approach leads to the Landau equation for the disturbance amplitude. In general form, it can be written as

$$\frac{\mathrm{d}|A|^2}{\mathrm{d}\tau} = \lambda_1 |A|^2 + \lambda_2 |A|^4 .$$

This equation includes the first terms in the amplitude expansion for the disturbance solution at $\mathrm{Re_L}$. The constants $\lambda_1 \in \mathbb{R}$ and $\lambda_2 \in \mathbb{R}$ are the Landau coefficients, which can be either positive or negative and are specific for a particular flow. The equation can be rewritten as

$$\frac{\mathrm{d}|A|^{-2}}{\mathrm{d}\tau} = -\lambda_1 |A|^{-2} - \lambda_2 ,$$

from which it is easy to obtain the general solution

$$|A|^{-2}(t) = |A_0|^{-2}\mathrm{e}^{-\lambda_1\tau} - \frac{\lambda_2}{\lambda_1}\left(1-\mathrm{e}^{-\lambda_1\tau}\right) ,$$

or

$$|A|^2(t) = \frac{|A_0|^2\mathrm{e}^{\lambda_1\tau}}{1+\frac{\lambda_2}{\lambda_1}|A_0|^2\left(1-\mathrm{e}^{\lambda_1\tau}\right)} ,$$

where A_0 is the initial value. Note that $\lambda_1 > 0$, when $\mathrm{Re}-\mathrm{Re_L} > 0$ and $\lambda_1 < 0$, when $\mathrm{Re}-\mathrm{Re_L} < 0$. Let us consider some possible variants depending on the signs of the coefficients:

$\lambda_1 > 0$, $\lambda_2 < 0$: The solution gives $A_0 \to 0$, as $\tau \to -\infty$ and steady state (amplitude saturation) at $A_\mathrm{f} = \sqrt{-\lambda_1/\lambda_2}$, as $\tau \to +\infty$ independent of A_0, i.e., the exchange of stabilities can occur.

$\lambda_1 < 0$, $\lambda_2 < 0$: $A_0 \to 0$ and $A_\mathrm{f} \to 0$, i.e., the flow is stable.

$\lambda_1 > 0$, $\lambda_2 > 0$: An unlimited growth occurs independent of A_0. The solution becomes infinite at $\tau = \ln[1+\lambda_1/(\lambda_2 A_0^2)]/\lambda_1$. In practice, this situation corresponds to a fast transition to turbulence.

$\lambda_1 < 0$, $\lambda_2 > 0$: If $A_0 < \sqrt{-\lambda_1/\lambda_2}$, then $A_\mathrm{f} \to 0$, i.e., the flow becomes stable. If, however, $A_0 > \sqrt{-\lambda_1/\lambda_2}$, then an unlimited growth occurs and the solution becomes infinite at $\tau = \ln[A_0^2/(A_0^2 - A_\mathrm{f}^2)]/\lambda_1$.

Therefore, two types of instability phenomena can be distinguished. If a nonzero steady state or turbulence can be reached by introducing an infinitesimal disturbance at $\mathrm{Re} > \mathrm{Re_L}$, such a system is called the *system with soft excitation*, (Fig. 1.5**a**). Otherwise flow instability can be triggered at $\mathrm{Re} < \mathrm{Re_L}$ with the initial disturbance of finite amplitude, that is the basic motion at these Reynolds numbers is *metastable*. Such a system is called the system with *hard excitation* (Fig. 1.5**b**). The nonzero

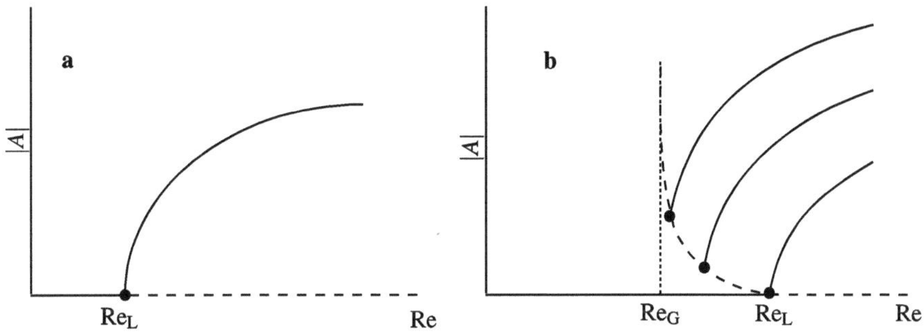

Fig. 1.5 General instability features based on analysis of the Landau equations: system with soft excitation (**a**) and system with hard excitation (**b**). Amplification curves (*solid lines*) and envelopes of the initial amplitudes destabilizing the flow (*dashed lines*) are outlined

solutions that appear at $\mathrm{Re} < \mathrm{Re_L}$ and $\mathrm{Re} > \mathrm{Re_L}$, are called *subcritical* and *supercritical*, respectively. In particular, it is known that the plane Couette flow and plane Poiseuiile flows considered in Chap. 3 are metastable, while the Blasius boundary layer considered in Chap. 4 represents a system with soft excitation.

1.2.3 Growth of disturbance energy

At first glance, it is not obvious that the global and conditional stabilities relate closely to the linear one. To show it, let us introduce and examine the equation of energy balance for a disturbed flow.

Let $\mathbf{U}(\mathbf{r})$ and $P(\mathbf{r})$ be distributions of velocity and pressure of a known stationary solution of the equations of motion (the Navier–Stokes equations) of an incompressible fluid

$$(\mathbf{U}\cdot\nabla)\mathbf{U} = -\nabla P + \frac{1}{\mathrm{Re}}\nabla^2\mathbf{U}\,, \tag{1.1a}$$

$$\nabla\cdot\mathbf{U} = 0\,, \tag{1.1b}$$

with natural boundary conditions $\mathbf{U}|_S = 0$ at boundaries S (walls and/or infinity). Let us impose a disturbance $\mathbf{u}(\mathbf{r},t)$ and $p(\mathbf{r},t)$, where $\mathbf{r}$ is the coordinate vector and t is the time, so that the resultant motion $\mathscr{U}(\mathbf{r}) = \mathbf{U}(\mathbf{r}) + \mathbf{u}(\mathbf{r},t)$, and $\mathscr{P}(\mathbf{r}) = P(\mathbf{r}) + p(\mathbf{r},t)$, also satisfy the Navier–Stokes equations and the boundary conditions

$$\frac{\partial\mathscr{U}}{\partial t} + (\mathscr{U}\nabla)\cdot\mathscr{U} = -\nabla\mathscr{P} + \frac{1}{\mathrm{Re}}\nabla^2\mathscr{U}\,, \tag{1.2a}$$

$$\nabla\cdot\mathscr{U} = 0\,. \tag{1.2b}$$

Substitution of (1.1) into (1.2) yields

$$\frac{\partial\mathbf{u}}{\partial t} + (\mathbf{U}\cdot\nabla)\mathbf{u} + (\mathbf{u}\cdot\nabla)\mathbf{U} + (\mathbf{u}\cdot\nabla)\mathbf{u} = -\nabla p + \frac{1}{\mathrm{Re}}\nabla^2\mathbf{u}\,, \tag{1.3a}$$

$$\nabla\cdot\mathbf{u} = 0\,. \tag{1.3b}$$

Equations (1.3) are fundamental hydrodynamic equations for disturbances.

The equation of energy balance for a disturbed flow, *the Reynolds–Orr equation*, which serves as a basis for the energy theory of stability, is one of the consequences of the above given equations. It is obtained by multiplication of (1.3a) by $\mathbf{u}$ and integration by parts over the volume $\mathscr{V}$ using the continuity equation (1.3b). The procedure yields:

$$\frac{dE_{\mathscr{V}}}{dt} = -\int_{\mathscr{V}} \mathbf{u}\cdot(\mathsf{S}\mathbf{u})d\mathscr{V} - \frac{2}{\mathrm{Re}}\int_{\mathscr{V}} \mathsf{s}:\mathsf{s}d\mathscr{V}\,, \tag{1.4}$$

where, using the Cartesian coordinates denoted as x_1, x_2, and x_3,

$$\mathsf{S}_{ij} = \frac{1}{2}\left(\frac{\partial U_i}{\partial x_j} + \frac{\partial U_j}{\partial x_i}\right),$$

$$\mathsf{s}_{ij} = \frac{1}{2}\left(\frac{\partial u_i}{\partial x_j} + \frac{\partial u_j}{\partial x_i}\right),$$

are the symmetric deformation (rate of strain) tensors of the basic laminar flow and the disturbances, respectively, so that using the summation convention $(\mathsf{S}\mathbf{u})_i = \mathsf{S}_{ij}u_j$, and $2\mathsf{s}:\mathsf{s} = 2\mathsf{s}_{ij}\mathsf{s}_{ij} = (\partial u_i/\partial x_j)^2$.

The first right-hand-side term of (1.4) describes the exchange of energy with the basic flow and the second term describes the energy dissipation due to viscosity. If the first term is positive, the energy is extracted from the basic flow. The dissipation term is always negative. The relative value of these two terms defines whether the disturbance energy decays or grows. The rate of these two terms at $dE_{\mathscr{V}}/dt = 0$ forms the global stability problem for $\mathrm{Re_E}$, below which any disturbance monotonically decays:

$$\frac{1}{\mathrm{Re_E}} = \max_{\mathbf{u}(\mathbf{r})\neq 0} -\frac{\int_{\mathscr{V}} \mathbf{u}\cdot(\mathsf{S}\mathbf{u})d\mathscr{V}}{2\int_{\mathscr{V}} \mathsf{s}:\mathsf{s}d\mathscr{V}}\,. \tag{1.5}$$

This is a variational problem to search for *kinematically allowable* velocities that maximize $1/\mathrm{Re_E}$—the rate of two quadratic forms (the Rayleigh quotient). It is known that such a maximum always exists and is positive.

Note that the described way to find $\mathrm{Re_E}$ is limited to a particular definition of the disturbance measure (its kinetic energy) and to fluids in which the basic flowfield is independent of the field of disturbances (i.e., the basic flow does not change when the disturbance energy grows). As seen, in this case, the nonlinear terms vanish in the Reynolds–Orr equation, and the instantaneous disturbance growth rate

$$\frac{1}{E_{\mathrm{V}}}\frac{dE_{\mathrm{V}}}{dt}$$

is independent of the initial disturbance amplitude. This indicates that the possible energy growth of any disturbances in *incompressible Newtonian fluids* described by (1.1) is related to linear energy growth mechanisms. In particular, it means that there are linear mechanisms for the transient growth of disturbances kinetic energy at 'subcritical' values of the Reynolds number, i.e., at $\mathrm{Re_E} < \mathrm{Re} < \mathrm{Re_L}$.

Additional material for this chapter is given in the multimedia file No. 1, which supplements the book, titled 'Concept of hydrodynamic stability.'

References

Betchov R, Criminale WO (1967) Stability of parallel flows. Academic, New York

Drazin PG, Reid WH (1981) Hydrodynamic stability. Cambridge University Press, Cambridge

Landau LD, Lifshitz EM (1959) Fluid mechanics. Pergamon, Oxford

Reynolds O (1883) An experimental investigation of the circumstances which determine whether the motion of water shall be direct or sinuous, and of the law of resistance in parallel channels. Philos Trans R Soc Lond A 174:935–982

Further Reading

van Dyke M (1982) An album of fluid motion. Parabolic Press, Stanford

Galdi GP, Padula M (1990) A new approach to energy theory in the stability of fluid motion. Arch Rat Mech Anal 110:187–286

Henningson DS, Reddy SC (1994) On the role of linear mechanisms in transition to turbulence. Phys Fluids A 6(3):1396–1398

Huerre P, Monkewitz PA (1990) Local and global instabilities in spatially developing flows. Ann Rev Fluid Mech 22:473–537

Joseph DD (1976) Stability of Fluid Motion, Springer Tracts on Natural Philosophy, vol 1. Springer–Verlag, Berlin

Schlichting H, Gersten K (2000) Boundary layer theory, 8th edn. Springer–Verlag, Berlin

Waleffe F (1995) Transition in shear flows. Nonlinear normality versus non-normal linearity. Phys Fluids 7(12):3060–3066

Chapter 2
Theoretical aspects

When studying the asymptotic stability of the basic flow, it is possible to simplify the problem essentially by reducing the consideration of nonlinear equations of motion to the analysis of linearized equations for disturbances. In this chapter, various aspects of the corresponding theory for the so-called parallel shear flows are expounded beginning from formulation of linear hydrodynamic stability problems in time and space for wavy (modal) disturbances. The classical Gaster's transformation between these two approaches is explained. The dual role of viscosity for flow instability is outlined. Then, relevance of oblique waves in instability is discussed. Finally, such special issues, important in the following chapters, as bi-orthogonality of modes and completeness of their set are introduced.

2.1 Formulation of linear hydrodynamic stability problems

Dropping the quadratic terms for perturbations in (1.3a) yields

$$\frac{\partial \mathbf{u}}{\partial t} + (\mathbf{U}\cdot\nabla)\mathbf{u} + (\mathbf{u}\cdot\nabla)\mathbf{U} = -\nabla p + \frac{1}{\mathrm{Re}}\nabla^2\mathbf{u}\,. \tag{2.1}$$

These equations, with relevant initial and boundary conditions as well as the continuity equation for disturbances (1.3b), form the *linear stability problem.* A general theoretical study even of these linearized equations is impossible at present time. Therefore, simplified models that reflect real-life phenomena well enough are usually considered.

For this purpose, it is important to choose an appropriate coordinate system, which reflects possible symmetry features of a flow. Frequently, a useful choice is a boundary and streamline-aligned coordinate system. Particularly, when the streamlines are parallel to each other, they should also be parallel to the boundaries, see Fig. 2.1. Then, for plane cases, the Cartesian coordinates x, y, and z; for axisymmetric cases, the cylindrical coordinates r, θ, and z; etc. are appropriate. Typical representatives of these so-called parallel shear flows are the Poiseuille flows (e.g.,

A.V. Boiko et al., *Physics of Transitional Shear Flows*,
Fluid Mechanics and Its Applications 98, DOI 10.1007/978-94-007-2498-3_2,

Fig. 2.1 A parallel shear flow

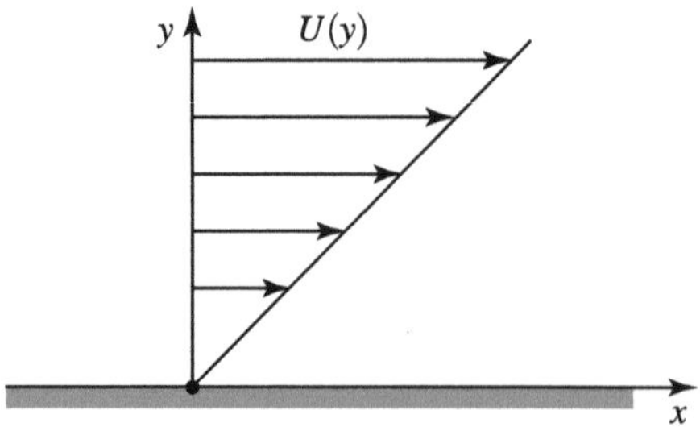

flows in a plane channel, round or elliptic pipe, rectangular duct, etc.) as well as some boundary layers (e.g., asymptotic suction profile over a permeable flat plate). Meanwhile, many other flows can be treated in the framework of the linear stability theory (after a proper justification) as locally parallel, in which case the streamline nonparallelism is neglected.

Let us begin with the plane case. The velocity vectors **U** and **u** in component-wise form in the Cartesian coordinates are

$$\mathbf{U} = (U(y), V = 0, W = 0)^T \,, \tag{2.2a}$$

$$\mathbf{u} = (u(x,y,z,t), v(x,y,z,t), w(x,y,z,t))^T \,, \tag{2.2b}$$

where x is the streamwise coordinate, y is the wall-normal coordinate, z is the transverse coordinate normal to x and y, and the superscript 'T' denotes transpose of the vectors.

Taking into account (2.2), the equations of motion (2.1) and the continuity equation can be written as

$$\frac{\partial u}{\partial t} + U\frac{\partial u}{\partial x} + v\frac{\partial U}{\partial y} = -\frac{\partial p}{\partial x} + \frac{\nabla^2 u}{\mathrm{Re}} \,, \tag{2.3a}$$

$$\frac{\partial v}{\partial t} + U\frac{\partial v}{\partial x} = -\frac{\partial p}{\partial y} + \frac{\nabla^2 v}{\mathrm{Re}} \,, \tag{2.3b}$$

$$\frac{\partial w}{\partial t} + U\frac{\partial w}{\partial x} = -\frac{\partial p}{\partial z} + \frac{\nabla^2 w}{\mathrm{Re}} \,, \tag{2.3c}$$

$$\frac{\partial u}{\partial x} + \frac{\partial v}{\partial y} + \frac{\partial w}{\partial z} = 0 \,. \tag{2.3d}$$

At the assumptions made above, these equations together with corresponding boundary and initial conditions describe completely the evolution of an arbitrary infinitesimal disturbance of a given basic flow in space and time. At solid surfaces, which we assume below, the no-slip boundary conditions (vanishing of disturbances) are imposed.

Note that there are six boundary conditions for velocity to close the problem, while the pressure at the boundaries is not specified. Instead, the conditions for the pressure are given implicitly through the continuity equation. Thus, the pressure can be eliminated yielding

$$\left(\frac{\partial}{\partial t}+U\frac{\partial}{\partial x}-\frac{\nabla^2}{\mathrm{Re}}\right)\left(\frac{\partial w}{\partial y}-\frac{\partial v}{\partial z}\right)+\frac{\partial U}{\partial y}\frac{\partial w}{\partial x} = 0\,, \qquad (2.4a)$$

$$\left(\frac{\partial}{\partial t}+U\frac{\partial}{\partial x}-\frac{\nabla^2}{\mathrm{Re}}\right)\left(\frac{\partial u}{\partial z}-\frac{\partial w}{\partial x}\right)+\frac{\partial U}{\partial y}\frac{\partial v}{\partial z} = 0\,, \qquad (2.4b)$$

$$\left(\frac{\partial}{\partial t}+U\frac{\partial}{\partial x}-\frac{\nabla^2}{\mathrm{Re}}\right)\left(\frac{\partial v}{\partial x}-\frac{\partial u}{\partial y}\right)+\frac{\partial U}{\partial y}\frac{\partial w}{\partial z}-\frac{\partial^2 U}{\partial y^2}v = 0\,, \qquad (2.4c)$$

where the Laplace operator

$$\nabla^2 = \frac{\partial^2}{\partial x^2}+\frac{\partial^2}{\partial y^2}+\frac{\partial^2}{\partial z^2}\,.$$

The equation for pressure can be also obtained by taking the divergence of the momentum equations, summing them up, and eliminating most of the terms with the help of the continuity equation that yields the following Poisson equation:

$$\nabla^2 p = -2\frac{\partial U}{\partial y}\frac{\partial v}{\partial x}\,.$$

Further successive elimination of variables makes it possible to form an equation for the single wall-normal velocity:

$$\left[\left(\frac{\partial}{\partial t}+U\frac{\partial}{\partial x}\right)\nabla^2-\frac{\partial^2 U}{\partial y^2}\frac{\partial}{\partial x}-\frac{1}{\mathrm{Re}}\nabla^4\right]v = 0\,. \qquad (2.5)$$

This equation, together with equation (2.4a) and the continuity equation (2.3), describes completely the disturbance flowfield.

The classical approach to solving such stability problems is the method of normal modes. Let us assume that the solution is separable, i.e., it can be expressed as a sum of elementary solutions (*normal modes*)[1], which have the form of plane waves

$$(u,v,w,p)^T = (\hat{u}(y),\hat{v}(y),\hat{w}(y),\hat{p}(y))^T \mathrm{e}^{\mathrm{i}(\boldsymbol{\kappa}\cdot\mathbf{r}-\omega t)}\,, \qquad (2.6)$$

where $\hat{u}$, $\hat{v}$, $\hat{w}$, $\hat{p}$ are the complex amplitude functions of the waves; κ is the *wavevector*, so that its magnitude $|\kappa| = \varkappa = \sqrt{\alpha^2+\beta^2}$, where α and β are the wavenumbers in the streamwise and transverse directions x and z, respectively; ω is the circular frequency (which is generally complex in case of instability in time). The absolute value of an amplitude function of the wave can be considered as its 'shape' or amplitude profile along the coordinate y, whereas the phase ψ, which also depends on y, denotes time shifts of the oscillations, so that, e.g., $\hat{u}(y) = |\hat{u}(y)|\mathrm{e}^{\mathrm{i}\psi y}$.

Some further kinematic definitions are useful. The quantity ω_{i} is the *growth rate* of the wave. The angle between the wavevector κ and the streamwise axis x is $\gamma = \arctan(\beta/\alpha)$. If $\gamma = 0$, the normal mode and the corresponding plane wave are two-dimensional (independent of z); otherwise the normal mode is three-dimensional, i.e., the wave is *oblique*. The complex vector of the *phase velocity* (the velocity of the wave crest in the direction of the wavevector) is determined by the expression

[1] Hence, in such reducing the initial-boundary-value problem, it is necessary to check the completeness of the obtained set of the modes (see Sect. 2.6 for further details).

$$\mathbf{c} = \mathbf{c}_\mathrm{r} + \mathrm{i}\mathbf{c}_\mathrm{i} = \frac{\kappa}{\varkappa}\frac{\omega}{\varkappa}\,, \tag{2.7}$$

where the unit vector $\kappa/\varkappa$ determines the direction of the vector $\mathbf{c}$ and the scalar values of the real and imaginary parts of $\omega/\varkappa$ determine the real phase velocity and the reduced growth rate, respectively. Hence, the projections of the phase velocity vector to the x and z axes are

$$c_\mathbf{x} = \frac{\alpha\omega}{\varkappa^2}\,, \quad c_\mathbf{z} = \frac{\beta\omega}{\varkappa^2}\,.$$

The decomposition (2.6) is equivalent to the transformation from the physical space of the Cartesian coordinates and time to the Fourier–Laplace spectral space of wavenumbers and frequencies. In certain configurations, the problem can be further simplified to consider separately symmetrical or antisymmetrical modes if the flow has a symmetry in the wall-normal direction as, e.g., in the plane channel flows.

Solving the corresponding set of homogeneous equations with homogeneous boundary conditions constitutes an *eigenvalue problem*, that means finding such values of α, β, and ω (with the Reynolds number Re and flow velocity U as basic parameters) at which there is a nontrivial solution $\hat{u}$, $\hat{v}$, $\hat{w}$, and $\hat{p}$. Such an implicit relation between α, β and ω can be characterized by the so-called characteristic (*dispersion*) relation written in general form as

$$\mathcal{F}(\alpha,\beta,\omega) = 0\,. \tag{2.8}$$

The elementary solutions (the normal modes) are the *eigenfunctions* of the corresponding spectral problem. From the viewpoint of physics, they can be interpreted as free hydrodynamic waves inside the shear flow appearing as a 'hydrodynamic resonator.'

Substituting (2.6) in (2.5) yields the famous *Orr–Sommerfeld equation*

$$\left[\mathrm{i}(\alpha U-\omega)\left(\mathcal{D}^2-\varkappa^2\right)-\mathrm{i}\alpha U_{yy}-\frac{1}{\mathrm{Re}}\left(\mathcal{D}^2-\varkappa^2\right)^2\right]\hat{v}=0\,, \tag{2.9}$$

where the symbol $\mathcal{D}$ and subscript 'y' denote the first derivative with respect to y. The Orr–Sommerfeld equation at homogeneous boundary conditions[2] $\hat{v} = \hat{v}_y = 0$ plays the role of the dispersion relation that constitutes the eigenvalue problem for the wall-normal velocity of the disturbance $\hat{v}$. Finally, by solving the Orr–Sommerfeld equation, we can make a step back and find, e.g., the transverse velocity $\hat{w}$ from

$$\left[\mathrm{i}(\alpha U-\omega)-\frac{1}{\mathrm{Re}}\left(\mathcal{D}^2-\varkappa^2\right)\right](\mathcal{D}\hat{w}-\mathrm{i}\beta\hat{v})+\mathrm{i}\alpha U_y\hat{w}=0\,, \tag{2.10}$$

and then the streamwise velocity $\hat{u}$ from the continuity equation

$$\mathrm{i}\alpha\hat{u}+\mathcal{D}\hat{v}+\mathrm{i}\beta\hat{w}=0\,. \tag{2.11}$$

Taking into account that we do not need to integrate to find $\hat{u}$, we can conclude that only two velocity components are required to represent the disturbance flowfield

[2] The condition $\hat{v}_y = 0$ following from the continuity equation is introduced to make the problem complete.

completely. For this purpose, instead of $\hat{w}$ or $\hat{u}$, sometimes their combination in the form of the normal vorticity

$$\tilde{\eta} = \frac{\partial u}{\partial z} - \frac{\partial w}{\partial x}$$

or

$$\hat{\eta} = \mathrm{i}(\beta\hat{u} - \alpha\hat{w})$$

is used. Comparing terms in (2.4b) with these definitions, we find that

$$\left[\mathrm{i}\,(\alpha U - \omega) - \frac{1}{\mathrm{Re}}\left(\mathcal{D}^2 - \varkappa^2\right)\right]\hat{\eta} = -\mathrm{i}\beta U_y \hat{v}\,. \tag{2.12}$$

This equation is known as the normal vorticity or the *Squire equation.*

2.1.1 Spectral formulation of stability

Let us assume that an elementary disturbance (a mode) develops in time, i.e., α and β are real, and $\omega = \omega_{\mathrm{r}} + \mathrm{i}\omega_{\mathrm{i}}$ is complex. If there are complex eigenvalues $\omega = \omega_{\mathrm{r}} + \mathrm{i}\omega_{\mathrm{i}}$ of the characteristic equation (2.8) such that $\omega_{\mathrm{i}} > 0$ at certain real α and β, the basic state is (linearly) unstable. If $\omega_{\mathrm{i}} < 0$ for all α and β, it is stable to small disturbances. In the case when $\omega_{\mathrm{i}} = 0$ at certain α and β, it is said that the system is *neutrally (indefinitely) stable*. The neutral disturbance can be of two types: if $\omega_{\mathrm{r}} = 0$, it is stationary; if $\omega_{\mathrm{r}} \neq 0$, it is periodic in time and represents a traveling wave.

The value of Re_{L} minimized over all frequencies at which the instability occurs is called the (first) critical Reynolds number of the spectral problem. If Re_{L} is finite, the equation $\omega_{\mathrm{i}}(\alpha,\beta,\mathrm{Re}) = 0$ parametrically determines a surface or several surfaces in space $(\alpha,\beta,\mathrm{Re})$ separating the region where the flow is stable from an instability region; it is called a boundary or *surface of neutral stability* (Fig. 2.2). As the neutral stability surface and the critical Reynolds number bound the parameter regions where the basic laminar flow is stable or unstable to small disturbances, their determination is the primary problem of the stability theory. Certainly, it is possible to formulate other problems, for example, to find the parameters of the dispersion relation at which the growth rate is maximal, etc.

Note that $\hat{v}$ plays the role of a forcing term in both (2.10) and (2.12). The homogeneous parts of these equations (i.e., assuming $\hat{v} = 0$) together with the corresponding homogeneous boundary conditions constitute eigenvalue problems which should be considered separately. For example, the solution of the homogeneous equation (2.12) gives a set of eigenvalues and eigenfunctions of the normal vorticity waves for the flow of interest. However, these modes are always stable. Multiplication of the homogeneous normal vorticity equation by its complex conjugate solution $\hat{\eta}^*$, integration over y, and subsequent separation of imaginary and real parts yield

$$\int_{y_1}^{y_2}\left[|\mathcal{D}\hat{\eta}|^2 + \left(\varkappa^2 + \omega_{\mathrm{i}}\mathrm{Re}\right)|\hat{\eta}|^2\right]\mathrm{d}y = 0\,,$$

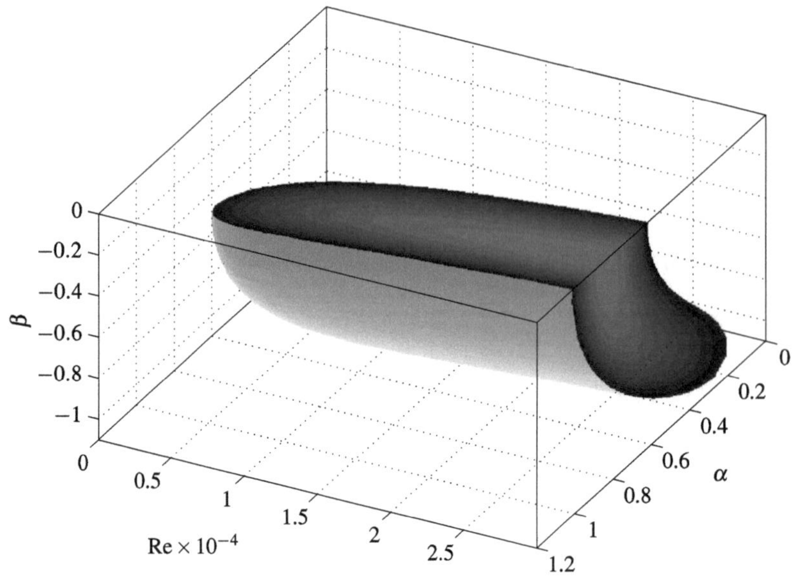

Fig. 2.2 Neutral stability surface with isocaps, which show cross-sectional views of the interior of the isosurfaces $\omega_i > 0$, for the plane Poiseuille flow (calculated by A. V. Boiko)

$$\int_{y_1}^{y_2} \left(U - \frac{\omega_r}{\alpha}\right) |\hat{\eta}|^2 \, dy = 0 \, .$$

The first of these equations can be satisfied only if the waves decay, i.e., $\omega_i < 0$. From the second equation, it follows directly that the propagation velocity of a vorticity wave lies between the minimum and maximum velocities of the basic flow.

Therefore, it is possible to study the asymptotic linear stability of a parallel flow by considering only the Orr–Sommerfeld equation[3]. Elementary solutions of the eigenvalue problem for the Orr–Sommerfeld equation with $\omega \neq 0$, i.e., traveling normal modes of the shear layer, are usually called *Tollmien–Schlichting waves* in honor of the pioneers in the boundary-layer stability W. Tollmien and H. Schlichting.

2.1.2 Inviscid instability mechanism

Difficulties with the general theoretical analysis of the Orr–Sommerfeld equation associated with the presence of the small parameter $(\alpha \mathrm{Re})^{-1}$ at the highest deriva-

[3] The streamwise and transverse velocities of the modes, however, are still significant in considering the initial-value problem of disturbance propagation.

tive can be reduced by considering the linear stability of a parallel shear flow at high Reynolds numbers. Then it seems natural to drop the viscous terms in (2.9). By putting the transverse wavenumber $\beta = 0$, we obtain the following Rayleigh equation for two-dimensional waves

$$\mathcal{D}^2\hat{v} - \alpha^2\hat{v} - \frac{U_{yy}}{U-c}\hat{v} = 0\,, \tag{2.13}$$

where $c = \omega/\alpha$ is the value of phase velocity, cf. (2.7). Since it is an ordinary differential equation of the second order, it is possible to satisfy only two of the four boundary conditions of the viscous problem—keeping those of zero wall-normal velocity at the boundaries and omitting the condition $\hat{v}_y = 0$. As practice shows, such simplification at high Reynolds numbers is justified, in particular, for free or separated shear flows when solid boundaries are absent or far from the shear layer of interest. At low Reynolds numbers, however, the influence of the boundaries is essential and the role of viscosity cannot be neglected, see, e.g., Chap. 7.

Note first of all that all coefficients in the Rayleigh equation are real; hence, its eigenvalues either are real (in which case the waves are neutrally stable) or occur in complex conjugate pairs. The solutions with $c_\text{i} > 0$ correspond to the instability waves, and, hence, are called Rayleigh or inviscid instability waves.

Considering the inviscid stability problem in time for two-dimensional waves, Lord Rayleigh (1880) proved some important general theorems. The first one (or the *inflection point criterion*) states that the necessary condition for inviscid instability is the presence of an inflection point in the basic flow velocity distribution $U(y)$. The *inflection point* is a point y_s such that $U_{yy}(y_\text{s}) = 0$. Later Tollmien (1935) proved that the condition is also sufficient for symmetric velocity distributions in channels and velocity distributions of the boundary-layer type.

The *second Rayleigh theorem* states that the velocity of propagation of neutral disturbances (i.e., when $c_\text{i} = 0$) in a shear layer is between the minimum U_min and maximum U_max flow velocities, i.e., $U_\text{min} < c_\text{r} < U_\text{max}$. The slice of the basic shear flow where the local velocity is equal to the phase velocity of the wave, i.e., $U(y_\text{c}) = c_\text{r}$, is called the *critical layer*, as it corresponds to a singularity in the Rayleigh equation (2.13) for the neutral disturbances, unless $y_\text{s} = y_\text{c}$. Hence, to get rid of the singularity in a general case, viscosity has to be taken into account in the critical layer.

Based on the inflection point criterion, nothing can be said about, e.g., the linear stability of a plane Couette flow with $U(y) = y$, where $U_{yy} = 0$ for all y. An additional necessary condition for instability of monotonous basic flow velocity profiles, the *criterion of the maximum vorticity* (i.e., the gradient U_y), was found later by Fjørtoft (1950), see Fig. 2.3.

Howard (1961) proved the so-called *semicircle theorem* for the bound of the growth rates for unstable Rayleigh waves. It appears that

$$\left[c_\text{r} - \frac{1}{2}(U_\text{max} + U_\text{min})\right]^2 + c_\text{i}^2 \le \left[\frac{1}{2}(U_\text{max} - U_\text{min})\right]^2 .$$

Hence, in the phase-velocity plane, all unstable modes lie inside a semicircle.

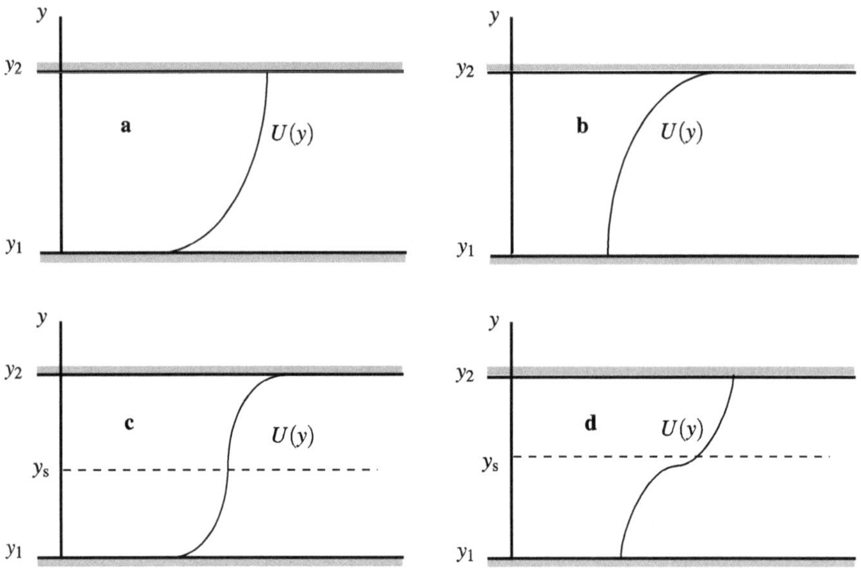

Fig. 2.3 Stability criteria: stable, $U_{yy} < 0$ (**a**); stable, $U_{yy} > 0$ (**b**); stable, $U_{s,yy} = 0$ in y_s, but $U_{yy}(U - U_s) \geq 0$ (**c**); probably unstable, $U_{s,yy} = 0$ and $U_{yy}(U - U_s) \leq 0$ (**d**)

2.1.3 Viscous instability mechanism

The equation (2.9) that was later called the Orr–Sommerfeld equation was derived independently by Orr (1907a,b) and Sommerfeld (1908). One of the first attempts to solve it belongs to Prandtl (1921, 1922), who studied stability of an idealized flow with a piecewise linear velocity distribution and came to an unexpected conclusion that viscosity can destabilize the flow. Now we know that many shear flows with velocity profiles without inflection are linearly unstable. Such instability is called *viscous instability*. Obviously, it can manifest itself only at finite Reynolds numbers. For example, Fig. 2.4 illustrates that the asymptotic behavior of the neutral

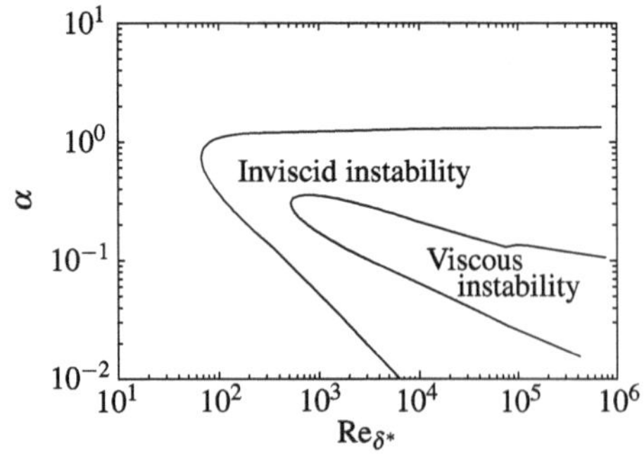

Fig. 2.4 Typical viscous (for Blasius boundary-layer profile) and inviscid (for pre-separation Hartree boundary-layer profile) neutral stability curves which are the slices of the neutral stability surfaces at $\beta = 0$ (calculated by A. V. Boiko)

stability curves for a boundary layer without the inflection point in the velocity profile interior differs from that with an inflection point, as Re approaches infinity. In the former case, it tends to zero; in the latter case, it becomes a constant value.

The physical mechanism of viscous instability can be illustrated by the following two-dimensional considerations. In a case of a two-dimensional normal mode, Eq. (1.4) can be reduced to the form

$$\frac{\mathrm{d}E}{\mathrm{d}t} = \int \tau_{12} U_y \mathrm{d}y - \frac{1}{\mathrm{Re}} \iint \left(\frac{\partial v}{\partial x} - \frac{\partial u}{\partial y} \right)^2 \mathrm{d}x\mathrm{d}y \,,$$

where $\tau_{12} = (\hat{v}\hat{u}^* - \hat{u}\hat{v}^*)/2$ is the Reynolds stress averaged in the streamwise direction x over one wavelength of the traveling wave. As can be seen, the disturbance kinetic energy increases only if the phase shift between $\hat{u}$ and $\hat{v}$ changes the sign of τ_{12}. The Orr–Sommerfeld equation shows that viscosity (as a part of the imaginary multiplier containing Re at the higher derivatives) introduces a phase shift compared to the inviscid case. The phases of $\hat{u}$ and $\hat{v}$ depending on the wall-normal distance are different, since the viscous terms for these velocities are not the same and $\hat{v}$ should satisfy the additional *viscous boundary condition* $\hat{v}_y = 0$. At a given velocity distribution $U(y)$, certain scalar parameters ω, α, β, and some finite values of Re, this effect can appear to be destabilizing and exceed the effect of energy dissipation due to viscosity.

2.2 Instability in space

In the framework of the eigenvalue approach, there is no means to guess a priori, which formulation—spatial or temporal—corresponds to a particular problem of interest in a given reference system. To identify how the disturbance develops, it is necessary to consider the impulse response of the system as illustrated in Fig. 1.2. This problem can be attacked analytically or numerically by considering the wave-packet group velocity (see Chap. 8).

Meanwhile, it is known from laboratory experiments that small-amplitude disturbances do grow downstream in many open near-wall shear layers (e.g., in attached boundary layers, channel flows, etc.) at least at the Reynolds numbers which are important in practice. Hence, we may assume even without a rigorous (and quite delicate) theoretical justification that it is more appropriate for such basic flows to consider the stability of waves with real frequencies ω and complex wavenumbers α and β. However, spatial instability is a process that differs significantly from instability in time. The initial-value problem for the wall-normal velocity disturbance as $v(x,y,z,t) = \tilde{v}(y,t)\mathrm{e}^{\mathrm{i}(\alpha x + \beta z)}$ obtained with the help of (2.3) is given by the parabolic partial differential equation

$$\left[\left(\frac{\partial}{\partial t} + \mathrm{i}\alpha U \right) (\mathcal{D}^2 - \varkappa^2) - \mathrm{i}\alpha U_{yy} \frac{1}{\mathrm{Re}} \left(\mathcal{D}^2 - \varkappa^2 \right)^2 \right] \tilde{v}(y,t) = 0 \,.$$

The spatial evolution problem for $v(x,y,z,t) = \tilde{v}(x,y,z)e^{-i\omega t}$, on the contrary, is described by the elliptic equation

$$\left[\left(-i\omega + U\frac{\partial}{\partial x}\right)\nabla^2 - U_{yy}\frac{\partial}{\partial x} - \frac{1}{\mathrm{Re}}\nabla^4\right]\tilde{v}(x,y,z) = 0\,.$$

The corresponding boundary-value problem requires the conditions at the downstream boundary (e.g., at the trailing edge of a flat plate or channel exit), i.e., there exist disturbances that propagate upstream. It is known from experimental and numerical studies, however, that such disturbances in parallel flows decay and can be neglected. Hence, one can parabolize the problem by imposing additional constrains to exclude all upstream-propagating solutions.

In mathematical studies of convective instability, the Reynolds number Re, the disturbance frequency ω, and one of the wavenumbers, usually β, are given, while α is obtained by solving the characteristic equation (2.8), which parametrically determines the eigenvalues of the spatial problem. For simple two-dimensional shear layers, the flow homogeneity with respect to the z coordinate ensures that the transverse wavenumbers β are real.

As in the case of instability in time, it is possible to introduce a real wavevector κ_r such that its magnitude and the angle between the direction κ_r and the axis x are

$$\varkappa_r = \sqrt{\alpha_r^2 + \beta_r^2} \quad \text{and} \quad \gamma_r = \arctan\left(\frac{\beta_r}{\alpha_r}\right),$$

respectively, so that $\kappa_r = \varkappa_r \exp(i\gamma_r)$. The phase velocity is then expressed by the formula

$$\mathbf{c} = \frac{\kappa_r}{\varkappa_r}\frac{\omega}{\varkappa_r}\,.$$

However, in contrast to the temporal problem, it is also possible to define the vector $\kappa_i = \varkappa_i \exp(i\gamma_i)$, where

$$\varkappa_i = \sqrt{\alpha_i^2 + \beta_i^2} \quad \text{and} \quad \gamma_i = \arctan\left(\frac{\beta_i}{\alpha_i}\right).$$

If the direction along the vector κ_i is denoted as $\mathbf{r}_i$ and the direction along the vector κ_r as $\mathbf{r}_r$, then the expression for the wave reads

$$\tilde{v} = \hat{v}(y)\exp(-\kappa_i \cdot \mathbf{r}_i)\exp[i(\kappa_r \cdot \mathbf{r}_r - \omega t)]\,.$$

Hence, $-\varkappa_i$ is the spatial amplification rate along the direction $\mathbf{r}_i$. In particular, if $\beta_i = 0$ and $\beta_r \neq 0$, the wave propagates at an angle $\gamma_r \neq 0$, but amplifies strictly downstream, because $\gamma_i = 0$.

Assuming that β is real, the condition $\alpha_i = 0$ provides the neutral stability curves $\omega = \omega(\beta, \mathrm{Re})$ for different β (Fig. 2.4) constituting together the neutral stability surface separating the regions of flow stability at $\alpha_i > 0$ and flow instability at $\alpha_i < 0$.

Due to differences in the mathematics of linear equations for eigenvalue problems, some of the general theorems of the inviscid temporal problem related to the real wavenumbers and the complex frequency given in Sect. 2.1.2 are not applicable to the problem of spatial amplification.

2.3 Gaster's transformation

Compared to the temporal instability case, the spatial instability problem is complicated as α enters the Orr–Sommerfeld equation nonlinearly which leads to a nonlinear (polynomial) eigenvalue problem. However, Gaster's transformation (Gaster, 1962) outlined below provides a useful approximate relation between the temporal and spatial growth rates for two-dimensional waves. This makes it possible to estimate the spatial growth rates from the temporal ones with minimized efforts.

Let us consider the dispersion relation in the form

$$\mathcal{F}(\alpha,\omega)=0\,.$$

Its total differential is

$$\mathrm{d}\mathcal{F}=\frac{\partial\mathcal{F}}{\partial\alpha}\mathrm{d}\alpha+\frac{\partial\mathcal{F}}{\partial\omega}\mathrm{d}\omega+\frac{\partial\mathcal{F}}{\partial\mathrm{Re}}\mathrm{dRe}=0\,.$$

By denoting the neutral stability curve parameters as α_0, ω_0, and Re_0, we obtain

$$\begin{aligned}
\mathrm{d}\omega|_{\alpha_0} &= -\mathrm{dRe}\frac{\partial\mathcal{F}}{\partial\mathrm{Re}}\Big/\frac{\partial\mathcal{F}}{\partial\omega}\,,\\
\mathrm{d}\alpha|_{\omega_0} &= -\mathrm{dRe}\frac{\partial\mathcal{F}}{\partial\mathrm{Re}}\Big/\frac{\partial\mathcal{F}}{\partial\alpha}\,,\\
\mathrm{d}\omega|_{\mathrm{Re}_0} &= -\mathrm{d}\alpha\frac{\partial\mathcal{F}}{\partial\alpha}\Big/\frac{\partial\mathcal{F}}{\partial\omega}=\left.\frac{\partial\omega}{\partial\alpha}\right|_{\mathrm{Re}_0}\mathrm{d}\alpha\,.
\end{aligned}$$

It follows that

$$\mathrm{d}\omega|_{\alpha_0}=\mathrm{d}\alpha|_{\omega_0}\frac{\partial\mathcal{F}/\partial\alpha}{\partial\mathcal{F}/\partial\omega}=-\left(\left.\frac{\partial\omega}{\partial\alpha}\right|_{\mathrm{Re}_0}\right)\mathrm{d}\alpha|_{\omega_0}=-c_\mathrm{g}\mathrm{d}\alpha|_{\omega_0}\,,$$

where c_g is called the group velocity of disturbances, which is a complex number. To perform the desired spatial-to-temporal growth rate transformation, we put $\Delta\alpha=-\mathrm{i}\alpha_\mathrm{i}$ and obtain the relation

$$\Delta\omega=-\mathrm{i}c_\mathrm{g}\alpha_\mathrm{i}\,.$$

At the opposite transformation, we put $\Delta\omega=\mathrm{i}\omega$, which yields

$$\Delta\alpha=\mathrm{i}\omega/c_\mathrm{g}\,.$$

The expression for c_g can be found, e.g., by finite differences, performing two numerical calculations for close values of α. A more rigorous approach described in Sect. 8.2, however, utilizes adjoint functions, which are considered in Sect. 2.5 and are important in some other aspects of linear stability.

2.4 Squire theorem

When $\beta=0$, the Orr–Sommerfeld equation (2.9) reduces to describe only two-dimensional waves. Written in the form

$$\left[\left(U - \frac{\omega_{2D}}{\alpha_{2D}}\right)(\mathcal{D}^2 - \alpha_{2D}^2) - U_{yy} + \frac{i}{\alpha_{2D}\mathrm{Re}_{2D}}\left(\mathcal{D}^2 - \alpha_{2D}^2\right)^2\right]\hat{v} = 0\,,$$

it clearly shows equivalence to (2.9) written in the form

$$\left[\left(U - \frac{\omega}{\alpha}\right)(\mathcal{D}^2 - \varkappa^2) - U_{yy} + \frac{i}{\alpha\mathrm{Re}}\left(\mathcal{D}^2 - \varkappa^2\right)^2\right]\hat{v} = 0$$

if to put

$$\frac{\omega_{2D}}{\alpha_{2D}} = \frac{\omega}{\alpha}\,,$$
$$\alpha_{2D}^2 = \varkappa^2\,,$$
$$\mathrm{Re}_{2D} = \mathrm{Re}\frac{\alpha}{\alpha_{2D}} = \mathrm{Re}\frac{\alpha}{\varkappa}\,.$$

Hence, in the case of the real wavenumbers α and β, the three-dimensional modes have effective Reynolds numbers that are smaller compared to the corresponding two-dimensional waves. Accordingly, if a two-dimensional parallel shear flow loses linear stability at a certain Reynolds number to a two-dimensional normal mode with real streamwise wavenumber α_{2D}, then it loses linear stability to a three-dimensional mode with the same wavevector magnitude κ at higher Reynolds number; see, e.g., Fig. 2.2. Hence, to study linear stability of a parallel flow in time, it is sufficient to confine attention to the two-dimensional waves. This statement constitutes the *Squire theorem*. In other words, to find $\mathrm{Re_L}$, it is possible to consider only two-dimensional waves and to build the *neutral stability curve* formed by values $\omega_i(\alpha, \beta = 0, \mathrm{Re}) = 0$ instead of the neutral stability surface as shown in Fig. 2.4.

Note that the result of the Squire theorem with respect to the critical Reynolds number $\mathrm{Re_L}$ is independent of either temporal or spatial stability of the basic flow considered, because both α_i and ω_i for neutral disturbances are zero by definition. However, cases when the oblique waves amplify faster than their two-dimensional counterparts are not excluded: under certain conditions they can be the main source for the transition to turbulence. For example, situations in boundary layers, considered in Chap. 4, with the downstream-growing Reynolds number are possible where the two-dimensional wave has already left the instability region, but some oblique waves still amplify.

Finally, for solving the initial-value problem, e.g., in studying the development of wave packets and transient disturbances (see Chaps. 8 and 9), the solution of the linear problem is only a starting point, since it is necessary to know not only the eigenvalues but also the eigenfunctions $\hat{u}$ $\hat{v}$, and $\hat{w}$, which are different for the cases of two-dimensional and oblique waves. Then, to describe the temporal development of an arbitrary initial disturbance under homogeneous boundary conditions, it is necessary to know a complete set of the waves constituting it. Moreover, in this case, the attempt to reduce the initial-value problem *only* to searching for the eigenvalues and eigenfunctions $\hat{v}$ is not physically justified, as the Orr–Sommerfeld operator is not self-adjoint, and its eigenfunctions are not orthogonal.

2.5 Adjoint problem and bi-orthogonality of normal modes

Let us consider the homogeneous linearized Navier–Stokes momentum equations and the continuity equation describing the wave dynamics in a shear flow (2.3) written in the form:

$$\frac{\partial \mathbf{u}}{\partial t} + \mathscr{L}\mathbf{u} + \nabla p = 0\,,$$
$$\nabla \cdot \mathbf{u} = 0\,,$$

with homogeneous boundary conditions, where $\mathscr{L}$ is the linearized Navier–Stokes operator, which is expressed in the Cartesian coordinates $x \equiv x_1, y \equiv x_2, z \equiv x_3$ as

$$(\mathscr{L}\mathbf{u})_i = U_j \frac{\partial u_i}{\partial x_j} + u_j \frac{\partial U_i}{\partial x_j} - \frac{1}{\mathrm{Re}} \frac{\partial^2 u_i}{\partial {x_j}^2}\,, \qquad i = 1,2,3\,,$$

where for brevity the summation convention is utilized. Using the Lagrange identity, one of the fundamental identities in the theory of ordinary differential equations, we can now define the adjoint linearized Navier–Stokes operator $\mathscr{L}^\dagger$ through the scalar products defined as

$$\mathbf{p} \cdot \mathbf{s} = \int_{y_1}^{y_2} \mathbf{ps} dy\,,$$

where the values y_1 and y_2 bound the domain of the vectors $\mathbf{p}$ and $\mathbf{s}$:

$$\begin{pmatrix} \mathbf{u}^\dagger \\ p^\dagger \end{pmatrix} \cdot \left[\begin{pmatrix} \frac{\partial}{\partial t} + \mathscr{L} & \nabla \\ \nabla \cdot & 0 \end{pmatrix} \begin{pmatrix} \mathbf{u} \\ p \end{pmatrix} \right] + \begin{pmatrix} \mathbf{u} \\ p \end{pmatrix} \cdot \left[\begin{pmatrix} \frac{\partial}{\partial t} + \mathscr{L}^\dagger & \nabla \\ \nabla \cdot & 0 \end{pmatrix} \begin{pmatrix} \mathbf{u}^\dagger \\ p^\dagger \end{pmatrix} \right] = \frac{\partial (\mathbf{u}^\dagger \cdot \mathbf{u})}{\partial t} + \nabla \cdot \mathbf{j}\,,$$

where $\mathbf{q} = (\mathbf{u}, p)^T$ and $\mathbf{q}^\dagger = (\mathbf{u}^\dagger, p^\dagger)^T$ are suitably differentiable fields, which do not necessarily satisfy (2.3),

$$\mathscr{L}^\dagger \mathbf{u}^\dagger = U_j \frac{\partial u_i^\dagger}{\partial x_j} - u_j^\dagger \frac{\partial U_j}{\partial x_i} + \frac{1}{\mathrm{Re}} \frac{\partial^2 u_i^\dagger}{\partial {x_j}^2}\,, \qquad i = 1,2,3\,,$$

(which can be obtained using integration by parts) and the vector $\mathbf{j}$ is the bilinear concomitant depending on both $\mathbf{q}$ and $\mathbf{q}^\dagger$ so that

$$\mathbf{j}_j(\mathbf{q}^\dagger, \mathbf{q}) = \mathscr{S}_{ij}^\dagger u_i + \mathscr{S}_{ij} u_i^\dagger\,, \quad i,j = 1,2,3\,, \tag{2.14}$$

where

$$\mathscr{S}_{ij} = \left(U_j - \frac{1}{\mathrm{Re}} \frac{\partial}{\partial x_j} \right) u_i + p\delta_{ij}\,, \quad \mathscr{S}_{ij}^\dagger = \frac{1}{\mathrm{Re}} \frac{\partial u_i^\dagger}{\partial x_j} + p^\dagger \delta_{ij}\,.$$

The corresponding adjoint equations by definition are related to the part in square brackets of the second term in the Lagrange identity

$$\frac{\partial \mathbf{u}^\dagger}{\partial t} + \mathscr{L}^\dagger \mathbf{u}^\dagger + \nabla p^\dagger = 0\,,$$
$$\nabla \cdot \mathbf{u}^\dagger = 0\,,$$

where $\mathbf{u}^\dagger = (u^\dagger, v^\dagger, w^\dagger)^T$, and $p^\dagger$ are the adjoint velocity and pressure, respectively, and homogeneous boundary conditions are assumed to zero the concomitant term.

Next, let us consider the normal modes, whose velocities and pressure are proportional to $\exp[\mathrm{i}(\alpha x-\omega t)]$, and their adjoints are proportional to $\exp[-\mathrm{i}(\alpha^\dagger x-\omega^\dagger t)]$ and assume that their set is complete to form a Hilbert space. Substituting the corresponding expressions into the Lagrange identity, assuming $\mathbf{U}=(U,0,0)^T$, and noticing that the left-hand side is identically zero, we obtain the following bi-orthogonality condition

$$(\alpha-\alpha^\dagger)\mathbf{e_x}\cdot\mathbf{j}-(\omega-\omega^\dagger)\mathbf{u}^\dagger\cdot\mathbf{u}=0\,,$$

where $\mathbf{e_x}$ is the unit vector codirectional with the x-axis. These are the required bi-orthogonality conditions. When considering, for example, the spatial stability of a shear flow, $\omega=\omega^\dagger$ is given. Hence, the equality is possible either if $\alpha=\alpha^\dagger$, in which case we can put $\mathbf{e_x}\cdot\mathbf{j}=1$ without loss of generality, or if $\alpha\neq\alpha^\dagger$, in which case $\mathbf{e_x}\cdot\mathbf{j}=0$.

In a similar way one can obtain the adjoint Orr–Sommerfeld equation for adjoint disturbances $\hat{v}^\dagger$, which takes the form

$$\mathscr{L}^\dagger_{\mathrm{OS}}\hat{v}^\dagger=0\,,$$

with the boundary conditions $\hat{v}^\dagger=\hat{v}^\dagger_y=0$. Here

$$\mathscr{L}^\dagger_{\mathrm{OS}}=-\mathrm{i}(\alpha U-\omega)\left(\mathcal{D}^2-\varkappa^2\right)-2\mathrm{i}\alpha U_y\mathcal{D}-\frac{1}{\mathrm{Re}}\left(\mathcal{D}^2-\varkappa^2\right)^2\,.$$

The orthogonality relation between the modes $\hat{v}_i$ and $\hat{v}^\dagger_j$ can be now obtained directly from the corresponding Lagrange identity:

$$\left(\hat{v}^\dagger_j,\mathscr{L}_{\mathrm{OS}}\hat{v}_i\right)-\left(\mathscr{L}^\dagger_{\mathrm{OS}}\hat{v}^\dagger_j,\hat{v}_i\right)=0\,.$$

Integrating by parts yields the following relation for the modes developing in time[4]:

$$\mathrm{i}(\omega_i-\omega^\dagger_j)\int_{y_1}^{y_2}\hat{v}^\dagger_j(\mathcal{D}^2-\varkappa^2)\hat{v}_i\mathrm{d}y=0\,.$$

Hence, finally,

$$\int_{y_1}^{y_2}\hat{v}^\dagger_j(\mathcal{D}^2-\varkappa^2)\hat{v}_i\mathrm{d}y=C\delta_{ij}\,,\tag{2.15}$$

where δ_{ij} is the Kronecker delta and C is an arbitrary constant.

2.6 Completeness of solutions for the Orr–Sommerfeld and Squire equations

As has already been mentioned in Sect. 2.1, to describe the development of an arbitrary disturbance in a particular flow, it is necessary to know the complete set of the

[4] Using a similar procedure, one can obtain the orthogonality relation for normal modes developing in space.

waves constituting the disturbance. The completeness theorem, i.e., the existence of the complete set of *discrete* eigenvalues with corresponding eigenfunctions for the linearized Navier–Stokes equations, was proved for the plane Poiseuille flow by Schensted (1960), and later for *any* parallel internal flow by Yudovich (1965) and independently by DiPrima and Habetler (1969). For external parallel flows, the completeness theorem in such a formulation cannot be justified. In that case, to obtain a complete set of eigenfunctions, as it has been shown by Salwen and Grosch (1981) and Zhigulev and Tumin (1987), it is necessary to supplement the set of discrete modes by a *continuous* spectrum: to consider the so-called generalized eigenfunctions, i.e., the eigenfunctions corresponding to the Orr–Sommerfeld and Squire equations, which do not decay at infinity, but rather are only bounded. Physically, these generalized eigenfunctions are required to describe the behavior of disturbances in the free stream (including infinity).

Let us specify some general features of the modes in a semibounded flow. The Orr–Sommerfeld equation has four linearly independent fundamental solutions. It is easy to establish their asymptotic behavior at $y \to \infty$. First, we present the Orr–Sommerfeld equation (2.9) in the form

$$\left[\left(\varkappa\overline{U}-\omega\right)\left(\mathcal{D}^2-\varkappa^2\right)-\varkappa\overline{U}_{yy}+\frac{\mathrm{i}}{\mathrm{Re}}\left(\mathcal{D}^2-\varkappa^2\right)^2\right]\hat{v}=0\,,$$

where $\overline{U} = \alpha/\varkappa U$ is the basic flow velocity projection to the direction of the wavevector κ. In this form it is equivalent to the Orr–Sommerfeld equation for two-dimensional waves with the effective basic flow velocity $\overline{U}$ and the streamwise wavenumber $\varkappa$.

For simplicity, we assume that $\overline{U} \to 1$ beyond a narrow shear layer at the wall. The Orr–Sommerfeld equation is reduced at $y \to \infty$ to

$$\left[\left(\varkappa-\omega\right)\left(\mathcal{D}^2-\varkappa^2\right)+\frac{\mathrm{i}}{\mathrm{Re}}\left(\mathcal{D}^2-\varkappa^2\right)^2\right]\hat{v}=0\,. \tag{2.16}$$

After the standard substitution $\hat{v} = \mathrm{e}^{qy}$, it becomes a biquadratic algebraic equation

$$\left(\varkappa-\omega\right)\left(q^2-\varkappa^2\right)+\frac{\mathrm{i}}{\mathrm{Re}}\left(q^2-\varkappa^2\right)^2=0\,.$$

This equation has four roots, so that the general solution of (2.16) is

$$\hat{v} = C_1\varphi_1 + C_2\varphi_2 + C_3\varphi_3 + C_4\varphi_4\,, \tag{2.17}$$

where

$$\varphi_1 = \mathrm{e}^{\varkappa y}, \quad \varphi_2 = \mathrm{e}^{-\varkappa y}, \quad \varphi_3 = \mathrm{e}^{\varpi y}, \quad \varphi_4 = \mathrm{e}^{-\varpi y}\,,$$

and $\varpi = \sqrt{\varkappa^2 + \mathrm{i}\mathrm{Re}(\varkappa-\omega)}$.

Let us first consider the case where $\varkappa$ and ϖ are complex and $\mathrm{Real}(\varkappa) < 0$, $\mathrm{Real}(\varpi) < 0$. Then the only allowable bounded solution

$$\hat{v} = C_1\varphi_1 + C_3\varphi_3$$

exhibits $\hat{v} = 0$ as $y \to \infty$. The boundary conditions at $y = 0$ yield

$$C_1\varphi_1(0)+C_3\varphi_3(0)=0\,,$$
$$C_1\varphi_{1,y}(0)+C_3\varphi_{3,y}(0)=0\,.$$

For the existence of nonzero C_1 and C_3, it is necessary that the determinant of the system be zero:

$$\varphi_1(0)\varphi_{3,y}(0)-\varphi_3(0)\varphi_{1,y}(0)=0\,. \tag{2.18}$$

Thus, the spectrum of these waves is discrete; i.e., at a fixed ω and Re there is a number of $\varkappa_n$ $(n=1,2,3,\ldots)$ satisfying (2.18). They can both grow or decay in the direction of wave propagation as well as be neutral. The waves are near-wall waves corresponding to that discussed in Sect. 2.1: their amplitudes approach zero at large y, i.e., they have a maximum (or maxima) near the wall (usually in the shear layer). The requirement of disappearance of the disturbances at infinity guarantees that the flow relaxes to its undisturbed state in the free stream.

It is seen that finite, but nonzero disturbances exist at $y\to\infty$ when either $\varkappa$ or ϖ is purely imaginary. Let $\varkappa=\mathrm{i}q$, $q>0$ (note that this is possible only when disturbances developing in space are considered). If $\mathrm{Real}(\varpi)<0$, the solutions limited at $y\to\infty$ take the form

$$\hat{v}=C_1\varphi_1+C_2\varphi_2+C_3\varphi_3\,.$$

The boundary conditions $\hat{v}(0)=\hat{v}_y(0)=0$ result in the equations

$$C_1\varphi_1(0)+C_2\varphi_2(0)+C_3\varphi_3(0)=0\,,$$
$$C_1\varphi_{1,y}(0)+C_2\varphi_{2,y}(0)+C_3\varphi_{3,y}(0)=0\,.$$

Based on them, two constants, e.g., C_2 and C_3, can be expressed through the third one which is an arbitrary constant. To be specific, let us choose $C_1=1$, then

$$\hat{v}=\varphi_1+C_2\varphi_2+C_3\varphi_3\,.$$

A particular form of C_2 and C_3 for the other y coordinates is determined by the basic flow velocity distribution through the Orr–Sommerfeld equation and the boundary conditions. The solution exists at any q and, consequently, corresponds to the continuous spectrum. In the case under consideration, we have

$$\lim_{y\to\infty}\hat{v}=\mathrm{e}^{\mathrm{i}qy}+C_2\mathrm{e}^{-\mathrm{i}qy}\,.$$

It is possible to show that, as $y\to\infty$, the wave appears as a disturbance of vorticity equal to zero and a finite disturbance of pressure. Therefore, such waves are called *pressure waves*. Furthermore, these waves are standing, as their phase velocities are purely imaginary. The case $\varkappa=\mathrm{i}q$, $q<0$ can be considered similarly. Depending on the sign at q, the wave amplitudes decrease or grow in the direction of the wavevector κ.

Let now $\varpi=\pm\mathrm{i}q$, $q>0$. Then the equation for the wavenumbers α is

$$\varkappa^2+\mathrm{i}\varkappa\mathrm{Re}+q^2-\mathrm{i}\omega\mathrm{Re}=0\,,$$

whence, under the assumption that $\mathrm{Re}\gg1$,

$$\varkappa_1\approx\omega+\mathrm{i}q^2/\mathrm{Re}\quad\text{and}\quad\varkappa_2\approx-\omega-\mathrm{i}\mathrm{Re}\,.$$

The first wave has the real part of the phase velocity coinciding with the effective free-stream velocity $\overline{U}$

$$c_{\mathrm{r}} = \mathrm{Real}\left(\frac{\omega}{\varkappa}\right) = 1$$

and is damped weakly with x. The second wave propagates with the same velocity, but upstream ($c_{\mathrm{r}} = -1$), and strongly decays in the direction of its propagation. Their amplitude functions are

$$\hat{v} = C_2\varphi_2 + C_3\varphi_3 + C_4\varphi_4\,, \quad \hat{v} = C_1\varphi_1 + C_3\varphi_3 + C_4\varphi_4\,.$$

As in the previous case, two homogeneous boundary conditions at the wall determine two arbitrary constants through the third one. Thus, the considered waves exist at any q and, consequently, also constitute a part of the continuous spectrum. It could be shown that they are *vorticity waves.*

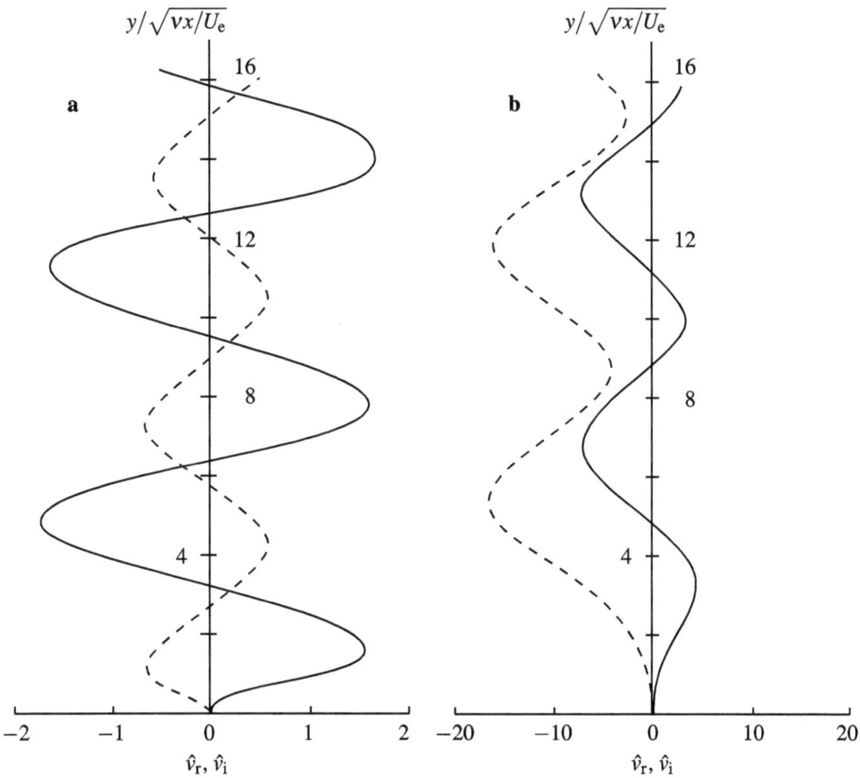

Fig. 2.5 Modes of the continuous spectrum for the Blasius boundary layer: pressure wave (**a**); vorticity wave (**b**). $\mathrm{Re_x} = 9 \times 10^4$, $\omega = 0.006$, $\beta = 0$, $q = 1$; (Zhigulev and Tumin, 1987)

The waves of the indicated classes constitute the complete set of solutions of the spatial Orr–Sommerfeld equation. In the temporal stability formulation, pressure waves are obviously absent, and the continuous spectrum consists only of vorticity waves. Figure 2.5 shows some examples of the real $\hat{v}_r$ and imaginary $\hat{v}_i$ parts of the eigenfunctions for pressure and vorticity waves obtained numerically.

Finally, the analysis of the homogeneous Squire equation shows that it also has a continuous spectrum of vorticity waves, the discrete spectrum being empty. Note that at $\alpha = 0$ the Orr–Sommerfeld equation reduces to

$$(\mathcal{D}^2 - \beta^2 - \mathrm{i}\omega\mathrm{Re})(\mathcal{D}^2 - \beta^2)\hat{v} = 0\,,$$

which also has no discrete solutions.

The observation that parallel flows are always stable to continuous spectrum components must be interpreted with caution. In particular, among these modes, there are weakly decaying low-frequency waves. Though such two-dimensional waves cannot penetrate deeply into the boundary layer, they can produce small quasi-stationary changes of the basic flow velocity distribution close to the shear-layer edge, where U_{yy} is small enough. This results in quasi-stationary inflections in the basic flow velocity profile capable of affecting the flow stability. Moreover, large-amplitude free-stream disturbances and the boundary-layer waves can interact nonlinearly. Additionally, as will be shown in Chap. 9, three-dimensional modes of the continuous spectrum can play a key role in the algebraic instability phenomenon in external flows.

Additional material for this chapter is given in the multimedia file No. 2, which supplements the book, titled 'Formulation of linear hydrodynamic stability problems.'

References

DiPrima RC, Habetler GJ (1969) A completeness theorem for nonself-adjoint eigenvalue problems in hydrodynamic stability. Arch Rat Mech Anal 34:218–227

Fjørtoft R (1950) Amplification of integral theorems in deriving criteria for instability for laminar flows and the baroclinic circular vortex. Geofus Publ 17(6):1–52

Gaster M (1962) A note on the relation between temporally-increasing and spatially-increasing disturbances in hydrodynamic stability. J Fluid Mech 14:222–224

Howard LN (1961) Note on a paper by John. W. Miles. J Fluid Mech 10:509–512

Orr WM (1907a) The stability or instability of the steady motions of a perfect liquid and of a viscous liquid. Part 1: A perfect liquid. Proc R Irish Acad A 27:9–68

Orr WM (1907b) The stability or instability of the steady motions of a perfect liquid and of a viscous liquid. Part 2: A viscous liquid. Proc R Irish Acad A 27:69–138

Prandtl L (1921) Bemerkungen über die Entstehung der Turbulenz. Z Angew Math Mech 1:431–436

Prandtl L (1922) Bemerkungen über die Entstehung der Turbulenz. Phys Z 23:19–25

Lord Rayleigh (1880) On the stability, or instability, of certain fluid motions. Proc Lond Math Soc 9:57–70

Salwen H, Grosch CE (1981) The continuous spectrum of the Orr–Sommerfeld equation. Part 2. Eigenfunction expansion. J Fluid Mech 104:445–465

Schensted IV (1960) Contributions to the theory of hydrodynamic stability. PhD thesis, University of Michigan

Sommerfeld A (1908) Ein Beitrag zur hydrodynamischen Erklärung der turbulenten Flüssigkeitsbewegungen. In: Atti IV Congr. Internaz. Mat., Acad. dei Lincei, Rome, vol 3, pp 116–124

Tollmien W (1935) Ein algemeines kriterium der instabilität laminarer geschwindigkeitsverteilungen. Nachr Ges Wiss Göttingen (Matematik) 1(5):79–114, translated as NACA TM 792, 1936

Yudovich VV (1965) On stability of stationary flows of viscous incompressible fluid. Dokl Akad Nauk 161(5):1037–1040

Zhigulev VN, Tumin AM (1987) Onset of turbulence. Dynamical theory of excitation and development of instabilities in boundary layers. Nauka. Sib. Otd., Novosibirsk, In Russian.

Further Reading

Lin CC (1955) The theory of hydrodynamic stability. Cambridge University Press, Cambridge

Prandtl L (1935) The mechanics of viscous fluids. In: Durand WF (ed) Aerodynamic Theory, vol 3, Springer–Verlag, Berlin

Saffman PG (1962) On the stability of a laminar flow to a dusty gas. J Fluid Mech 13:120–128

Sattinger DH (1970) The mathematical problem of hydrodynamic stability. J Math Mech 20(9):797–817

Schlichting H, Gersten K (2000) Boundary layer theory, 8th edn. Springer–Verlag, Berlin

Squire HB (1933) On the stability for three-dimensional disturbances of viscous fluid between parallel walls. Proc R Soc Lond A 142:621–628

Part II
Generic problems

Chapter 3
Instability of plane parallel flows

In this chapter the essentials of the linear-stability analysis and corresponding experimental results for such truly parallel internal flows as the plane Couette and the plane Poiseuille flow are presented. For an interested reader, a set of simple MATLAB functions and exercises are provided helping in understanding various physical and computational aspects of linear stability for shear flows.

3.1 Plane Couette flow

A shear flow with a linear velocity profile, i.e., the so-called plane Couette flow occurs, e.g., between two parallel walls moving with equal absolute velocities U_a in the opposite directions, see Fig. 3.1. The flow is linearly stable at $Re_L \to \infty$. Furthermore, numerical experiments show that it is also stable at any finite Reynolds number.

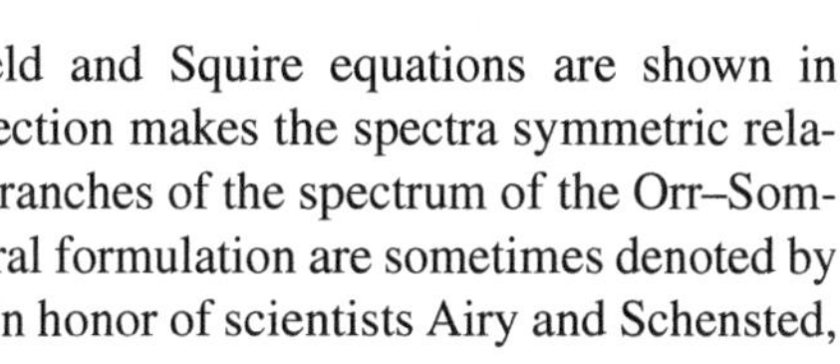

Fig. 3.1 The plane Couette flow

Typical spectra of the Orr–Sommerfeld and Squire equations are shown in Fig. 3.2. The absence of a preferential direction makes the spectra symmetric relative to the axis $\omega_r = 0$ (or $\alpha_r = 0$). Three branches of the spectrum of the Orr–Sommerfeld and Squire equations in the temporal formulation are sometimes denoted by the letters A for $|\omega_r| \to 1$ and S for $\omega_r = 0$ in honor of scientists Airy and Schensted, respectively.

Hence, the simplicity of the basic flow velocity profile is in this case misleading, as the linear stability of the flow rises the question on some other routes of

A.V. Boiko et al., *Physics of Transitional Shear Flows*, Fluid Mechanics and Its Applications 98, DOI 10.1007/978-94-007-2498-3_3,

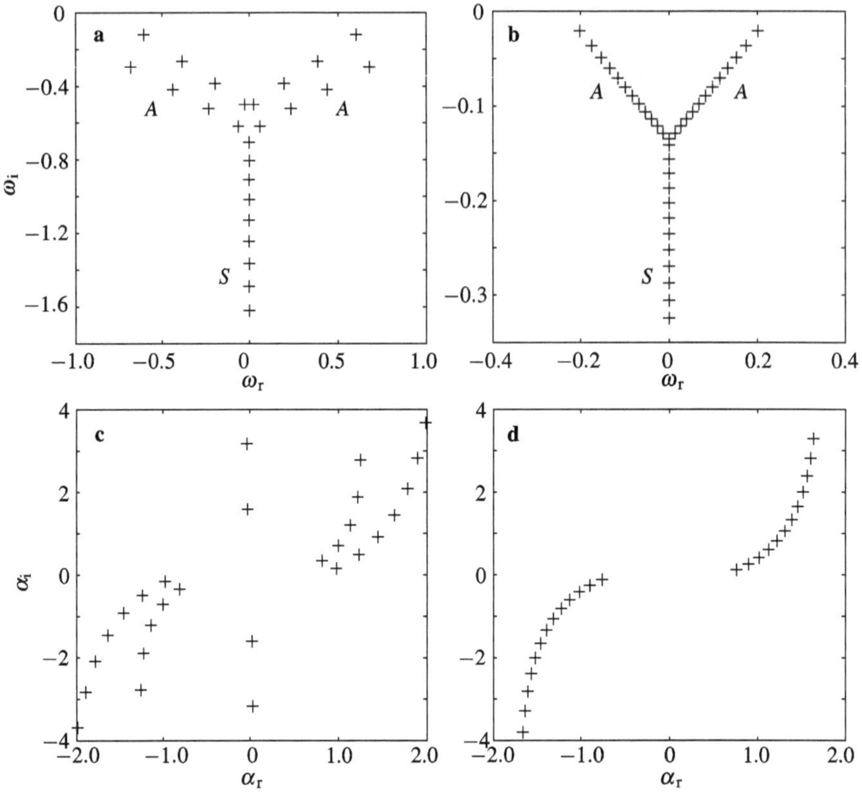

Fig. 3.2 Leading eigenvalues in temporal formulation for the Orr–Sommerfeld (**a**) and Squire (**b**) equations for a plane Couette flow at $\mathrm{Re}_h = U_a h/\nu = 1000$, $\alpha = 1$, $\beta = 0$, $\beta = 0$. The same for spatial formulation (**c**, **d**) at $\mathrm{Re}_h = 1000$, $\omega = 0.6$, $\beta = 0$

its turbulization. Evidently, the experimentally observed transition to turbulence in laboratory Couette flow can be related to imperfection of the real flow, in which case inflections in the basic flow velocity profile can cause the inviscid instability, or to some nonlinear phenomena. However, even the ideal plane Couette flow loses its monotonic and conditional stabilities at finite Reynolds numbers, i.e., the kinetic energy of some disturbances at higher Reynolds numbers can experience a growth. A theoretical and numerical description of transient disturbances due to the loss of the monotonic stability is given in Chap. 9.

3.2 Plane Poiseuille flow

A plane Poiseuille flow ia a classical example of a parallel flow subject to viscous linear hydrodynamic instability. The first theoretical results on the stability of a

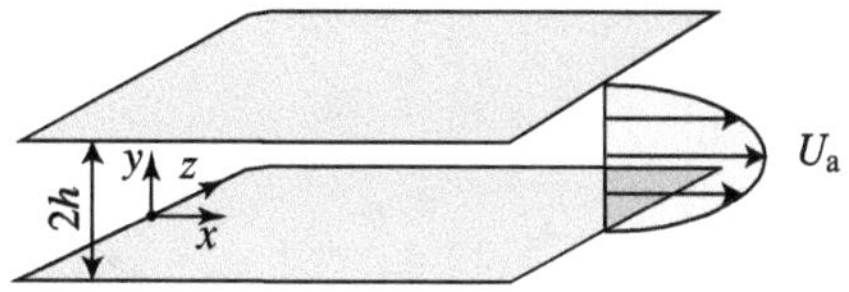

Fig. 3.3 The Poiseuille flow in a plane channel

plane Poiseuille flow were presented by Heisenberg (1924), who calculated the position of a part of the neutral stability curve using an asymptotic method. Lin (1955) advanced further the asymptotic analysis by finding that instability occurs above a critical Reynolds number approximately equal to $\mathrm{Re}_{\mathrm{L},h} = U_a h \nu \approx 5300$ at $\alpha \approx 1$. The Reynolds number is based here on the maximum flow velocity in the channel U_a and on the half distance between the walls[1] $h/2$ (Fig. 3.3). Since that time, $\alpha = 1$ has been widely used as a test value. A general theoretical proof of absolute stability for the Poiseuille flow at $\mathrm{Re} \gg 1$ (as well as for the boundary-layer velocity profiles without the inflection point) was done first by Iordanskii and Kulikovskii (1965). Hence, the flow instability considered in this section is of convective nature.

3.2.1 Numerical results

Thomas (1953) for the first time solved the problem using a numerical method on a computer and found $\mathrm{Re}_{\mathrm{L},h} = 5780$ at $\alpha = 1.02$. Orszag (1971) applied a very effective spectral method utilizing the fast convergence of the Chebyshev polynomial approximation to solving the instability problem. He obtained $\mathrm{Re}_{\mathrm{L},h} = 5772.22$ at $\alpha = 1.02056$. The spectral approach appeared to be highly valuable for the study of linear stability. It produces matrices of a moderate size, which provide an approximation of the set of leading eigenvalues and corresponding eigenfunctions in one run with high accuracy.

The leading parts of the temporal and spatial spectra of the Orr–Sommerfeld and Squire equations are shown in Fig. 3.4. In the temporal approach, the eigenvalues are located along three main branches denoted usually as A ($\omega_r \to 0$), P ($\omega_r \to \infty$), and S ($\omega_r \approx 2/3$) in honor of Airy, Pekeris and Schensted, respectively. The number of modes on a branch depends on the values of Re, α, and β. For the given set of parameters, there is an unstable disturbance with $\omega_i > 0$. Note also that the symmetry of the basic flow velocity profile leads to separation of the normal modes into two subsets with symmetric (even) and antisymmetric (odd) eigenfunctions $\hat{v}$ with respect to the axis $y = 0$. The corresponding modes are denoted by different symbols in Fig. 3.4.

Typical results of the linear stability calculation on the basis of the Orr–Sommerfeld equation are shown in Fig. 3.5. The outer lines represent the neutral stability curve. The curve can be interpreted in terms of multiple-valued functions as consisting of two branches. Branch I, shown in Fig. 3.5**a**, is also called the lower branch of

[1] Sometimes $\overline{\mathrm{Re}} = 2/3\mathrm{Re}$, based on the mean basic flow velocity, is used.

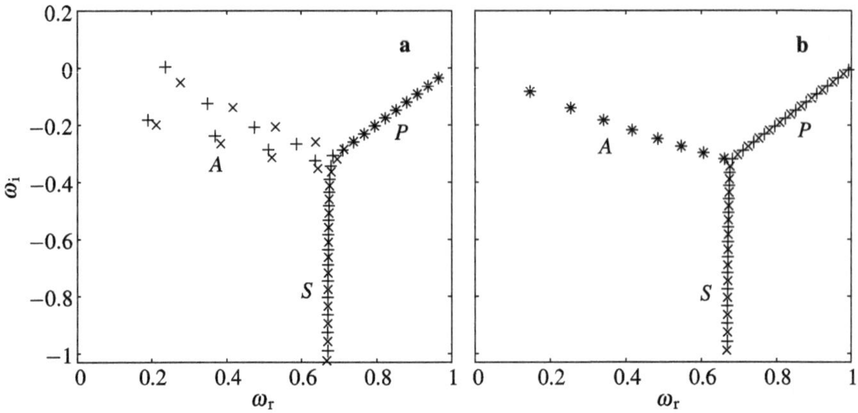

Fig. 3.4 Leading temporal eigenvalues for the Orr–Sommerfeld (**a**) and Squire (**b**) equations obtained by the pseudospectral method with 120 Chebyshev polynomials. The leading even (+) and odd (×) modes are shown. $\mathrm{Re}_h = 10000$, $\alpha = 1$, $\beta = 0$ (calculated by A. V. Boiko)

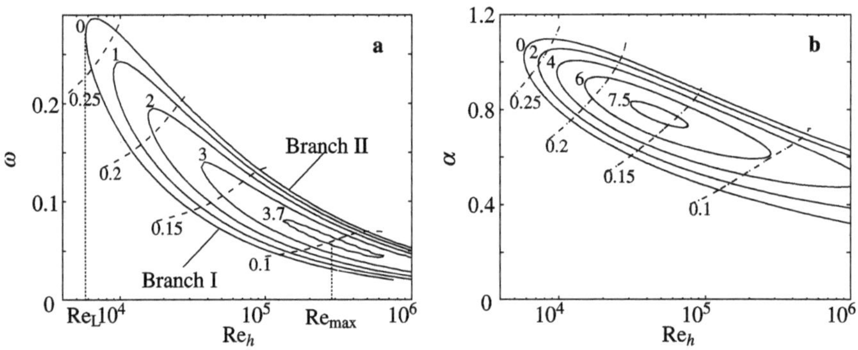

Fig. 3.5 Real parts of phase velocities $c_r = \omega/\alpha_r$ (*dashed lines*) as well as neutral stability curves and isolines of spatial $-\alpha_i \times 10^2$ (**a**) and temporal $\omega_i \times 10^3$ (**b**) growth rates (*solid lines*) for two-dimensional Tollmien–Schlichting waves in the plane Poiseuille flow (calculated by A. V. Boiko)

the neutral stability curve. Branch II is called the upper branch. The same definitions are valid for the neutral stability curve in Fig. 3.5**b**.

The neutral stability curves are perhaps the most important (but not unique) parameters describing the instability in the Re–α or Re–ω-planes. In particular, the area inside the curves corresponds to amplifying disturbances. The isolines or values with nonzero growth rates and the Reynolds numbers of the maximum amplification Re_{max} shown in Fig. 3.5**a** can also be of interest.

3.2.2 Experimental linear stability investigations

At first glance, the linear instability of a plane channel flow is quite easy to describe, since it is strictly parallel and bounded. This statement seems true for the numerical description, but it is not so straightforward as regards experimental studies. For a long time there was a lack of empirical results suitable for comparison. This is explained by certain difficulties at accurate experimental modeling of the Poiseuille flow between two infinite plates and the fact that the flow is only metastable at $Re < Re_L$. In the majority of early studies, the inflow conditions, the basic flow uniformity, and the turbulence level were insufficiently controlled, which produced large scattering of the transition Reynolds number Re_T.

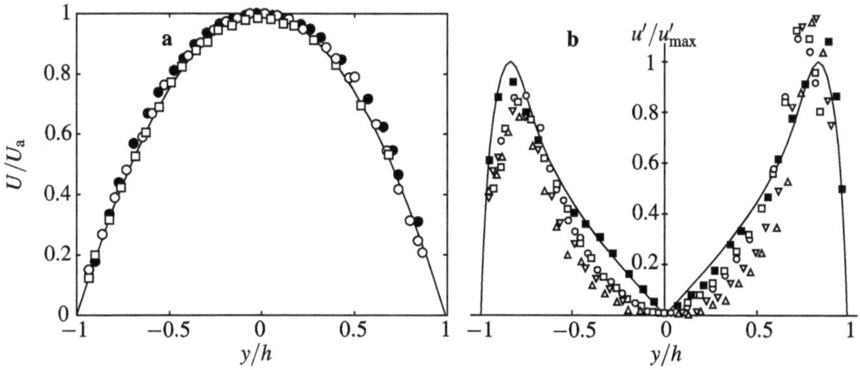

Fig. 3.6 Mean velocity profiles for the plane Poiseuille flow (Kozlov and Ramazanov, 1982) (**a**) and profiles of velocity oscillations $\omega = 0.27$ (**b**): $Re_h = 3000$ (○), 4000 (□), 5000 (▽), 6000 (△) (Kozlov and Ramazanov, 1982), 4000 (■) (Nishioka et al, 1975); calculations at $Re_h = 4000$ (*solid line*); $u'_{max} = \max_y[u'(y)]$ (Kozlov and Ramazanov, 1982)

Finally, the experiments of Nishioka et al (1975) and Kozlov and Ramazanov (1982) were able to confirm definitively the applicability of the linear stability theory for the plane Poiseuille flow. In these studies, channels with large width-to-height ratios and smooth elongated contractors at the channel entrance were used. Special care was taken to reduce the levels of background disturbances inside the channels to 0.05% (Nishioka) and 0.1% of U_a (Kozlov and Ramazanov), and, thus, to keep the flow laminar to $Re_{T,h} = 8000$ and 7000, respectively. The experiments allowed a detailed study of spatial development of disturbances generated in a control manner at various frequencies and over a wide range of Reynolds numbers ($Re_h = 3000$–8000) as well as obtaining points at the neutral stability curve. Figure 3.6**a** with the experimental mean and theoretical basic flow velocity profiles for $Re_h = 3000$, 4000, and 5000 assures that the flow in the channel is close to the plane Poiseuille flow.

The profiles of excited small-amplitude disturbances measured as the root-mean-square of the periodic signal amplitude (denoted throughout the book by a prime in contrast to theoretical values) are shown in Fig. 3.6**b**. The small asymmetry probably results from weak distortions introduced by a disturbance generator and a hot-wire probe used to measure the flow velocity in the experiment. At the channel centerline, a 180° phase jump occurs, which indicates the antisymmetric nature of the streamwise velocity oscillations (corresponds to a symmetric shape of wall-normal disturbance velocity).

Similarly, Fig. 3.7**a** shows the amplification rate $-\alpha_{\rm i}$ as a function of the frequency. Figure 3.7**b** shows the neutral stability curve. The symbols corresponding to the growing perturbations are filled; those of decaying waves are empty.

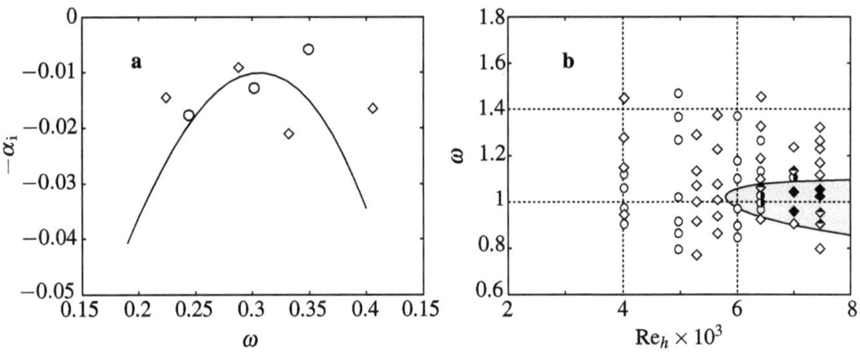

Fig. 3.7 Amplification rate as a function of frequency for the plane Poiseuille flow at $\mathrm{Re}_h = 4000$ (**a**): experiments (*symbols*) and linear stability theory (*solid line*). Stability boundary of plane Poiseuille flow for small disturbances (**b**): growing disturbances (*filled symbols*); nearly neutral disturbances (*half-filled symbols*); decaying disturbances (*open symbols*); linear stability theory (*solid line*). Data of Nishioka et al (1975) (◊) and Kozlov and Ramazanov (1982) (○)

As a whole, the experiments showed that a laminar plane Poiseuille flow with a proper control of disturbances at the channel entrance can be implemented up to Reynolds numbers $\mathrm{Re}_h = 7000$–8000. Small-amplitude waves do not experience streamwise growth up to the critical Reynolds number $\mathrm{Re_L}$, as is well-predicted by the linear stability theory.

3.3 Method of linear stability calculations

Below we describe the collocation technique for solving the Orr–Sommerfeld equation as applied to the plane Poiseuille flow. It is based on interpolation of the solution to be found by trial polynomial functions as

$$u \approx \sum_{i=1}^{n} \tilde{u}_i \phi_i \,,$$

where $\tilde{u}_i$ are the values of the function u in a set of interpolation points referred to as collocation points or nodes (knots) and ϕ_i are the known or basis functions, whose set is complete. The interpolation implies that $\phi_k = 1$ at the node k and $\phi_k = 0$ at the other nodes (but not necessarily between them). Such basis functions are specified by the elementary interpolation Lagrange polynomials.

For solving problems with nonperiodic solutions, the roots of the second-kind Chebyshev polynomials (Gauss–Lobatto nodes) $\varsigma_i = \cos[\pi i/(n-1)]$, where $i = 0,\ldots,n-1$, are frequently used as the interpolation nodes. If u is infinitely differentiable in the interval $y \in [-1;\,1]$, then the method provides high accuracy of the result with the use of a relatively small number of nodes.

The matrix of the discrete analog of the first derivative $\mathbf{D}_n$ is defined as

$$(\mathbf{D}_n u)_i = \sum_{j=0}^{n-1} (D_n)_{ij} u_j\,, \quad i = 0,\ldots,n-1\,,$$

where

$$(\mathbf{D}_n)_{ij} = \begin{cases} \frac{\bar{c}_i}{\bar{c}_j}\frac{(-1)^{i+j}}{\varsigma_i - \varsigma_j}\,, & i \neq j\,, \\ \frac{-\varsigma_j}{2\left(1-\varsigma_j^2\right)}\,, & 1 \le i = j \le n-2\,, \\ \frac{2(n-1)^2+1}{6}\,, & i = j = 0\,, \\ -\frac{2(n-1)^2+1}{6}\,, & i = j = n-1\,, \end{cases}$$

and

$$\bar{c}_j = \begin{cases} 2\,, \; j = 0, n-1\,, \\ 1\,, \; j = 1,\ldots,n-2\,. \end{cases}$$

Higher-order derivative matrices can be obtained either by matrix multiplications or by more accurate (due to a smaller number and proper ordering of operations) recurrent relations.

The Orr–Sommerfeld equation (2.9) is then approximated as

$$\left[\operatorname{diag}(\alpha U - \omega)\left(\mathbf{D}_n^2 - \varkappa^2 \mathbf{I}_n\right) - \operatorname{diag}(\alpha \mathbf{D}_n^2 U) + \frac{\mathrm{i}}{\mathrm{Re}}\left(\mathbf{D}_n^2 - \varkappa^2 \mathbf{I}_n\right)^2\right]\hat{v} = 0\,, \quad (3.1)$$

where $\mathbf{I}_n$ is the identity matrix of order $n \times n$. To solve the equation, we need to account for the double conditions $\hat{v} = \mathbf{D}_n \hat{v} = 0$ at both boundaries. However, this is not a unique procedure, which is also quite delicate. Improper application of the boundary conditions can lead to appearance of the so-called *spurious modes* exhibiting apparent flow instability, whose origin is, however, purely numeric.

The easiest task is the application of the Dirichlet boundary conditions $\hat{v} = 0$, which is accomplished by removing boundary rows and columns of the resulting matrices. One of the tricks to apply the Neumann boundary conditions $\mathbf{D}_n \hat{v} = 0$ consists in modifying $\mathbf{D}_n^4$ in (3.1). To this end, let us consider the polynomial $\tilde{q}$ of degree $\le n+2$ interpolating $\hat{v}$ at y_j, so that $\tilde{q}(\pm 1) = \tilde{q}_y(\pm 1) = 0$. The latter condition can be written as

$$\hat{v}_j = \tilde{q}(y_j) = \left(1 - y_j^2\right) q(y_j)\,,$$

where q is a polynomial of degree less than or equal to n with $q(\pm 1) = 0$. Differentiating the expression four times yields

$$\mathcal{D}^4\hat{v}_j = \mathcal{D}^4\tilde{q}(y_j) = \left(1-y_j^2\right)\mathcal{D}^4 q(y_j) - 8(y_j)\mathcal{D}^3 q(y_j) - 12\mathcal{D}^2 q(y_j)\,.$$

Hence, taking into account that $q(y_j) = v_j/(1-y_j^2)$, the matrix of the discrete analog of the highest derivative is modified accordingly:

$$\mathbf{D}_n^4 \to \left[\mathrm{diag}\left(1-y_j^2\right)\mathbf{D}_n^4 - 8\mathrm{diag}(y_j)\mathbf{D}_n^3 - 12\mathbf{D}_n^2\right]\mathrm{diag}\left(\frac{1}{1-y_j^2}\right)\,.$$

The first four collocation derivatives, coordinates of the Gauss–Lobatto nodes and integration metric coefficients (weights) are returned in a single container by MATLAB function `chder1` given below. The weights, which will be used later in this book, are given for the so-called Clenshaw–Curtis quadrature, which is the formula of the optimal order based on the fixed set of the Chebyshev nodes.

```
function D=chder1(n)
% CHDER1 returns 1st 4 derivative matrices nxn in
% D{1:4}, coordinates of the Gauss-Lobatto nodes in
% D{5} and integration metric w to find
% Int[f(y)dy]=Sum[f(y)*w] and scalar products in D{6}.

D{1}=gallery('chebspec',n,0);
D{2}=D{1}*D{1};
D{3}=D{1}*D{2};
D{4}=D{1}*D{3};

Phi=pi*(0:n-1)'/(n-1);
D{5}=cos(Phi);

% Modify the 4th derivative
Y=1-D{5}.^2;
D{4}=(diag(Y)*D{4}-diag(8*D{5})*D{3}...
     -12*D{2})*diag([0;1../Y(2:n-1);0]);

% Calculate the weight
i=2:n-1;
k=1:floor(n/2)-1;
[D{6}(1),D{6}(n)]=deal(1/((n-1)^2-mod(n,2)));
D{6}(i)=...
   2*(1-2*sum(cos(2*Phi(i)*k)/diag(4*k.^2-1),2)-...
   cos((n-1)*Phi(i))/((n-1)^2-1)*mod(n,2))/(n-1);
```

Function `PoiseuilleT` implements the solution algorithm to solve the problem of linear stability of the plane Poiseuille flow in time with the help of the QZ-algorithm. One of the features in solving eigenvalue problems numerically is accurate resolution of only relatively smooth functions when using only limited number of nodes. In the case under consideration, they usually correspond to the leading eigenvalues. Hence, the Tollmien–Schlichting wave is sorted out by tracing the convergence of solutions to fixed values, as the matrix dimension grows.

```
function [e,v]=PoiseuilleT(a,b,Re,U,D)
% POISEUILLE solves the Orr-Sommerfeld equation
% in temporal formulation for plane Poiseuille flow.
%     a  - streamwise wavenumber
%     b  - transverse   wavenumber
%     Re - Reynolds number
%     U  - basic velocity profile
%     D  - derivatives
%     e  - vector of calculated eigenvalues
%     v  - matrix of calculated eigenvectors

I=eye(length(D{1})); % Identity matrix
K2=(a^2+b^2)*I;      % Squared wavevector length

A=a*diag(U)*(D{2}-K2)+2*a*I+...
  1i/Re*(D{4}-2*K2*D{2}+K2^2);
B=D{2}-K2;

A=A(2:end-1,2:end-1);% Dirichlet boundary conditions
B=B(2:end-1,2:end-1);

[v,e]=eig(B\A);
e=diag(e);
```

When considering the instability in space, the real frequency ω is given, and (3.1) is a nonlinear eigenvalue problem for α. The problem is nonlinear in the sense that the Orr–Sommerfeld equation can be written as a polynomial, rather than a linear function of α:

$$P_4(\alpha)\hat{v} = \left(\mathbf{C}_0\alpha^4 + \mathbf{C}_1\alpha^3 + \mathbf{C}_2\alpha^2 + \mathbf{C}_3\alpha^1 + \mathbf{C}_4\right)\hat{v} = 0\,,$$

where the matrix coefficients are

$$\begin{aligned}
\mathbf{C}_0 &= \mathbf{I}_n\,,\\
\mathbf{C}_1 &= \mathrm{iRediag}\,(U)\,,\\
\mathbf{C}_2 &= -\mathrm{iRe}\omega\mathbf{I}_n - 2\left(\mathbf{D}_n^2 - \beta^2\mathbf{I}_n\right)\,,\\
\mathbf{C}_3 &= \mathrm{iRe}\left[\mathrm{diag}\left(\mathbf{D}_n^2 U\right) - \mathrm{diag}(U)\left(\mathbf{D}_n^2 - \beta^2\mathbf{I}_n\right)\right]\,,\\
\mathbf{C}_4 &= \left(\mathbf{D}_n^4 - 2\beta^2\mathbf{D}_n^2 + \beta^4\mathbf{I}_n\right)^2 + \mathrm{iRe}\omega\left(\mathbf{D}_n^2 - \beta^2\mathbf{I}_n\right)\,.
\end{aligned}$$

There are several ways to attack the problem. Probably the simplest one consists in transforming it to a linear problem of larger dimension by forming the so-called companion matrices, which are not actually unique. One of their implementations is given below:

$$\left[\begin{pmatrix} -\mathbf{C}_1 & -\mathbf{C}_2 & -\mathbf{C}_3 & -\mathbf{C}_4 \\ \mathbf{I}_n & 0 & 0 & 0 \\ 0 & \mathbf{I}_n & 0 & 0 \\ 0 & 0 & \mathbf{I}_n & 0 \end{pmatrix} - \alpha \begin{pmatrix} \mathbf{C}_0 & 0 & 0 & 0 \\ 0 & \mathbf{I}_n & 0 & 0 \\ 0 & 0 & \mathbf{I}_n & 0 \\ 0 & 0 & 0 & \mathbf{I}_n \end{pmatrix}\right] \begin{pmatrix} \alpha^3 \hat{v} \\ \alpha^2 \hat{v} \\ \alpha^1 \hat{v} \\ \hat{v} \end{pmatrix} = 0 \,.$$

This equation constitutes a generalized eigenvalue problem for α, which can be solved completely by the QZ-algorithm. Function `PoiseuilleS` implements the solution algorithm for the problem of linear stability of the plane Poiseuille flow. However, as the problem dimension is quadruplicated with respect to the problem of stability in time, it is often more efficient to find a subset of the leading eigenvalues including the Tollmien–Schlichting wave by a variant of the iterative Arnoldi algorithm.

```
function [e,v]=PoiseuilleS(o,b,Re,U,D)
% POISEUILLE solves the Orr-Sommerfeld equation
% in spatial formulation for plane Poiseuille flow.
%    o  - frequency
%    b  - transverse wavenumber
%    Re - Reynolds number
%    U  - basic velocity profile
%    D  - derivatives
%    e  - vector of calculated eigenvalues
%    v  - matrix of calculated eigenvectors

I=eye(length(D{1})); % Identity matrix
D2=D{2}-b^2*I;       % Useful expression

C0=I;
C1=diag(1i*Re*U);
C2=-1i*o*Re*I-2*D2;
C3=1i*Re*(diag(D{2}*U)-diag(U)*D2);
C4=(D{4}-2*b^2*D{2}+b^4*I)+1i*o*Re*D2;

C0=C0(2:end-1,2:end-1);% Dirichlet boundary conditions
C1=C1(2:end-1,2:end-1);
C2=C2(2:end-1,2:end-1);
C3=C3(2:end-1,2:end-1);
C4=C4(2:end-1,2:end-1);

[v,e]=polyeig(C4,C3,C2,C1,C0);
```

Additional material for this chapter is given in the multimedia file No. 3, which supplements the book, titled ‘Instability of plane parallel flows.’

Exercises

3.1. Write a program for numerical solution of the Orr–Sommerfeld equation for the plane Poiseuille flow at given values of α, β, and Re, which finds separately even and odd modes. Hint: for simplicity, use even number of nodes. Compare the results of the calculations with those given in Fig. 3.4**a**. How do the results change if $\beta \neq 0$?

3.2. Write a program for numerical solution of the homogeneous normal vorticity equation (2.12) (with a zero right-hand side) for the plane Poiseuille flow at given values of α, β, and Re. Compare the results of the calculations with those given in Fig. 3.4**b**.

3.3. Compare the leading eigenvalues obtained with the programs in Exercises 3.1 and 3.2 with those shown in Fig. 3.4 at $n = 40, 70, 100, 150$. How does the spectrum change with growth of the matrix dimension?

3.4. Design and implement algorithms to obtain and draw the growth rates (both in temporal and spatial formulation) and phase velocities of the leading normal modes for the plane Poiseuille flow as shown in Fig. 3.5. Hint: use MATLAB function `contour` to draw the isolines.

References

Heisenberg W (1924) Über Stabilität und Turbulenz von Flüssigkeitsströmen. Ann Phys 74:577–627, translated as NACA TM 1291, 1951

Iordanskii SV, Kulikovskii AG (1965) On absolute stability of some plane parallel flows at high Reynolds numbers. Zh Èksper Teoret Fiz 49(10):1326, In Russian

Kozlov VV, Ramazanov SP (1982) An experimental investigation of the stability of Poiseuille flow. Fluid Mech – Soviet Res 11(4):41–46

Lin CC (1955) The theory of hydrodynamic stability. Cambridge University Press, Cambridge

Nishioka M, Iida S, Ichikawa Y (1975) An experimental investigation of the stability of plane Poiseuille flow. J Fluid Mech 72:731–751

Orszag SA (1971) Accurate solution of the Orr–Sommerfeld stability equation. J Fluid Mech 50:689–703

Thomas LH (1953) The stability of plane Poiseuille flow. Phys Rev 91:780–783

Further Reading

Davies SJ, White CM (1928) An experimental study of the flow of water in pipe of rectangular section. Proc R Soc Lond A 19:92–107

Narayanan MA, Narayana T (1967) Some studies on transition from laminar to turbulent flow in a two-dimensional channel. Z Angew Math Phys 18:642–650

Patel VC, Head MR (1969) Some observations on skin friction and velocity profiles in fully developed pipe and channel flows. J Fluid Mech 38:181–201

Sherlin GC (1960) Behaviour of isolated disturbances superimposed on laminar flow in a rectangular pipe. J Res N B S, Sec A, Phys and Chem 64A:281–289

Trefethen LN (2000) Spectral methods in Matlab. Society for Industrial and Applied Mathematics, Philadelphia, USA

Chapter 4
Instability of the flat-plate boundary layer

The linear stability of the Blasius boundary layer which is a quasi-parallel rather than a strictly parallel flow is addressed in this chapter. At high Reynolds numbers, the wall-normal velocity V is small and the growth of the boundary layer is slow compared to distances characteristic for the laminar–turbulent transition induced by small disturbances. Hence, it is not too much surprising that the parallel-flow approximation works quite satisfactorily for two-dimensional waves observed in experiments.

4.1 Historical notes

For the first time Tollmien (1929), and then Schlichting (1933, 1935) were able to calculate the neutral stability curve for the Blasius boundary layer in the parallel-flow approximation. The least critical Reynolds number for the flat-plate boundary layer, gained from modern numerical solutions of the linear stability problem, is

$$\mathrm{Re}_{\mathrm{L},\delta^*} = \frac{U_\mathrm{e}\delta^*}{\nu} \approx 520\,,$$

where the displacement thickness is defined as

$$\delta^* = \int_0^\infty \left(1 - \frac{U(y)}{U_\mathrm{e}}\right) \mathrm{d}y$$

and U_e is the local external-flow velocity. For the Blasius boundary layer, $\delta^* = 1.72\sqrt{\nu x/U_\mathrm{e}} \approx 1.72x/\sqrt{\mathrm{Re}_\mathbf{x}}$, so that $\mathrm{Re}_{\delta^*} \approx 1.72\sqrt{\mathrm{Re}_\mathbf{x}}$.

The boundary-layer growth, see Fig. 4.1, leads to a specific behavior of instability waves during their propagation downstream. Let us consider a disturbance with dimensionless frequency $\omega = 2\pi f \delta^*/U_\mathrm{e}$, where f is a corresponding dimensional frequency. Due to a continuous change of the characteristic boundary layer scale δ^* with x, ω also experiences a continuous change. This is the reason for introducing the so-called frequency parameter

A.V. Boiko et al., *Physics of Transitional Shear Flows*,
Fluid Mechanics and Its Applications 98, DOI 10.1007/978-94-007-2498-3_4,

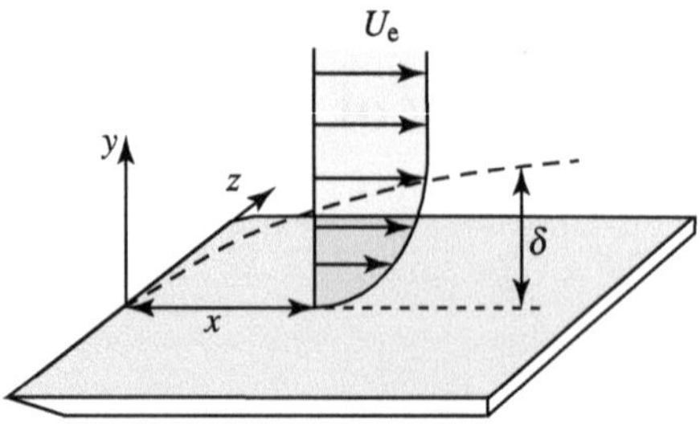

Fig. 4.1 Boundary layer over a flat plate

$$F = \frac{2\pi f \nu}{U_e^2} = \frac{\omega}{\mathrm{Re}_{\delta^*}}.$$

The dimensionless frequency parameter F is independent of the local boundary-layer thickness. Downstream propagation of the disturbance corresponds in the Re–F-plane to a continuous displacement to higher Reynolds numbers along a selected value of F. Thus, the disturbance is first damped and then, upon reaching branch I of the neutral stability curve, begins to grow. It grows to branch II and then decays again. Such behavior is the first evidence of 'nonparallel flow effects,' which we consider later.

The first experiments proving the correctness of the linear instability concept were carried out by Schubauer and Skramstad (1948) using a unique for that time experimental facility, setup, and advanced techniques, which later became widely accepted.

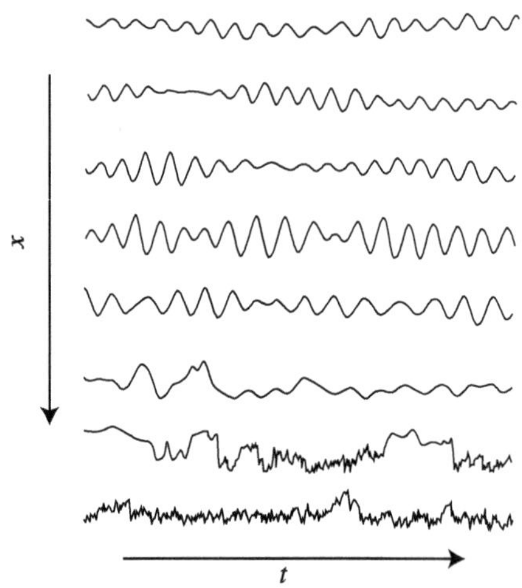

Fig. 4.2 Oscilloscope traces of streamwise fluctuations of streamwise velocity in a flat-plate boundary layer obtained at an arbitrary amplitude scale by Schubauer and Skramstad (1948)

Reducing the turbulence level in the test section of a subsonic wind tunnel down to Tu = 0.01–0.03%[1], they detected quasi-harmonic wave motions preceding the laminar–turbulent transition in the flat-plate boundary layer (Fig. 4.2). The hot-wire data were obtained for various streamwise pressure gradients by placing the flat plate at different angles of attack to promote or attenuate the transition. The flow stability characteristics were studied with the help of small-amplitude periodic disturbances excited in a controlled manner in the laminar boundary layer. After several trials they chose the vibrating-ribbon technique: an alternating current passing through a metal strip stretched in transverse direction inside the boundary layer forced its vibration in a field of a magnet located beneath the test surface. Using this method it appeared to be possible to excite normal modes of the boundary layer of a given frequency and study their characteristics.

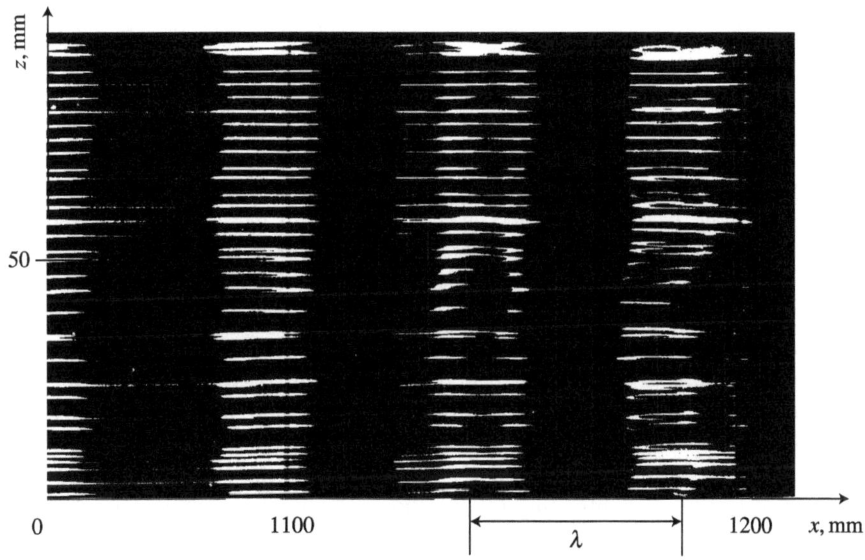

Fig. 4.3 Smoke visualization of the ribbon response in a flat-plate boundary layer in the region of Tollmien–Schlichting wave development; λ is the wave wavelength (photo by V. V. Kozlov)

In the linear normal-mode approach, the ribbon response at a certain distance from the source, where the parallel-flow approximation is applicable, can be expanded into a set of Orr–Sommerfeld modes. In the Blasius boundary layer, as well as in many other subsonic flows, all modes, except for the leading discrete eigenvalue, strongly decay with the streamwise coordinate x at any Reynolds number. Therefore, at a distance from the source comparable to the characteristic wavelength

[1] The free-stream turbulence level is determined here and below as the value $\mathrm{Tu} = u'/U_e$, as measured near the leading edge of a body.

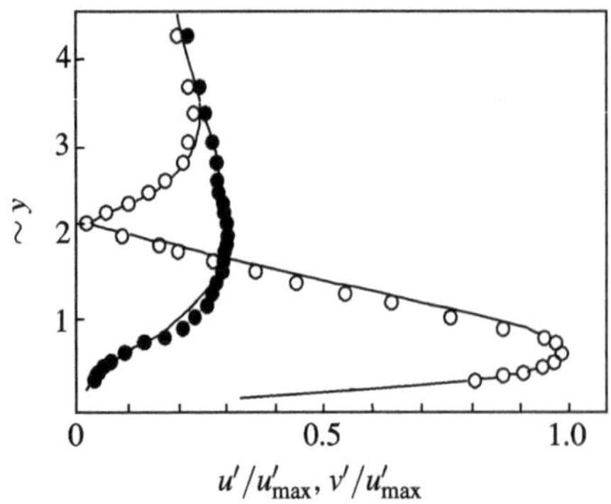

Fig. 4.4 Typical profiles of streamwise u' (○) and wall-normal v' (•) velocities of a two-dimensional Tollmien–Schlichting wave excited by the vibrating-ribbon technique in the Blasius boundary layer. The velocities are normalized to the internal velocity maximum u'_{max}. The *solid lines* show corresponding calculation using the linear stability theory for parallel flows (Westin, 1997)

of the induced oscillations, only the lading (Tollmien–Schlichting) mode can 'survive' (Fig. 4.3).[2]

A typical experimental distribution of the Tollmien–Schlichting wave amplitude measured in the Blasius boundary layer, together with the corresponding curves obtained with the linear stability theory for parallel flows, is shown in Fig. 4.4. The wave has two characteristic maxima in the streamwise velocity u', separated by a phase 'jump.' The phase jump is observed in accordance with the continuity equation (2.11) at the amplitude maximum of the wall-normal velocity v'.

Schubauer and Skramstad (1948) were able to localize both branches of the neutral stability curve by measuring the streamwise disturbance velocity at a fixed distance from the wall (Fig. 4.5). They also inspected the disturbance amplitude distribution in the wall-normal direction. A comparison of the obtained data with the theoretical results of Schlichting (1935) showed qualitative agreement. The data confirmed excellently the validity of the linear theory of hydrodynamic stability and formed a classical basis for the majority of further experimental and theoretical studies of the laminar–turbulent transition in boundary layers and other quasi-parallel flows.

Wortmann (1953, 1964) and later Kozlov and Babenko (1978) investigating boundary-layer stability in water determined the amplitudes of velocity oscillations at various frequencies and reconstructed some part of the neutral stability curve. Ross et al (1970) published experimental results that included detailed quantitative data on both the position of the neutral stability curve and the amplitude growth, as well as on the amplitude profiles of the waves at several Reynolds numbers. At the same time, calculations of the Blasius boundary-layer stability by Jordisson (1970) refined the position of the neutral stability curve obtained with asymptotic methods by Lin (1955), which was accepted earlier as the reference data.

[2] Opposite examples are given, e.g., by the boundary layer over a flat-plate compliant and a concave walls: in some situations, several linearly unstable modes can appear in that cases.

Nevertheless, the quantitative agreement of the experimental and numerical results was obtained only at low frequencies of the boundary-layer perturbations. In experiments the oscillations grew up to $F = 400 \times 10^{-6}$, though the calculations indicated that all disturbances above $F \approx 250 \times 10^{-6}$ should decay. A number of issues were left open concerning the correspondence between the temporal and spatial instability approaches, the nonparallel flow effects, the behavior of naturally occurring wave packets, and the influence of free-stream turbulence upon transition. We address the nonparallel flow effects below in Sect. 4.3 and the other above mentioned points in the next chapters.

4.2 Solution of the Orr–Sommerfeld equation for the boundary layers

The basic flow velocity distribution enters the Orr–Sommerfeld equation as a parameter function. Removing the requirement that the basic flow velocity profile is an exact solution of the stationary equations of motion, it is possible to consider it as a model function of y, thereby making it possible to apply the linear stability analysis to various types of quasi-parallel flows. Such reasoning can be also applied for flows described in other coordinate systems, e.g., for axisymmetric configurations. The correctness of the approach must, however, be verified in each particular case.

Theoretical self-similar profiles of the basic flow velocity across the boundary layer over a flat plate at an angle of attack are obtained as a result of solving the Falkner–Skan equation:

$$\Psi_{\eta\eta\eta} + \Psi_{\eta\eta}\Psi + \beta_{\mathrm{H}}\left[1 - (\Psi_\eta)^2\right] = 0\,, \tag{4.1}$$

with the boundary conditions

$$\begin{aligned} \eta = 0 &: \quad \Psi = \Psi_\eta = 0\,, \\ \eta \to \infty &: \quad \Psi_\eta \to 1\,, \end{aligned}$$

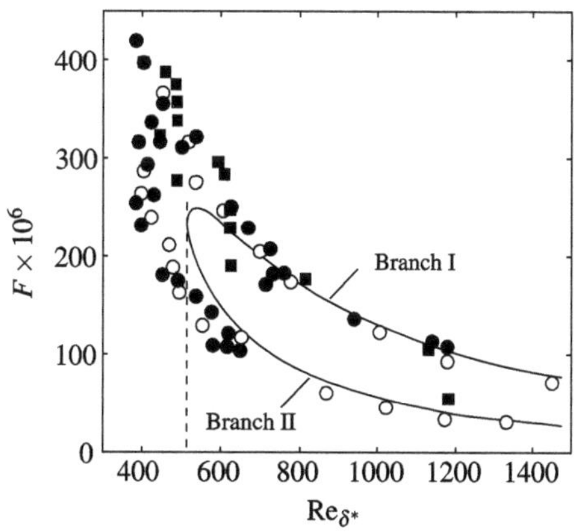

Fig. 4.5 Neutral stability data for the Blasius boundary layer of Schubauer and Skramstad (1948) (■), Ross et al (1970) (○), Kachanov et al (1979) (•); numerical calculation using linear stability theory (*solid line*); location of the theoretical critical Reynolds number $\mathrm{Re}_{\mathrm{L},\delta^*} \approx 520$ (*dashed line*)

where the subscript 'η' denotes the first derivative with respect to the self-similar wall-normal coordinate

$$\eta = y\sqrt{\frac{U_\mathrm{e}}{(2-\beta_\mathrm{H})\nu x}} = \frac{y}{x}\frac{\mathrm{Re}_\eta}{\sqrt{2-\beta_\mathrm{H}}} \,. \tag{4.2}$$

Here $\mathrm{Re}_\eta = \sqrt{\mathrm{Re}_\mathbf{x}} = \sqrt{U_\mathrm{e}x/\nu}$, $U_\mathrm{e} = Qx^{\beta_\mathrm{H}/(2-\beta_\mathrm{H})}$, Q is a constant, and β_H is the Hartree pressure gradient parameter. The function Ψ is a dimensionless stream function, so that

$$\frac{U}{U_\mathrm{e}} = \Psi_y ,$$
$$\frac{V}{U_\mathrm{e}} = \frac{1}{\mathrm{Re}_\eta}\frac{(1-\beta_\mathrm{H})\eta\Psi_y - \Psi}{\sqrt{2-\beta_\mathrm{H}}} \,.$$

When $\beta_\mathrm{H} > 0$, the flow is accelerated downstream, and when $\beta_\mathrm{H} < 0$, it is retarded and the streamwise velocity profiles have an inflection point. In the latter case, owing to the nonlinearity of the equation, there exist two solutions in the range $-0.199 \lesssim \beta_\mathrm{H} < 0$, one of which (without back flow) is called the Hartree solution and the other one (with back flow) is called the Stewartson solution (Stewartson, 1954). At $\beta_\mathrm{H} \lesssim -0.199$, solutions resembling velocity profiles of near wall jets with overflow in the middle of boundary layer can exist (Libby and Liu, 1967). At zero angle of attack, $\beta_\mathrm{H} = 0$, the Falkner–Skan equation (4.1) reduces to the Blasius boundary-layer equation

$$\Psi_{yyy} + \Psi_{yy}\Psi = 0\,, \quad \Psi(0) = \Psi_y(0) = 0\,, \quad \Psi_y(y \to \infty) \to 1\,.$$

The procedure of the numerical solution of the Orr–Sommerfeld equation for the boundary layers is, in general, analogous to that described in Sect. 3.3 for the plane Poiseuille flow. However, as the Chebyshev polynomials are defined on a finite interval $\varsigma \in [-1;1]$, in the framework of this method one has to perform a coordinate transformation from a physical to a numerical domain using, e.g., the algebraic function

$$\eta = L\frac{1+\varsigma}{b-\varsigma} \,,$$

where L is a scale factor and $b = 1 + 2L/\eta_\mathrm{max}$ is useful to adjust the distribution of nodes in the interval. For simplicity, the physical domain is assumed here to be truncated at a large distance from the wall η_max. MATLAB function `mapping` performs such a transformation.

Additionally, as there is no analytic solution for the Falkner–Skan equation, it must be found numerically. This can be done by a number of techniques. Among them, there are two general approaches consisting either of a direct solution of the problem as the boundary-value problem (which is rather stiff) or its transformation to an initial-value problem preserving the boundary conditions at the wall, followed by its iterative solution with the false position method to satisfy the condition at infinity. In both cases, it is useful to transform the third-order differential equation to a set of the first-order differential equations as

```
function D=mapping(a,b,D)
% MAPPING algebraic mapping of [0;2*a/(b-1)] to [-1;1]
% using ETA=a*(1+Z)/(b-Z):
% a,b - mapping parameters;
% D   - the container returned by CHDER1.

f_zz=2/(b+1)/a;
f_z =f_zz.*(D{5}-b);
f   =f_z.*(D{5}-b)/2;

z_ee  =f_z.*f;
fz_ee =f.*z_ee;
z_eee =f_z.*z_ee+f_zz.*f.^2;
z_eeee=f_z.*z_eee+3*f_zz.*fz_ee;

z_ee  =diag(z_ee);
fz_ee =diag(fz_ee);
z_eee =diag(z_eee);
z_eeee=diag(z_eeee);

D{1}=diag(f    )*D{1};
D{2}=diag(f.^2)*D{2}+    z_ee      *D{1};
D{3}=diag(f.^3)*D{3}+3*fz_ee      *D{2}+ z_eee*D{1};
D{4}=diag(f.^4)*D{4}+6* z_ee*diag(f.^2)*D{3}+...
     (4*z_eee*diag(f)+3* z_ee.^2) *D{2}+z_eeee*D{1};
D{5}=a*(1+D{5})./(b-D{5});
D{6}=D{6}./sqrt(f');
```

$$\frac{d}{dy}\begin{pmatrix}\Psi_0\\ \Psi_1\\ \Psi_2\end{pmatrix}=\begin{bmatrix}\Psi_1\\ \Psi_2\\ \beta_H\left(\Psi_1^2-1\right)-\Psi_0\Psi_2\end{bmatrix},$$

where $\Psi_0=\Psi$, $\Psi_1=\Psi_y$ and $\Psi_2=\Psi_{yy}$.

To solve it using the false position method, the 'initial' conditions $\Psi_0(0)=\Psi_1(0)=0$ are fixed at $\eta=0$. Then, by varying $\Psi_2(0)$, the value of $|\Psi_1(\infty)-1|$ is minimized. To solve the corresponding boundary-value problem, one can use off-the-shelf software packages. For example, the MATLAB function `bvp4c` can be used for it. Function `FS` implements the latter approach. The results of the numerical calculation of linear stability of the Blasius boundary layer to two-dimensional Tollmien–Schlichting waves are given in Fig. 4.6.

Note that the homogeneous boundary conditions at $\eta_{\max}$ used in solving the Orr–Sommerfeld equation in the above approach deteriorate the continuous spectrum. Instead, to find a spectrum mode accurately, it is necessary to use more general asymptotic boundary conditions, which can be derived from (2.17), (see, e.g., Maslowe and Spiteri, 2001). Function `Contin` implements the corresponding solution algorithm for disturbances developing in time.

```
function [U,V]=FS(y,b)
% FS solves the Falkner-Skan equation.
% y - variable step descending coordinates,
%     for which the solution is required;
% b - Hartree pressure parameter;
% U - streamwise  velocity profile.
% V - wall-normal velocity profile.

% initial conditions require y is in ascending order
y=y(end:-1:1);

% provide initial guess
solinit=bvpinit(y,[0 1 0]);

% register Jacobians to accelerate convergence
opts=bvpset('FJacobian' ,@stwjac,...
            'BCJacobian',@stwBCjac);

% solve the equation
sol=bvp4c(@stw,@stwbc,solinit,opts,b);

% if fractional steps were used implicitly by bvp4c,
% restore the solution on the original nodes
sol=deval(sol,y);

% extract the velocities and restore original order
U=sol(2,end:-1:1).';
V=((1-b)*y.*U-sol(1,end:-1:1).')/sqrt(2-b);
% -----------------------------------------------------
function dydt=stw(x,y,b)
% Description of the Falkner-Skan equation.
dydt=[y(2); y(3); b*(y(2).^2-1)-y(1).*y(3)];
% -----------------------------------------------------
function bc=stwbc(ya,yb,b)
% Description of the boundary conditions.
bc=[ya(1); ya(2); yb(2)-1];
% -----------------------------------------------------
function JF=stwjac(x,y,b)
% Jacobian for the FS-equation.
JF=[0 1 0; 0 0 1; -y(3) 2*b*y(2) -y(1)];
% -----------------------------------------------------
function [Ja,Jb]=stwBCjac(ya,yb,b)
% Jacobians for the boundary conditions:
% Ja - at the wall; Jb - in infinity.
Ja=[1 0 0; 0 1 0; 0 0 0];
Jb=[0 0 0; 0 0 0; 0 1 0];
```

4.3 Nonparallel flow effects

For a long time, the discrepancy of results from the linear stability theory and experiments in the region of low Reynolds numbers and high frequencies was attributed

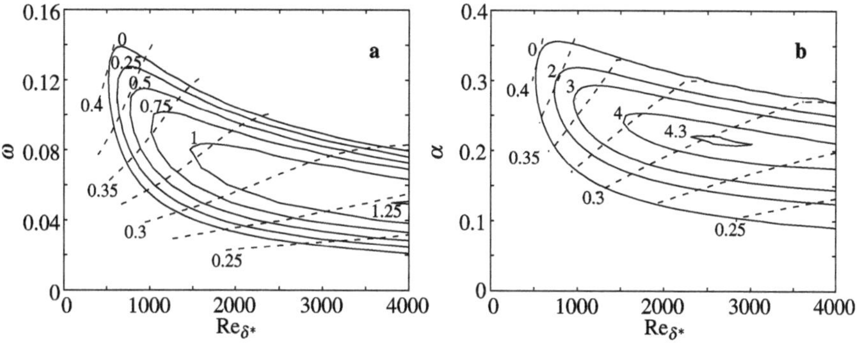

Fig. 4.6 Real parts of phase velocities $c_{\mathrm{r}} = \omega/\alpha_{\mathrm{r}}$ (*dashed lines*) as well as neutral stability curves and isolines of spatial $-\alpha_{\mathrm{i}} \times 10^2$ (**a**) and temporal $\omega_{\mathrm{i}} \times 10^3$ (**b**) growth rates (*solid lines*) for two-dimensional Tollmien–Schlichting waves in the Blasius boundary layer; $\mathrm{Re}_{\delta^*} = 1.72\mathrm{Re}_\eta$ (calculated by A. V. Boiko)

to the boundary-layer growth with the streamwise coordinate. Later it became clear that the main reason for the discrepancy has a little relation to the nonparallel flow effects. However, the divergence of the boundary-layer streamlines is still of great significance to the development of both oblique waves or wave packets in the Blasius boundary layer.

4.3.1 Outline of theoretical approaches to account for nonparallel effects

The nonparallel flow effects can be divided into the parametric variation of the Reynolds number with the streamwise coordinate, resulting in downstream changes of the velocity profiles and wavenumbers of the instability waves, and the effect of the basic flow divergence, i.e., presence of both the wall-normal velocity $V(y)$ and the streamwise velocity gradient $\partial U/\partial x$.

The first of these effects can be treated parametrically with the quasi-parallel flow approximation. A sequence of parallel-flow calculations with parametric variation of the Reynolds number gives the wavenumbers $\alpha(x)$. Then the imaginary part (the growth rate) α_{i} is related locally to the amplitude of, e.g., the streamwise disturbance velocity $A(x)$ as $\mathrm{d}A/\mathrm{d}x = -\alpha_{\mathrm{i}}(x)A$, so that

$$A(x) = \exp\left(-\int_{x_0}^{x} \alpha_{\mathrm{i}}(\chi)\mathrm{d}\chi\right) .$$

Corrections due to the divergence of basic flow streamlines require a special treatment. This can be done on the basis of either asymptotic or perturbation approaches, parabolized stability equations (PSEs), or with the help of direct numerical simulation (DNS). The latter technique consisting in the numerical integration of the full Navier–Stokes equations has been intensively developing over the past two or three

```
function v=Contin(a,b,Re,q,Y,U0,U2)
% CONTIN finds a continuous mode of the OS equation.
% a      - streamwise wavenumber;
% b      - transverse wavenumber;
% q      - continuous spectrum parameter;
% Re     - Reynolds number;
% Y      - vector of nodes;
% U0,U2 - basic velocity U0(Y) and U0''(Y);
% v      - amplitude function (profile) of the mode.

% useful constants
o=complex(a,(q^2-a^2-b^2)/Re); k=sqrt(a^2+b^2);

% use vectorization and analytic Jacobians
opt=bvpset('Vectorized','on','BCJacobian',@OSBCjac);

solinit=bvpinit(Y,[1 0 1 0],0); % initial guess
v=deval(bvp4c(@OS,@OSbc,solinit,...
              opt,a,k,o,Re,q^2,U0',U2',Y'),Y);
v=v(1,:).';
% -------------------------------------------------------
function dydt=OS(x,y,C,a,k,o,Re,q2,U0,U2,Y)
% The OS equation.

U0=interp1(Y,U0,x,'cubic'); U2=interp1(Y,U2,x,'cubic');
Z=a*U0-o; k2=k^2; k4=k^4;
dydt=[y(2,:); y(3,:); y(4,:);
(1i*Re*(k2*Z-a*U2)-k4).*y(1,:)+(2*k2+1i*Re*Z).*y(3,:)];
% -------------------------------------------------------
function bc=OSbc(ya,yb,C,a,k,o,Re,q2,U0,U2,Y)
% Boundary conditions (C is unknown):
% v(0)=0, v'(0)=0, v (y)=1,
% v''  (y)-q^2v(y)=  C(k^2-q^2)exp(-ky),
% v'''(y)-q^2v(y)=-kC(k^2-q^2)exp(-ky).

bc=[ya(1); ya(2); yb(1)-1;
    yb(3)-q2*yb(1)-  C*(k^2-q2)*exp(-k*max(Y))
    yb(4)-q2*yb(2)+k*C*(k^2-q2)*exp(-k*max(Y))];
% -------------------------------------------------------
function [Ja,Jb,Jp]=OSBCjac(ya,yb,C,a,k,o,Re,q2,U0,U2,Y)
% Jacobians for the boundary conditions:

Ja=[1,0,0,0; 0,1,0,0; 0,0,0,0;   0,0,0,0;   0,0,0,0];
Jb=[0,0,0,0; 0,0,0,0; 1,0,0,0; -q2,0,1,0; 0,-q2,0,1];
Jp=[0; 0; 0; -1; k*C];
```

decades. In this case, both the variations of the basic flow and the flow nonlinearity are automatically taken into account. One of the main difficulties of using the full Navier–Stokes equations for the hydrodynamic stability analysis is complexity of their numerical implementation. Thus, the other techniques outlined below are frequently used.

4.3.1.1 Perturbation methods

The stability of the growing boundary layer can be studied with perturbation methods. For the disturbance vector $\mathbf{q} = (u,v,w,p)^T$ that slowly varies with x, one can write in Wentzel–Kramers–Brillouin-approximation (or WKB) sense

$$\mathbf{q} = \bar{\mathbf{q}}(X,y,\varepsilon)\exp\left(\mathrm{i}\int_{x_0}^{x}\alpha(\varepsilon,\chi)\mathrm{d}\chi + \mathrm{i}\beta z - \mathrm{i}\omega t\right),$$

where $X = \varepsilon x$ is a 'slow' streamwise coordinate; both the disturbance wavenumber α and the disturbance amplitude function $\bar{\mathbf{q}}$ vary slowly downstream; $\varepsilon = \mathcal{O}(1/\mathrm{Re}_\delta) \ll 1$ is a small parameter to account for the basic flow variations and δ is a boundary layer scale. Similarly, basic flow velocities take the form

$$U(x,y) = U(X,y,\varepsilon)\,, \quad V(x,y) = V(X,y,\varepsilon)\,.$$

As a variant, it is possible to use two parameters, one of which accounts for the basic flow velocity variations and the other for the changes in the disturbance amplitude function. Expansion of the quantities depending on ε into a Taylor series and collection of terms of the same orders of magnitude yield a chain of systems of linear equations, which can be written in compact form as

$$\mathscr{L}_0\bar{\mathbf{q}}_0 = 0\,, \quad \mathscr{L}_0\bar{\mathbf{q}}_1 = \mathscr{L}_1\bar{\mathbf{q}}_0\,, \ldots$$

with homogeneous boundary conditions. The first of them is exactly the linearized Navier–Stokes equations in the parallel-flow approximation, the second one takes into account the first-order nonparallel flow effects. The sequence of systems of equations can be continued to include into consideration higher-order effects. Hence, the main difference between perturbation theories of this sort—as far as the expansion is correctly chosen—is the number of small parameters and the number of terms taken into account.

The Fredholm alternative requires that the higher-order inhomogeneous terms be orthogonal to the lower-order terms, that is, e.g., for the first-order correction,

$$\mathscr{L}_0^\dagger(\mathscr{L}_1\bar{\mathbf{q}}_0) = 0\,.$$

Using an explicit integral form of this equation, it is possible to extract the required nonparallel correction for the streamwise wavenumber including its imaginary part. In general, the results of these theories lead to the conclusion that the nonparallel flow effects destabilize the boundary layer in the region of low Reynolds numbers to high-frequency disturbances.

Also, the studies show a substantial influence of weak nonparallel flow effects on the development of small disturbances. In particular, a strong dependence of the growth rates on the wall-normal coordinate due to downstream changes of the disturbance amplitude profile was found. This leads to a considerable dependence

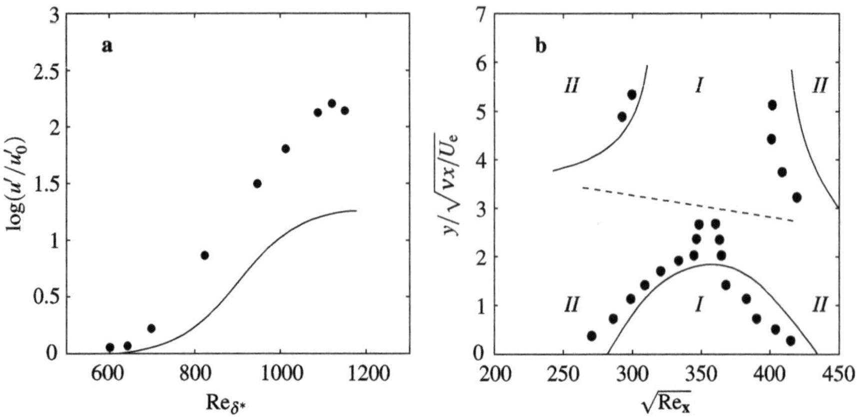

Fig. 4.7 Tollmien–Schlichting wave amplification curves for $F = 110 \times 10^{-6}$: experiment (•) of Kachanov et al (1979) and linear stability theory (*solid line*) of Levchenko et al (1975) (**a**). Location of neutral points for streamwise velocity disturbance in the $\mathrm{Re_x}$–η-plane for $F = 200 \times 10^{-6}$: experimental data (•) of Kachanov et al (1979); calculations (*solid lines*) of Tumin and Shepelev (1980); minimum $|\bar{u}|$ displacement (*dashed line*); growth of disturbances (*I*); decay of disturbances (*II*) (**b**)

of the stability characteristics on the experimental procedure of their determination (Fig. 4.7) and to an ambiguity in defining the neutral stability curve in calculations.

4.3.1.2 Parabolized stability equations

Excessive complexity and cost of DNS for stability applications as well as intricacy of the perturbation method resulted in the development of a relatively cheap and accurate (if applied wisely) approach: the parabolized stability equations (PSEs). When considering development of disturbances in space, the approximation enables a substantial elimination of elliptic components in the Navier–Stokes equations, so that the equations obtained can be solved by downstream marching. PSEs is a relatively simple approach, which can be adopted to more complicated problems such as the boundary-layer receptivity (Chap. 10), the nonlinearity (Chap. 12) and complex flow geometries (Chaps. 5 and 6).

The approximation procedure starts with substitution of the basic flow velocity distribution $\mathbf{U} = (U(x,y), V(x,y), 0)^T$ into the linearized Navier–Stokes equation, which leads to general stability equations for a nonparallel flow. The vector of disturbance flow variables $\mathbf{q} = (u, v, w, p)^T$ in the equations is written in the form

$$\mathbf{q} = \bar{\mathbf{q}}(x,y)\exp\left(\mathrm{i}\int_{x_0}^{x}\alpha(\chi)\mathrm{d}\chi + \mathrm{i}\beta z - \mathrm{i}\omega t\right), \tag{4.3}$$

where α is a complex wavenumber. It is assumed that both the basic flow and the disturbance experience only slow downstream changes in their velocity distributions in wall-normal directions and, hence, the terms containing the second derivatives in

x and their multiples are considered negligible and are dropped. This leads to the following system of differential equations with respect to $\bar{\mathbf{q}}$ and α:

$$\begin{array}{llll}
[\mathrm{i}\alpha\bar{u}+\mathcal{D}\bar{v}+\mathrm{i}\beta\bar{w}] & & +[\bar{u}_x] & =0\,,\\
[\mathscr{L}_{\mathrm{Sq}}\bar{u}+U_y\bar{v}+\mathrm{i}\alpha\bar{p}] & +[U_x\bar{u}+V\mathcal{D}u] & +\left[\left(U-\frac{2\mathrm{i}\alpha}{\mathrm{Re}}\right)\bar{u}_x+\bar{p}_x\right]-\left[\frac{\mathrm{i}}{\mathrm{Re}}\alpha_x\bar{u}\right] & =0\,,\\
[\mathscr{L}_{\mathrm{Sq}}\bar{v}+\mathcal{D}\bar{p}] & +[V_y\bar{v}+V\mathcal{D}\bar{v}] & +\left[\left(U-\frac{2\mathrm{i}\alpha}{\mathrm{Re}}\right)\bar{v}_x\right]-\left[\frac{1}{\mathrm{Re}}\alpha_x\bar{v}\right] & =0\,,\\
[\mathscr{L}_{\mathrm{Sq}}\bar{w}+\mathrm{i}\beta\bar{p}] & +[V\mathcal{D}\bar{w}] & +\left[\left(U-\frac{2\mathrm{i}\alpha}{\mathrm{Re}}\right)\bar{w}_x\right]-\left[\frac{1}{\mathrm{Re}}\alpha_x\bar{w}\right] & =0\,,
\end{array}$$

where the subscripts 'x' and 'y' mean differentiating in x and y, respectively, and

$$\mathscr{L}_{\mathrm{Sq}}=\mathrm{i}(\alpha U-\omega)-\frac{1}{\mathrm{Re}}\left(\mathcal{D}^2-\varkappa^2\right)\,.$$

The left-hand sides of the equations of motion contain four groups of terms in brackets. The first of them corresponds to the linearized Navier–Stokes equations for parallel flows. The second contains the terms taking into account the downstream modification of the basic flow. The streamwise derivatives of the amplitude functions $\bar{\mathbf{q}}$ belong to the third group. Finally, the last group includes the terms taking into account the downstream changes of the streamwise wavenumber α.

The resulting equations can be written compactly as

$$\mathscr{L}\bar{\mathbf{q}}+\mathscr{M}\bar{\mathbf{q}}_x+\alpha_x\mathscr{N}\bar{\mathbf{q}}=0\,,$$

where the operators $\mathscr{L}$, $\mathscr{M}$, and $\mathscr{N}$ contain only derivatives in y and

$$\mathscr{M}=-\mathrm{i}\frac{\partial\mathscr{L}}{\partial\alpha}\,,\quad \mathscr{N}=\frac{1}{2}\frac{\partial\mathscr{M}}{\partial\alpha}=-\frac{\mathrm{i}}{2}\frac{\partial^2\mathscr{L}}{\partial\alpha^2}\,.$$

It follows from (4.3) that the streamwise oscillations and growth of the disturbance can be included both in $\bar{\mathbf{q}}$ and α. This ambiguity is removed by applying some auxiliary compatibility conditions on the variation of $\bar{\mathbf{q}}$ to attribute most of the disturbance growth to α. To measure the disturbance growth, the amplification rate defined as A,

$$\sigma=(\mathrm{d}A/\mathrm{d}x)/A\,,\quad A(x)=\max_y[u'(x,y)]\,,$$

is frequently used in laboratory experiments, where u' is the root-mean-square amplitude of the disturbance streamwise velocity. In calculations, another auxiliary condition is usually utilized:

$$\bar{\mathbf{v}}\cdot\frac{\partial\bar{\mathbf{v}}}{\partial x}=0\,,$$

where $\mathbf{v}=(u,v,w)^T$, ensuring that the kinetic energy of the disturbances changes due to the exponential factor.

A special analysis as well as numerical experiments indicate that the PSEs are not completely parabolic, but rather weakly elliptical. This residual ellipticity manifests itself as a strong numerical instability when a sufficiently small step size is used for downstream marching. It is proposed either to enlarge the step to avoid instability or to remove the $\partial p/\partial x$-term in the streamwise momentum equation. Naturally, these modifications deteriorate the accuracy of the results and can prevent resolution of rapid variations of the shape functions in some situations.

For the Blasius boundary layer, however, comparisons with DNS results show that the PSEs provide better accuracy of the stability solutions compared to that obtained in the parallel-flow approximation (Bertolotti et al, 1992). Nevertheless, the nonparallel flow effects in this case are small and cannot explain the experimentally found stability characteristics at low Reynolds numbers and large frequencies.

4.3.2 Modern view on the place and role of nonparallel effects in the Blasius boundary layer

The difference between the theoretical and experimental results on the Blasius boundary-layer stability is related both to details of the experimental setup and to the chosen amplitude measure. The wind-tunnel study of the development of Tollmien–Schlichting waves in a boundary layer has many traps. One of the main problems is to suppress environmental perturbations, in particular, the level of free-stream turbulence. Though the external disturbances are extremely important for excitation of 'natural' Tollmien–Schlichting waves,[3] it is desirable to reduce them as much as possible when studying the development of *controlled* boundary-layer oscillations.

In the boundary layer above an ideal flat plate, there is no streamwise pressure gradient. In an experiment, a plate has a finite thickness and a pressure gradient at the leading edge. Therefore, the boundary layer differs substantially from the Blasius one in that region, which can be quite extended downstream. The pressure gradient at the leading edge depends on the plate thickness, edge shape, and position of the stagnation point. Thinner plates are preferable as they allow reduction of the Reynolds number near the beginning of the gradientless region. The stagnation point position can be controlled, for example, by a trailing edge flap on the plate.

Nonparallel flow effects require a selection of a stability criterion both in experiment and in theory. Unfortunately, such a physical measure of linear stability as the disturbance kinetic energy is rarely used. Schubauer and Skramstad (1948) in their pioneer study simply traversed the hot-wire probe at a fixed distance from the wall. Ross et al (1970) and Kachanov et al (1979) were the first to pay attention to a downstream displacement of the near-wall maximum in the streamwise disturbance velocity profile caused by the nonparallel flow development. They applied a modified measurement technique: the development of the disturbances was traced along the line of equal velocity (that is, along a fixed self-similar boundary-layer coordinate) close to the maximum of the streamwise velocity of the oscillations. Additionally, the influence of the distance to the wall on the position of the experimentally determined neutral stability curve and on the spatial amplification was studied. Various ways of defining the disturbance measure were also discussed by Bouthier (1973), Gaster (1974) and Saric and Nayfeh (1977). Despite these improvements in

[3] The terms *natural waves* and *natural transition* hereinafter denote the variety of ways of the transition obtained in various experiments without a controlled introduction of disturbances to a flow that are additional to the background ones.

the methodology of experiment, however, the growth of oscillations was once again found to occur up to the frequency parameter $F = 400 \times 10^{-6}$ (Fig. 4.5).

Experiments on flat-plate boundary-layer stability, in which special care was taken to minimize difference of the basic flow from the Blasius one, were conducted by Klingmann et al (1993). The plate used in that study had a leading edge whose shape was specially calculated with the help of the two-dimensional potential flow theory to take into account the real design of the wind-tunnel test section. The plate was equipped with a trailing edge flap to regulate the pressure distribution at the leading edge. Thus, it became possible to remove the pressure drop near the leading edge and to reduce substantially the size of the pressure gradient region. Controlled instability waves were excited by the vibrating-ribbon technique.

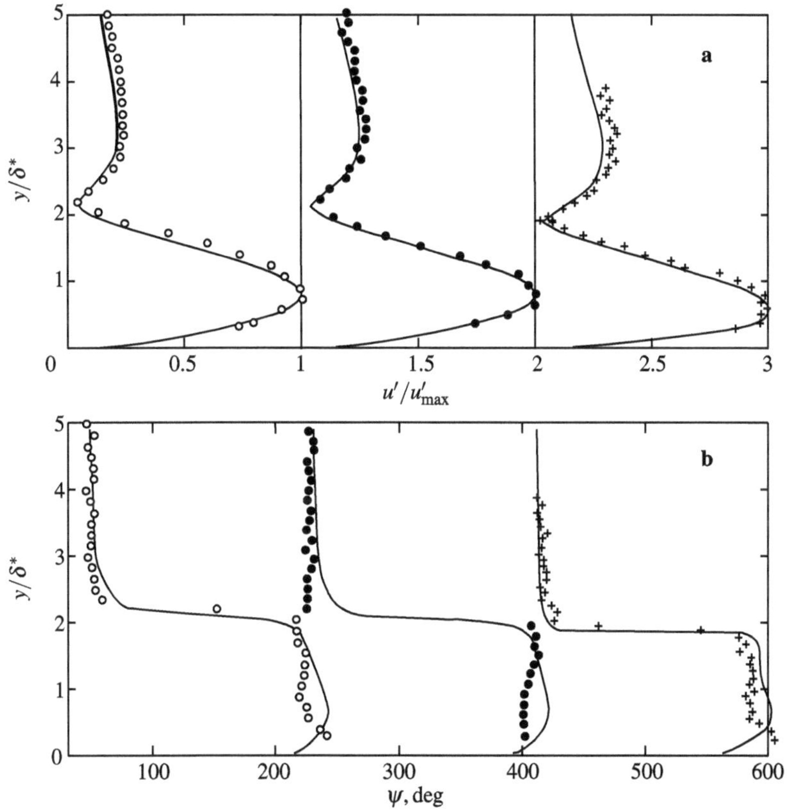

Fig. 4.8 Amplitude (**a**) and phase (**b**) distributions of Tollmien–Schlichting waves at $F = 250 \times 10^{-6}$: $\mathrm{Re}_{\delta^*} = 343$ (∘); $\mathrm{Re}_{\delta^*} = 396$ (•); $\mathrm{Re}_{\delta^*} = 574$ (+); corresponding results of parallel linear stability theory (*solid lines*) (Klingmann et al, 1993)

The disturbance streamwise amplitude and phase profiles measured at three different distances from the leading edge are shown in Fig. 4.8. The Tollmien–Schlichting wave amplitude is normalized to its inner maximum. Characteristic features of the profiles that are typical for the Tollmien–Schlichting waves predicted by the linear stability theory are the inner and the outer amplitude maxima as well as the phase 'jump' near the amplitude minimum. Generally, the profile varies as the wave moves downstream. The phase jump and the amplitude maxima displace monotonically to smaller y/δ^*, indicating that the wave evolution is not self-similar in the boundary-layer coordinates, as has been discussed earlier.

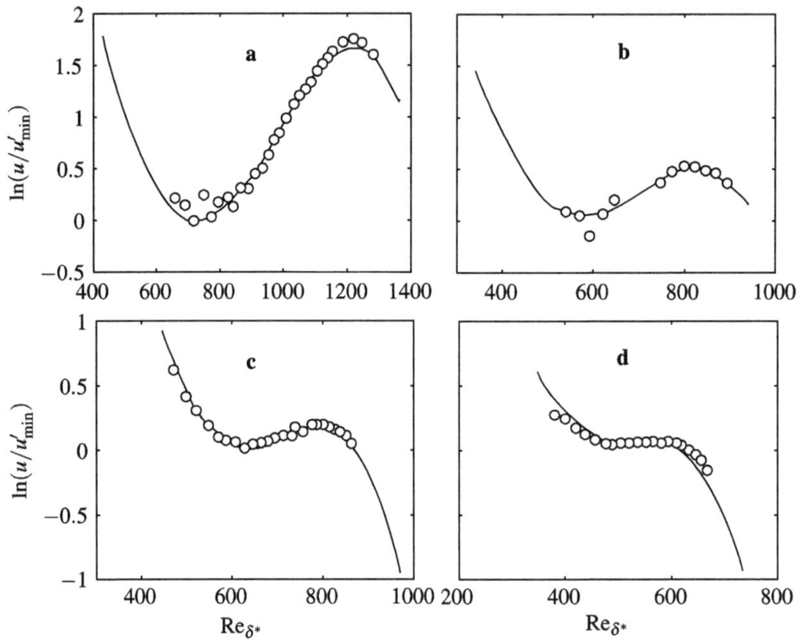

Fig. 4.9 Downstream behavior of the Tollmien–Schlichting wave amplitudes at $F \times 10^6 = 100$ (**a**), 162 (**b**), 200 (**c**) and 250 (**d**): experimental data (○); results of the linear stability theory for parallel flows (*solid lines*) (Klingmann et al, 1993)

The development of the disturbance amplitude with Re measured at the inner Tollmien–Schlichting wave maximum is compared in Fig. 4.9 with the results of the linear stability theory. The wave excited upstream of branch I of the neutral stability curve decays at first, then is amplified in the instability region, and decays again behind branch II. At low frequencies, the amplitude evolution is predicted quite well by the theory; at high frequencies, the experimental amplitude curves experience a little deviation from that calculated by the theory, but much less than in the previous experimental studies. In particular, at $F = 250 \times 10^{-6}$, the experimental data show a small amplification in the region about $\mathrm{Re}_{\delta^*} = 520$, whereas the linear stability theory for parallel flows predicts a monotonic decay of the Tollmien–Schlichting wave. The experimental and theoretical neutral stability curves are shown in Fig. 4.10. It can be seen that the experimental results are in good agreement with the nonparallel linear stability analysis of Gaster (1974). For low frequencies, the difference between the theoretical curves and experimental data is negligible.

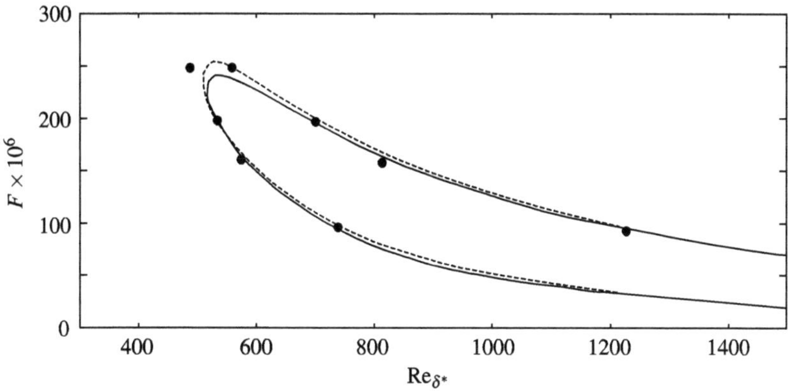

Fig. 4.10 Neutral stability curves: experimental data obtained at the inner amplitude maximum by Klingmann et al (1993) (•); parallel linear stability theory (*solid line*) and nonparallel linear stability theory of Gaster (1974) (*dashed line*)

These results show that the parallel linear stability theory predicts well the distributions of amplitudes, wavenumbers, growth rates, and neutral stability of the two-dimensional Tollmien–Schlichting waves. The parallel-flow approximation allows acquisition of quite precise results for the disturbance growth rates, except for the uppermost part of the neutral stability curve, where it is necessary to take into account the effect of downstream growth of the boundary layer, the effect being rather small. The discrepancy between the preceding experimental results and the predictions of the parallel linear stability theory arises mainly from inaccurate experimental simulation of the ideal Blasius boundary layer or, equivalently, from neglecting the actual properties of the boundary layer in the calculations. The results are most sensitive in the region of low Reynolds numbers near the leading edge of the plate where the pressure distribution is difficult to control with standard setups.

Additional material for this chapter is given in the multimedia file No. 4, which supplements the book, titled 'Instability of boundary layers.'

Exercises

4.1. Write a MATLAB function to find the Hartree solution of the Falkner–Skan equation using the false position method. Hint: $\Psi_2(0) \geq 0$ for the Hartree solution. Compare the speed and convergence of this technique and the technique used in function `FS` at different values of the pressure gradient parameter β_{H}.

4.2. For pre-separation Hartree boundary-layer profile ($\beta_{\mathrm{H}} = -0.198$) try to reproduce the neutral stability curve shown in Fig. 2.4.

4.3. Perform the analysis of the continuous spectrum of the Squire equations for a boundary layer using the approach of Sect. 2.6 as a template. Write a MATLAB function to calculate the eigenfunctions for modes of the continuous spectrum of the Squire equations for the Blasius boundary layer in temporal formulation.

4.4. Write a MATLAB function to calculate the eigenfunctions for modes of the continuous spectrum of the Orr–Sommerfeld and Squire equations for the Blasius boundary layer in spatial formulation. For the Orr–Sommerfeld equation, provide separate calculations for the pressure and vorticity modes. Compare the results with shown in Fig. 2.5.

4.5. How does the position of the nearest to the wall amplitude maximum of the eigenfunction obtained in Exercises 3.1–3.3 change when $\beta \gg \alpha$?

4.6. Modify the program from Exercise 3.4 to obtain and draw the growth rates (both in temporal and spatial formulation) and phase velocities of the leading normal modes for the Blasius boundary layer as shown in Fig. 4.6.

4.7. Modify the programs from Exercise 4.6 to obtain and draw eigenfunctions of the leading discrete mode (the Tollmien–Schlichting wave) as shown in Fig. 4.8. How do the amplitude and phase profiles of the eigenfunctions change in the region of instability at $\beta = 0$ and $\beta \approx \alpha$?

References

Bertolotti FP, Herbert T, Spallart PR (1992) Linear and nonlinear stability of the Blasius boundary layer. J. Fluid Mech 242:441–474

Bouthier M (1973) Stabilité linéaire des écoulements presque parallèles. Part 2. La couche limite de Blasius. J Mécanique 12(1):75–95

Gaster M (1974) On the effects of boundary-layer growth on flow stability. J Fluid Mech 66:465–480

Jordisson R (1970) The flat plate boundary layer. Part 1. Numerical integration of the Orr–Sommerfeld equation. J Fluid Mech 43:801–811

Kachanov YS, Kozlov VV, Levchenko VY (1979) The development of small-amplitude oscillations in a laminar boundary layer. Fluid Mech – Soviet Res 8(2):152–156

Klingmann BGB, Boiko AV, Westin KJA, Kozlov VV, Alfredsson PH (1993) Experiments on the stability of Tollmien–Schlichting waves. Eur J Mech B/Fluids 12(4):493–514

Kozlov LF, Babenko VV (1978) Experimental investigations of boundary layer. Naukova dumka, Kiev. In Russian.

Levchenko VY, Volodin AG, Gaponov SA (1975) Stability characteristics of boundary layers. Nauka. Sib. Otd., Novosibirsk. In Russian.

Libby PA, Liu TM (1967) Further solutions of the Falkner-Skan equation. AIAA J 5(5):1040–1042

Lin CC (1955) The theory of hydrodynamic stability. Cambridge University Press, Cambridge

Maslowe SA, Spiteri RJ (2001) The continuous spectrum for a boundary layer in a streamwise pressure gradient. Phys Fluids 13(5):1294–1299

Ross JA, Barnes FH, Burns JG, Ross MAS (1970) The flat plate boundary layer. Part 3. Comparison of theory with experiment. J Fluid Mech 43:819–832

Saric WS, Nayfeh AH (1977) Nonparallel stability of boundary layers with pressure gradients and suction. In: AGARD-CP–224 Laminar–Turbulent Transition, Copenhagen, pp 6–1621

Schlichting H (1933) Zur Entstehung der Turbulenz bei der Plattenströmung. In: Math. Phys. Klasse, Nachr. Ges. Wiss. Göttingen, Göttingen, pp 181–208

Schlichting H (1935) Amplitudenverteilung und Energiebilanz der kleinen Störungen bei der Plattenströmung. In: Math. Phys. Klasse, Fachgruppe I, vol 1, Nachr. Ges. Wiss. Göttingen, Göttingen, pp 47–78

Schubauer GB, Skramstad HK (1948) Laminar-boundary layer oscillations and transition on a flat plate. TN 909, NACA

Stewartson K (1954) Further solutions of the Falkner–Skan equation. Proc Camb Phil Soc 50:454–465

Tollmien W (1929) Über die Entstehung der Turbulenz. 1. Mitteilung. In: Math. Phys. Klasse. Nachr. Ges. Wiss. Göttingen, Göttingen, pp 21–44, translated as NACA TM 609, 1931.

Tumin AM, Shepelev VE (1980) Numerical analysis of disturbance development in incompressible flat plate boundary layer. Chisl Met Mekhan Splosh Sredy 1(3):141–152, In Russian.

Westin KJA (1997) Laminar–turbulent boundary layer transition influenced by free stream turbulence. TRITA-MEK TR 1997:10, Royal Institute of Technology, Stockholm, Doctoral Thesis.

Wortmann FX (1953) Eine Methode zur Beobachtung und Messung Wasserströmung mit Tellur. Z Angew Phys 5(6):200–206

Wortmann FX (1964) Experimental investigation of vortex occurence at transition in unstable boundary-layers. AFOSR 64–1280

Further Reading

Benney DJ, Rosenblatt S (1964) Stability of spatially varying and time dependent flows. Phys Fluids 7(8):1385

Boiko AV, Westin KJA, Klingmann BGB, Kozlov VV, Alfredsson PH (1994) Experiments in a boundary layer subjected to free stream turbulence. Part 2. The role of TS-waves in the transition process. J Fluid Mech 281:219–245

Bouthier M (1971) Stabilité linéaire des écoulements presque parallèles des échelles multiples. Compt Rendus Acad Sci, Sèr A 273:1101–1104

Bouthier M (1972) Stabilité linéaire des écoulements presque parallèles. Part 1. J Mécanique 11(4):599–621

Bouthier M (1973) Stabilité linéaire des écoulements presque parallèles. Part 2. La couche limite de Blasius. J Mécanique 12(1):75–95

van Dyke M (1964) Perturbation methods in Fluid Mechanics. Academic, New York

Fasel H (1976) Investigation of the stability of boundary layers by a finite-difference model of the Navier–Stokes equations. J Fluid Mech 78:355–383

Fasel H, Konzelmann U (1990) Non-parallel stability of a flat-plate boundary layer using the complete Navier–Stokes equations. J Fluid Mech 221:311–347

Haj-Hariri H (1994) Characteristics analysis of the parabolized stability equations. Stud Appl Math 92:41–53

Herbert T (1997) Parabolized stability equations. Ann Rev Fluid Mech 29:245–283

Herbert T, Bertolotti FP (1987) Stability analysis of nonparallel boundary layers. Bull Amer Phys Soc 32(10):2079

Itoh N (1986) The origin and subsequent development in space of Tollmien–Schlichting waves in a boundary layer. Fluid Dyn Res 1:119–130

Kachanov YS, Kozlov VV, Levchenko VY, Maksimov VP (1979) Transformation of external disturbances into the boundary layer waves. In: Proc. Sixth Intern. Conf. on Numerical Methods in Fluid Dyn., Springer–Verlag, Berlin, Lecture Notes in Physics, 19, pp 299–307

Li F, Malik MR (1997) Spectral analysis of parabolized stability equations. Computers & Fluids 23(3):279–297

Murdock JW (1977) A numerical study of nonlinear effects on boundary layer stability. AIAA J 15(8):1167–1173

Nayfeh AH (1973) Perturbation methods. Wiley-Interscience, New York

Nayfeh AH, Saric WS, Mook DT (1974) Stability of non-parallel flows. Arch Mech 26(3):401–406

Saric WS (1990) Low-speed experiments: Requirements for stability measurements. In: Hussaini MY, Voight RG (eds) Instability and Transition, Springer–Verlag, Berlin, ICASE/NASA LaRC Series, vol 1, pp 162–172

Saric WS (1994) Low-speed boundary-layer transition experiments. In: Corke TC, Erlebacher G, Hussaini MY (eds) Transition, Experiments, Theory & Computations. Oxford University Press, Oxford, pp 1–113

Volodin AG (1973) Stability of plane boundary layer with account of nonparallelity. Izv Sib Otd Akad Nauk SSSR, Ser Tekhn Nauk 8(2):14–17

Chapter 5
Instabilities of plane flows over curvilinear surfaces

In this chapter, we consider the linear stability of some flows over a curvilinear wall, which have close relations to the plane shear layers considered in previous chapters: the Couette flow between concentric rotating cylinders, the flow in a curved channel, and the self-similar boundary layer over a curved wall.

5.1 Influence of curvature on the basic flow

Let us first consider the effect of curvature on the velocity profiles of the basic flows, as the velocity profile enters the linearized Navier–Stokes equations as a parameter and, e.g., appearance of an inflection in the profile can lead to a crucial change of the flow stability characteristics. In applications, the surface curvature can experience intricate downstream variations. Therefore, in this section, starting from general equations, we restrict our attention only to internal and external flows with small curvature.

5.1.1 Equations of motion in cylindrical coordinates

In cylindrical coordinates (θ, r, z), the Navier–Stokes momentum equations and the continuity equation take the form

$$\frac{\mathrm{d}U}{\mathrm{d}t} + \frac{UV}{r} = -\frac{1}{r\rho}\frac{\partial P}{\partial \theta} + \nu\left(\nabla^2 U - \frac{U}{r^2} + \frac{2}{r^2}\frac{\partial V}{\partial \theta}\right), \tag{5.1a}$$

$$\frac{\mathrm{d}V}{\mathrm{d}t} - \frac{U^2}{r} = -\frac{1}{\rho}\frac{\partial P}{\partial r} + \nu\left(\nabla^2 V - \frac{V}{r^2} - \frac{2}{r^2}\frac{\partial U}{\partial \theta}\right), \tag{5.1b}$$

$$\frac{\mathrm{d}W}{\mathrm{d}t} = -\frac{1}{\rho}\frac{\partial P}{\partial z} + \nu\nabla^2 W, \tag{5.1c}$$

A.V. Boiko et al., *Physics of Transitional Shear Flows*,
Fluid Mechanics and Its Applications 98, DOI 10.1007/978-94-007-2498-3_5,

$$\frac{1}{r}\frac{\partial U}{\partial \theta}+\frac{\partial V}{\partial r}+\frac{V}{r}+\frac{W}{z}=0\,, \tag{5.1d}$$

where

$$\frac{\mathrm{d}}{\mathrm{d}t}=\frac{\partial}{\partial t}+\frac{U}{r}\frac{\partial}{\partial \theta}+V\frac{\partial}{\partial r}+W\frac{\partial}{\partial z}\,,\quad \nabla^2=\frac{1}{r^2}\frac{\partial^2}{\partial \theta^2}+\frac{\partial^2}{\partial r^2}+\frac{1}{r}\frac{\partial}{\partial r}+\frac{\partial^2}{\partial z^2}\,,$$

and U, V, and W are the azimuthal, radial, and axial velocities, respectively, so that $\mathbf{U}=(U,V,W)^T$.

5.1.1.1 Description of the basic flow between rotating cylinders

The basic flow being taken in the form $U=U(r)=r\Omega(r)$, $V=W=0$, and $P=P(r)$, equations (5.1) give

$$\frac{U^2}{r}=\frac{1}{\rho}\frac{\partial P}{\partial r}\,,$$

$$\nu\frac{\partial}{\partial r}\left(\frac{\partial}{\partial r}+\frac{1}{r}\right)U=0\,.$$

Then, the general solution has the form

$$U(r)=C_1 r+\frac{C_2}{r}\,,$$

where C_1 and C_2 are constants determined by boundary conditions as

$$C_1=\frac{\Omega_2 R_2^2-\Omega_1 R_1^2}{R_2^2-R_1^2}\,, \tag{5.2a}$$

$$C_2=R_1^2 R_2^2\frac{\Omega_1-\Omega_2}{R_2^2-R_1^2}\,, \tag{5.2b}$$

(5.2c)

where R_1 and R_2 are the radii and Ω_1 and Ω_2 the angular velocities of the inner and outer cylinders, respectively.

In the narrow-gap approximation, i.e., assuming that the gap between the cylinders is small,

$$\frac{R_2}{R_1}\approx 1\,,$$

the second term in the expression for the basic flow velocity can be neglected. That is, the velocity is changing linearly across the narrow gap as in the plane Couette flow. Hence, the centrifugal effects on the basic flow velocity in this case are completely neglected.

5.1.1.2 Description of the basic flow in a curved channel

The basic flow being taken in the form $U = U(r)$ and $V = W = 0$, the Navier–Stokes equations are reduced to yield

$$\frac{U^2}{r} = \frac{1}{\rho}\frac{\partial P}{\partial r},$$

$$\nu\frac{\partial}{\partial r}\left(\frac{\partial}{\partial r} + \frac{1}{r}\right)U = \frac{1}{r\rho}\frac{\partial P}{\partial \theta},$$

where $\partial P/\partial \theta$ is a constant. Then, the general solution has the form

$$U(r) = \frac{1}{\rho\nu}\frac{\partial P}{\partial \theta}\left(r\ln r + C_1 r + \frac{C_2}{r}\right), \qquad (5.3)$$

where C_1 and C_2 are constants determined by boundary conditions.

Let us assume that the channel walls are formed by two concentric cylinders with the inner and outer radii R_1 and R_2, respectively, in which case the no-slip boundary conditions require

$$C_1 = -\frac{R_2^2\ln R_2 - R_1^2\ln R_1}{R_2^2 - R_1^2},$$

$$C_2 = \frac{R_1^2 R_2^2}{R_2^2 - R_1^2}\ln\frac{R_2}{R_1}.$$

In the narrow-gap approximation, the velocity profile of the basic flow can be expressed as

$$U(r) = -\frac{d^2}{8\rho\nu R_1}\frac{\partial P}{\partial \theta}\left(1 - 4y^2\right),$$

where new variables

$$y = \frac{r - R_0}{d}, \quad R_0 = \frac{R_1 + R_2}{2}, \quad \text{and} \quad d = R_2 - R_1,$$

where $y \in [-1/2; +1/2]$ were introduced. The profile appears to be parabolic as in the plane Poiseuille flow and the centrifugal effects are also negligible.

5.1.2 Description of the flow in boundary layers over curvilinear surfaces

It is usually convenient to perform calculations of the basic boundary-layer flows in the so-called natural coordinates fitted to the surface, Fig. 5.1.

Let us introduce a dimensionless parameter of surface curvature as $k(x) = 1/R(x)$, where $R(x)$ is the local radius of curvature, normalized to a characteristic length l also involved into the Reynolds number, and denote $\bar{h} = 1/(1 + ky)$. Then, the equations of continuity and momentum take the form

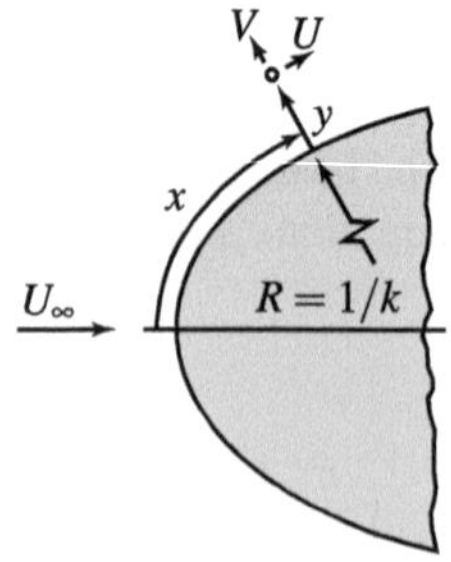

Fig. 5.1 Natural coordinate system for plane flows

$$
\begin{aligned}
U_x + (V/\bar{h})_y &= 0\,, \\
\bar{h}UU_x + VU_y + k\bar{h}UV &= -\bar{h}P_x + \frac{1}{\mathrm{Re}}\left(k^2U_{xx} + U_{yy} - y\bar{h}^3k_xU_x\right. \\
&\quad \left. + k\bar{h}U_y - k^2\bar{h}^2U + 2k\bar{h}^2V_x + k_x\bar{h}^3V\right)\,, \\
\bar{h}UV_x + VV_y - k\bar{h}U^2 &= -\ P_y + \frac{1}{\mathrm{Re}}\left(k^2V_{xx} + V_{yy} - y\bar{h}^3k_xV_x\right. \\
&\quad \left. + k\bar{h}V_y - k^2\bar{h}^2V - 2k\bar{h}^2U_x - k_x\bar{h}^3U\right)\,,
\end{aligned}
$$

where the subscripts denote corresponding directional derivatives, $k > 0$ corresponds to convex surfaces and $k < 0$ corresponds to concave ones.

Under the assumption that $k \sim \mathcal{O}(1)$ and $k_x \sim \mathcal{O}(l/\delta)$, where l and δ are a characteristic streamwise length and a characteristic shear layer thickness, respectively, so that $\delta/l \ll 1$, the momentum equations are reduced to the following boundary-layer equations:

$$
\begin{aligned}
UU_x + VU_y + kUV &= -P_x + \frac{1}{\mathrm{Re}}\left(U_{yy} + kU_y\right)\,, \\
kU^2 &= P_y
\end{aligned}
$$

with the ordinary no-slip boundary conditions at the wall $U = V = 0$. The need of a smooth transition from the boundary layer to the free stream with its local velocity U_e dictates the following conditions:

$$
y \to \infty: \quad U \to U_\mathrm{e}\mathrm{e}^{-ky}, \quad U_y \to -kU_\mathrm{e}\mathrm{e}^{-ky}\,,
$$

which can be obtained by considering the vorticity behavior:

$$
-\frac{\partial U}{\partial y} - kU = 0\,.
$$

If the curvature changes according to

$$
k = \overline{\Omega}\sqrt{\frac{1}{2-\beta}\frac{U_\mathrm{e}}{\nu x}}\,,
$$

and the free-stream velocity is given by

$$
U_\mathrm{e} = Qx^{\beta_\mathrm{H}/(2-\beta_\mathrm{H})}\,,
$$

where Q is a constant, then the boundary layer is self-similar in the wall-normal coordinates η (4.2), and the corresponding generalized Falkner–Skan equation takes the form

$$\Psi_{\eta\eta\eta\eta} + \overline{\Omega}\Psi_{\eta\eta\eta} + \Psi_{\eta\eta\eta}\Psi + \overline{\Omega}\Psi_{\eta\eta}\Psi - (2\beta_{\mathrm{H}} - 1)\left[\Psi_{\eta\eta}\Psi_{\eta} + \overline{\Omega}(\Psi_{\eta})^2\right] = 0\,,$$

with the boundary conditions

$$\Psi = \Psi_{\eta} = 0 \quad \text{at} \quad \eta = 0; \quad \Psi_{\eta} \to \mathrm{e}^{-\overline{\Omega}\eta}, \quad \Psi_{\eta\eta} \to -\overline{\Omega}\mathrm{e}^{-\overline{\Omega}\eta}\,, \quad \text{as} \quad \eta \to \infty\,,$$

where the subscript 'η' denotes the first derivative with respect to η. In case of the small curvature, the curvature can be neglected completely, hence, recovering the Falkner–Skan equation for plane self-similar boundary layers.

5.2 Hydrodynamic instability at curvilinear surfaces

With respect to the Tollmien–Schlichting instability in the plane-parallel-flow approximation, the changes implanted by the presence of a small curvature can be taken into account by a corresponding modification of the linearized equations of motion. Special studies show, however, that the modification of the plane equations provides only a limited correction to the plane solutions.

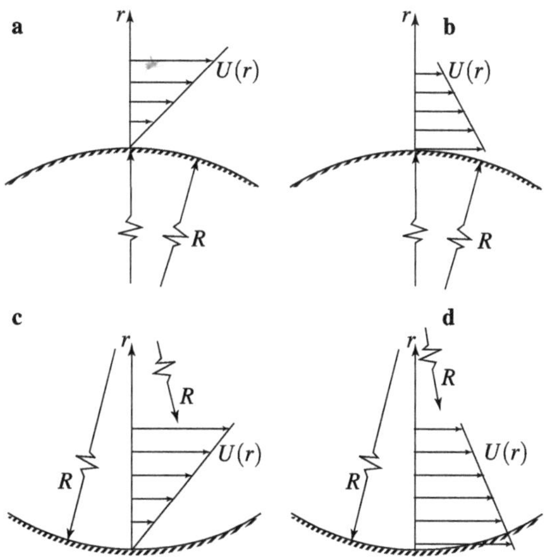

Fig. 5.2 Application of the Rayleigh circulation criterion: stable (**a**), unstable (**b**), unstable (**c**), stable (**d**)

Meanwhile, the surface curvature can cause another instability invoked by imbalance between pressure and centrifugal forces. In contrast to the Tollmien–Schlichting instability, this instability can appear even in the inviscid limit. A criterion for its onset was given by Lord Rayleigh (1916). Considering a flow in a cylindrical coordinate system (θ, r, z) and the velocity profile of the form $\mathbf{U} = (U(r), V = 0, W = 0)^T$, he found that the necessary condition for stability to axisymmetric disturbances is

$$\frac{d\Gamma^2}{dr} > 0\,,$$

where $\Gamma = rU$ is the circulation. This can be illustrated as follows. In a steady state, the pressure gradient balances the centrifugal force $\rho U^2/r$. If a fluid ring changes its radial position from r_1 to r_2 $(r_1 < r_2)$, then the centrifugal force becomes equal to $\rho(r_1U_1/r_2)^2/r_2 = \rho(r_1U_1)^2/r_2^3$. This value is larger than the pressure gradient $\rho U_2^2/r$ when $(r_1U_1)^2 > (r_2U_2)^2$ or $\Gamma_1^2 > \Gamma_2^2$. Hence, the ring moves further out of r_1, that is, such a flow is unstable. Otherwise, if $(r_1U_1)^2 < (r_2U_2)^2$ or $\Gamma_1^2 < \Gamma_2^2$, the ring tends to its original position, i.e., the flow is stable.

Examples, illustrating application of the criterion are shown in Fig. 5.2. In the Couette flow between rotating cylinders, when only the inner disk rotates, the case depicted in Fig. 5.2**b** can be realized. Corresponding instability is called the Taylor instability, as it was considered for the first time by Taylor (1923). Analogous instability in a curved channel is called the Dean instability (Dean, 1928). Both basic flows are internal with fixed distances between streamlines (as in plane parallel flows). However, the Couette flow is closed and the curved channel flow is open. It leads to predominance of absolute centrifugal instability in the Couette flow and convective centrifugal instability in the channel flow. The case depicted in Fig. 5.2**c** is characteristic for external flow with no-slip boundary conditions at the wall. Corresponding instability is called the Görtler instability (Görtler, 1941). Special numerical and experimental studies show that it is of convective nature in most practical cases. It becomes absolute, e.g., at very large curvatures and wavelengths, that is mainly of academic interest.

It was shown by Synge (1933) that the Rayleigh circulation criterion is also a sufficient condition for stability of swirling flows to axisymmetric disturbances. However, non-axisymmetric (e.g., helical) disturbances, which are not covered by the Rayleigh arguments, can be amplifying in the inviscid limit, even if the criterion of flow stability holds. On the contrary, viscosity has a stabilizing influence on centrifugal instability; at low enough Reynolds numbers, a limit of instability should be reached.

Laboratory experiments indicate that centrifugal instability tends in many cases to form stationary vortices directed along the basic flow. It appears that they, rather than the Tollmien–Schlichting waves, play the primary role during the transition to turbulence at, for example, turbine blades and compressors, walls of some nozzles, lower wing surfaces, curved channels in heat exchangers, etc. Note also that, for the instability to appear, the curvature of streamlines, rather than the proximity to the wall is important. Hence, it can, in principle, occur in a vicinity of the stagnation point or at flow separation, where the streamline curvature is significant.

5.2.1 Taylor problem

In the inviscid limit, the Rayleigh circulation criterion specified for the flow under consideration (Fig. 5.3) indicates that it is stable to axisymmetric disturbances if

$$\frac{\mathrm{d}\,[rU(r)]^2}{\mathrm{d}r} = \frac{\mathrm{d}(C_1 r^2 + C_2)^2}{\mathrm{d}r} = 4C_1 r^2 U(r) > 0\,, \tag{5.4}$$

or

$$\left(\Omega_2 R_2^2 - \Omega_1 R_1^2\right) U(r) > 0\,,$$

where $R_2 > R_1$ is assumed. This happens everywhere in the flow if the cylinders rotate in the same direction and the circulation of the outer cylinder is larger than that of the inner one. In the Ω_1–Ω_2-plane, the neutral stability condition corresponds to a ray from the coordinate origin $\Omega_2 \sim \Omega_1 R_1^2/R_2^2$, called the *Rayleigh line*. The region of stability is to the right of it and the region of instability is, correspondingly, to the left (Fig. 5.4). When the cylinders rotate in opposite directions, it is easy to show that the Rayleigh circulation criterion is always violated at the inner cylinder and satisfied at the outer. These two regions of the flow are separated by the so-called *nodal radius* on which the angular velocity of the basic flow is equal to zero.

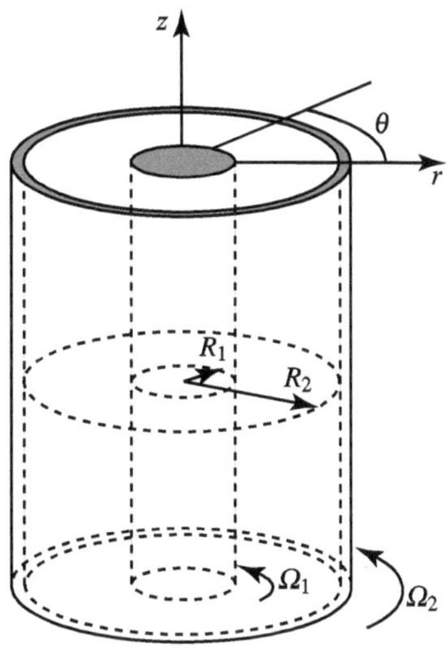

Fig. 5.3 Co-rotating concentric cylinders

The role of viscosity for this essentially inviscid instability is only damping. Hence, the instability is expected when viscosity is relatively small if the cylinders rotate in different directions or, otherwise, somewhere in a region to the left of the Rayleigh line. Assuming small velocity and pressure disturbances in the form $\mathbf{u} = (u, v, w)^T$ and p, respectively, and performing a linearization procedure similar to that in Sect. 2.1, the equations of motion (5.1) yield the equations

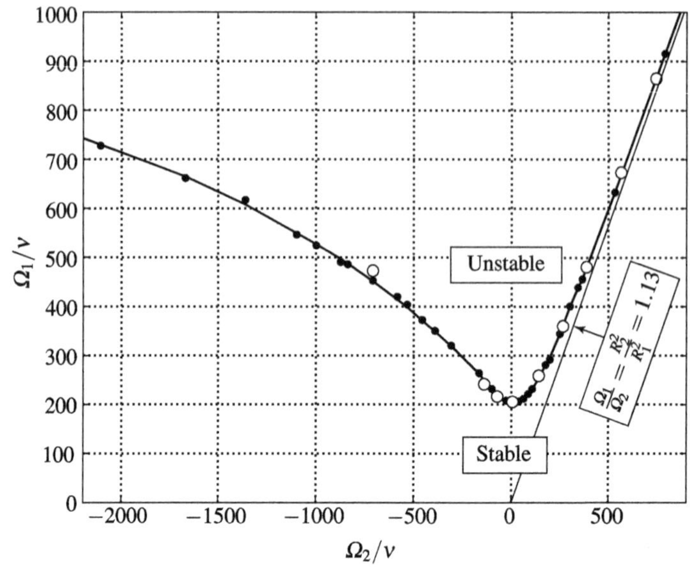

Fig. 5.4 Comparison of observed (•) and calculated (○) velocities at which instability first appears: $R_1 = 3.80\,\text{cm}$, $R_2 = 4.035\,\text{cm}$ (Taylor, 1923)

$$\frac{\partial u}{\partial t} + \frac{U}{r}\frac{\partial u}{\partial \theta} + \left(\frac{\partial U}{\partial r} + \frac{U}{r}\right) v = -\frac{1}{r}\frac{\partial p}{\partial \theta} + \frac{1}{\text{Re}}\left(\nabla^2 u - \frac{u}{r^2} + \frac{2}{r^2}\frac{\partial v}{\partial \theta}\right), \tag{5.5a}$$

$$\frac{\partial v}{\partial t} + \frac{U}{r}\frac{\partial v}{\partial r} - 2\frac{U}{r}u = -\frac{\partial p}{\partial r} + \frac{1}{\text{Re}}\left(\nabla^2 v - \frac{v}{r^2} - \frac{2}{r^2}\frac{\partial u}{\partial \theta}\right), \tag{5.5b}$$

$$\frac{\partial w}{\partial t} + \frac{U}{r}\frac{\partial w}{\partial \theta} = -\frac{\partial p}{\partial z} + \frac{1}{\text{Re}}\nabla^2 w, \tag{5.5c}$$

$$\frac{\partial v}{\partial r} + \frac{v}{r} + \frac{1}{r}\frac{\partial u}{\partial \theta} + \frac{w}{z} = 0, \tag{5.5d}$$

which are scaled with respect to the characteristic velocity U_{ch} and length l ($\text{Re} = U_{\text{ch}} l/\nu$). Assuming the axisymmetric behavior of the disturbances, the equations can be simplified further yielding (on eliminating the pressure), the system of two equations

$$\left[\frac{\partial}{\partial t} - \frac{1}{\text{Re}}\left(\nabla^2 - \frac{1}{r^2}\right)\right]\left(\nabla^2 v - \frac{v}{r^2}\right) = 2\frac{U}{r}\frac{\partial^2 u}{\partial z^2}, \tag{5.6a}$$

$$\left[\frac{\partial}{\partial t} - \frac{1}{\text{Re}}\left(\nabla^2 - \frac{1}{r^2}\right)\right] u = -\left(\frac{\partial U}{\partial r} + \frac{U}{r}\right) v, \tag{5.6b}$$

with the boundary conditions $u = v = \partial v/\partial r = 0$ at the walls. Using the normal-mode approach, in which the elementary disturbances take the form

$$u(r,z,t) = \hat{u}(r)\mathrm{e}^{\mathrm{i}(\beta z - \omega t)}, \quad v(r,z,t) = \hat{v}(r)\mathrm{e}^{\mathrm{i}(\beta z - \omega t)},$$

yields

$$\left[\mathrm{i}\omega + \frac{1}{\mathrm{Re}}\left(\hat{\Delta} - \frac{1}{r^2}\right)\right]\left(\hat{\Delta} - \frac{1}{r^2}\right)\hat{v} = 2\beta^2 \frac{U}{r}\hat{u}\,,$$

$$\left[\mathrm{i}\omega + \frac{1}{\mathrm{Re}}\left(\hat{\Delta} - \frac{1}{r^2}\right)\right]\hat{u} = \left(U_r + \frac{U}{r}\right)\hat{v}\,,$$

where the subscript 'r' denotes the first derivative with respect to r and

$$\hat{\Delta} = \frac{\mathrm{d}^2}{\mathrm{d}r^2} + \frac{1}{r}\frac{\mathrm{d}}{\mathrm{d}r} - \beta^2\,.$$

Finally, in the case of a narrow gap, the terms factored by $1/r^2$ and $1/r(\mathrm{d}/\mathrm{d}r)$ are small compared to the others and can be neglected. Recalling that $U = \Omega r$, where $\Omega \approx C_1$ is an effective circular velocity defined in (5.2a), we obtain

$$\left[\mathrm{i}\omega + \frac{1}{\mathrm{Re}}\left(\frac{\mathrm{d}^2}{\mathrm{d}r^2} - \beta^2\right)\right]\hat{u} = 2\Omega\hat{v}\,,$$

$$\left[\mathrm{i}\omega + \frac{1}{\mathrm{Re}}\left(\frac{\mathrm{d}^2}{\mathrm{d}r^2} - \beta^2\right)\right]\left(\frac{\mathrm{d}^2}{\mathrm{d}r^2} - \beta^2\right)\hat{v} = 2\beta^2\Omega\hat{u}\,.$$

The last set of equations can be rewritten as

$$\left(\frac{\mathrm{d}^2}{\mathrm{d}r^2} - \beta^2 - \sigma\right)\hat{u} = \tilde{v}\,, \tag{5.7a}$$

$$\left(\frac{\mathrm{d}^2}{\mathrm{d}r^2} - \beta^2 - \sigma\right)\left(\frac{\mathrm{d}^2}{\mathrm{d}r^2} - \beta^2\right)\tilde{v} = -\beta^2 \mathrm{Ta}\Omega\hat{u}\,, \tag{5.7b}$$

where

$$\tilde{v} = -2\Omega\mathrm{Re}\hat{v}\,, \quad \sigma = -\mathrm{i}\omega\mathrm{Re}$$

and the customary Taylor number $\mathrm{Ta} = -4\Omega\mathrm{Re}^2$ is introduced instead of the Reynolds number. The reason of it is clear if we compare the expression with condition (5.4) separating the regions of stability and instability in the inviscid case. The choice of σ instead of ω is stipulated by the observation that the flow is stable to any axisymmetric disturbance periodic in time, i.e., the principle of exchange of stabilities holds in this case. Hence, instability leads to growth of stationary toroidal (Taylor) vortices, which are unstable, when $\sigma > 0$, σ playing the role of growth rate. The numerical technique developed for solving the Orr–Sommerfeld equation in Chap. 3 can be easily adapted to (5.7).

The set of equations (5.7) can be combined by eliminating $\tilde{v}$ to obtain one six-order differential equation

$$\left(\frac{\mathrm{d}^2}{\mathrm{d}r^2} - \beta^2 - \sigma\right)^2\left(\frac{\mathrm{d}^2}{\mathrm{d}r^2} - \beta^2\right)\hat{u} = -\beta^2 \mathrm{Ta}\Omega\hat{u}$$

with the boundary conditions

$$\hat{u} = \left(\frac{\mathrm{d}^2}{\mathrm{d}r^2} - \beta^2 - \sigma\right)\hat{u} = \frac{\mathrm{d}}{\mathrm{d}r}\left(\frac{\mathrm{d}^2}{\mathrm{d}r^2} - \beta^2 - \sigma\right)\hat{u} = 0\,. \tag{5.8}$$

Then, the neutral stability is determined by the condition

$$\left(\frac{\mathrm{d}^2}{\mathrm{d}r^2}-\beta^2\right)^3\hat{u}=-\beta^2\mathrm{Ta}\Omega\hat{u}\,.$$

In dimensionless form,

$$\Omega=\frac{\Omega_2 R_2^2-\Omega_1 R_1^2}{(R_2+R_1)U_{\mathrm{ch}}}\,.$$

The choice of the length and velocity scales l and U_{ch} is arbitrary. Let, e.g., $l=R_2-R_1$ and $U_{\mathrm{ch}}=\Omega_1 l$. Then, the explicit expression for the Taylor number is

$$\mathrm{Ta}=-4\Omega\frac{\Omega_1^2 l^4}{\nu^2}=\left[2\frac{\Omega_1(R_2-R_1)^2}{\nu}\right]^2\frac{\bar{\eta}^2-\bar{\mu}}{1-\bar{\eta}^2}=\left(2\frac{\Omega_1 R_1^2}{\nu}\right)^2\left(\frac{1}{\bar{\eta}}-1\right)^4\frac{\bar{\eta}^2-\bar{\mu}}{1-\bar{\eta}^2}\,,$$

where

$$\bar{\mu}=\frac{\Omega_2}{\Omega_1},\quad \bar{\eta}=\frac{R_1}{R_2}<1\,.$$

As seen, the denominator of the last factor in the expression for Ta is always positive, while the sign of the numerator, and that of the Taylor number, depends on the ratios $\bar{\mu}$ and $\bar{\eta}$.

When $\bar{\mu}\lesssim-0.78$, helical disturbances become amplifying at smaller Taylor numbers than axisymmetrical ones. In that case, one has to consider the complete eigenvalue problem starting from equations (5.5) and assuming the normal modes in the form

$$\begin{aligned}u(r,z,t)&=\hat{u}(r)\mathrm{e}^{\mathrm{i}(n\theta+\beta z-\omega t)}\,,\\ v(r,z,t)&=\hat{v}(r)\mathrm{e}^{\mathrm{i}(n\theta+\beta z-\omega t)}\,,\\ w(r,z,t)&=\hat{w}(r)\mathrm{e}^{\mathrm{i}(n\theta+\beta z-\omega t)}\,,\end{aligned}$$

where the discrete azimuthal wavenumber or the azimuthal wave index n is integer. It is interesting to note that, at first, the leading mode at $\bar{\mu}\approx-0.78$ has $n=\pm1$, but, as $\bar{\mu}$ becomes even smaller, the azimuthal wavenumber of the leading mode takes higher values in the sequence $n=\pm2,\pm3,\ldots$.

5.2.2 Dean problem

The Rayleigh circulation criterion applied to the basic flow in a narrow curved channel (Fig. 5.5) indicates that the flow is locally stable in the inviscid limit in the interval $-1/2<y<0$ and unstable at $0<y<1/2$. The viscous equations (5.6) and the corresponding boundary conditions (5.8) are valid for the general velocity distribution in the channel (5.3) as well as for the flow in a narrow gap. In the last case, after the replacement $\tilde{v}=(3U_{\mathrm{m}}d^2/R_1\nu)\beta^2\hat{v}$, where the basic flow velocity

$$U(r)=\frac{3}{2}U_{\mathrm{m}}\left(1-4y^2\right),\quad U_{\mathrm{m}}=-\frac{d^2}{12\rho\nu R_1}\frac{\partial P}{\partial\theta},\quad \frac{\partial P}{\partial\theta}=\mathrm{const.}\,,$$

is introduced, the traditional form of the equations for normal modes is recovered:

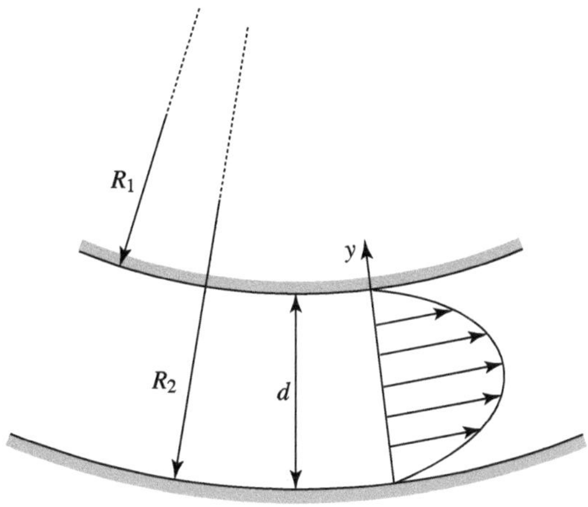

Fig. 5.5 Flow in a curved channel

$$\left(\frac{\mathrm{d}^2}{\mathrm{d}r^2} - \beta^2 - \sigma\right)\hat{u} = -\beta^2\bar{\Lambda}y\tilde{v}\,, \tag{5.9a}$$

$$\left(\frac{\mathrm{d}^2}{\mathrm{d}r^2} - \beta^2 - \sigma\right)\left(\frac{\mathrm{d}^2}{\mathrm{d}r^2} - \beta^2\right)\tilde{v} = \left(1 - 4y^2\right)\hat{u}\,, \tag{5.9b}$$

with the boundary conditions $\hat{u} = \tilde{v} = \mathrm{d}\tilde{v}/\mathrm{d}r = 0$, where

$$\bar{\Lambda} = \frac{36\mathrm{Re}^2 d}{R_1} = 36\mathrm{De}^2\,, \quad \mathrm{Re} = \frac{U_{\mathrm{m}} d}{\nu}\,,$$

and De is the Dean number. Note that the principle of exchange of stabilities cannot be proven for this flow. However, due to numerous experimental observations that the instability in curved channels is dominated by stationary vortices, and the instability being assumed convective, it is customary to put $\sigma = 0$, thus, reducing the number of parameters:

$$\left(\frac{\mathrm{d}^2}{\mathrm{d}r^2} - \beta^2\right)\hat{u} = -\beta^2\bar{\Lambda}y\tilde{v}\,, \tag{5.10a}$$

$$\left(\frac{\mathrm{d}^2}{\mathrm{d}r^2} - \beta^2\right)\left(\frac{\mathrm{d}^2}{\mathrm{d}r^2} - \beta^2\right)\tilde{v} = \left(1 - 4y^2\right)\hat{u}\,. \tag{5.10b}$$

The numerical technique developed for solving the Orr–Sommerfeld equation in Chap. 3 can be easily adapted to (5.10). The calculated critical parameters for this flow are De = 35.92 at $\beta = 3.95$. The perturbations originating due to the instability are called the Dean, or sometimes, the Görtler vortices because of their similarity to the vortex structure of boundary layers, which are described in the next section in more detail.

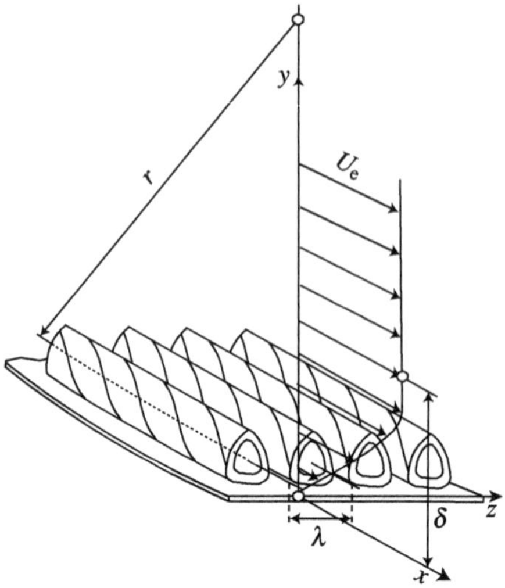

Fig. 5.6 Görtler vortices on a concave wall

5.2.3 Görtler problem

The Görtler instability is wide-spread in boundary layers over curved concave walls (Fig. 5.6). It is often responsible for amplification of streamwise counter-rotating vortices whose growth can lead to a premature laminar–turbulent transition in a wide range of free-stream speeds. For the first time, it was probably shown in experiments of Clauser and Clauser (1937) that the laminar boundary layers over concave walls become turbulent at lower Reynolds numbers than those over flat or convex surfaces. A relation of the early flow turbulization to the Görtler instability was demonstrated by Liepmann (1943). The first direct experimental evidence of the existence of stationary Görtler vortices was obtained by china-clay surface visualization by Gregory and Walker (1956) and by hot-wire measurements by Tani (1962). The experiments and subsequent analyses showed that the instability has a convective character.

Later numerous flow visualizations and detailed measurements (see, e.g., Ito, 1987) helped to improve significantly the understanding of this phenomenon and to evaluate its influence on the laminar–turbulent transition in boundary layers. The majority of corresponding studies were devoted to stationary Görtler vortices, though the principle of exchange of stabilities is not applicable in this case and non-stationary vortices are also observed in real transitional flows. Moreover, even for stationary Görtler vortices, no quantitative agreement between the experimental and theoretical linear stability characteristics has been obtained until recently.

Görtler (1941) adopted the quasi-parallel-flow approximation and used the normal-mode analysis to show that instability in the form of stationary vortices can occur in boundary layers over concave surfaces when a certain parameter (the Görtler number, akin of the Dean number for curved channels),

$$\text{Gö}_\delta = \text{Re}_\delta\sqrt{\frac{\delta}{R}},$$

exceeds a critical value. We define the Reynolds number here as $\text{Re}_\delta = U_e\delta/\nu$, where the length δ characterizes the boundary-layer thickness, U_e is the free-stream velocity at the boundary-layer edge, ν denotes the kinematic viscosity of the fluid, and R is the radius of wall curvature. The critical Görtler number changes with the disturbance transverse wavenumber β (normalized usually by the length scale δ). The dependence $\text{Gö}(\beta)$ forms a neutral stability curve defined in the Gö–β-plane.

In contrast to the flow along a flat plate, it appeared that the quasi-parallel-flow approximation has serious shortcomings in this case. For example, it leads to a non-physical conclusion that the critical Görtler number behaves as $\text{Gö}_L \to 0$ when $\beta \to 0$. Hall (1982, 1983) showed that the reduction of the problem under consideration to a set of homogeneous ordinary differential equations is not justified for values of $\text{Gö} \lesssim \mathcal{O}(1)$ and $\beta \lesssim \mathcal{O}(1)$. The neglect of the downstream variation of the vortex shape (i.e., consideration of the wall-normal amplitude function as independent of x and prehistory of its development) is invalid in this case. To account for the prehistory, one has to formulate the problem as a set of partial differential equations and solve the corresponding parabolic initial-boundary-value problem. Corresponding numerical tests show that the behavior of stationary vortices at small Görtler numbers depends substantially on initial conditions and determination of a unique neutral stability curve loses practical significance. In such a case, the most accurate approach, which allows one to trace the development of vortices from very beginning, consists in formulating and solving the receptivity problem (see Chap. 10 for details) to provide proper initial conditions.

Let us denote the disturbance velocity-vector components corresponding to the Cartesian coordinate system and pressure as $u(x,y,z,t)$, $v(x,y,z,t)$, $w(x,y,z,t)$, and $p(x,y,z,t)$, and the basic flow velocity as $\mathbf{U} = (U(x,y), V(x,y), 0)^T$. To be consistent with previous theoretical studies, in deriving the governing equations all these variables are made non-dimensional using the so-called boundary-layer scaling, which differs from the experimental one in using the Reynolds number $\text{Re}_0 = \sqrt{U_0 x_0/\nu}$ instead of local Re_{δ^*} based on the boundary-layer displacement thickness δ^*. We assume that the excitation of disturbances occurs locally at $x = x_0 > 0$ from inside the boundary layer, therefore x is scaled by x_0, and the streamwise velocities U and u are normalized with respect to the free-stream velocity U_0 at x_0. The wall-normal and transverse coordinates y and z as well as the wall radius of curvature are made dimensionless with respect to x_0/Re_0, while the wall-normal and transverse velocities V, v, and w are scaled with U_0/Re_0. The pressure is scaled with $\rho U_0^2/\text{Re}_0^2$ and time t is scaled with x_0/U_0.

Based on flow visualizations, it is believed usually that stationary vortices dominate in the flows in the presence of the Görtler instability. This is plausible quite often, but cannot be rigorously justified for all practical cases. Non-stationary vortices can be initiated by low-frequency vortical perturbations of incoming flows and seem to appear frequently in many practical situations (see, e.g., Schultz and Volino, 2003). Such perturbations are typical, for example, in boundary layers over curved blades of turbomachines. Consequently, the apparent predominance of stationary

Görtler vortices in numerous experiments might be explained, e.g., by a stronger boundary-layer receptivity to these disturbances, by larger dimensionless amplitudes of surface roughness, or free-stream nonuniformities in comparison with amplitudes of surface vibrations or free-stream velocity fluctuations, rather than by their largest amplification rates.

Keeping it in mind, let us consider stability of a two-dimensional boundary layer over a concave wall with a radius of curvature $R \gg 1$ at the high-Reynolds-number limit. In this case, the continuity equation and Navier–Stokes momentum equations linearized about the steady two-dimensional basic flow reduce to

$$\begin{aligned} u_x + v_y + w_z &= 0\,, \\ u_t + Uu_x + U_x u + Vu_y + U_y v &= u_{yy} + u_{zz}\,, \\ v_t + Uv_x + V_x u + Vv_y + V_y v + 2\text{Gö}^2 Uu + p_y &= v_{yy} + v_{zz}\,, \\ w_t + Uw_x + Vw_y + p_z &= w_{yy} + w_{zz}\,, \end{aligned}$$

where the subscripts 't' 'x', 'y', and 'z', means differentiating in t, x, y, and z, respectively. The equations are parabolic and, consequently, require initial conditions at x_0 and appropriate boundary conditions at the wall and in the free stream.

Let us assume that the disturbances are harmonic in time. Next, since the basic flow is independent of the transverse direction, it is possible to consider solutions periodic in z with a transverse wavenumber $\beta = 2\pi/\lambda_z$ and assume solutions of the form

$$\begin{pmatrix} u \\ v \\ w \\ p \end{pmatrix} = \begin{pmatrix} \bar{u}(x,y)\cos\beta z \\ \bar{v}(x,y)\cos\beta z \\ \bar{w}(x,y)\sin\beta z \\ \bar{p}(x,y)\cos\beta z \end{pmatrix} \mathrm{e}^{-\mathrm{i}\omega t}\,.$$

After simple transformations, this yields the conservative form of the equations being convenient for their numerical solution by marching

$$\bar{u}_x + \bar{v}_y + \beta\bar{w} = 0\,, \tag{5.11a}$$

$$(U\bar{u})_x + V\bar{u}_y + U_y\bar{v} = \bar{u}_{yy} - \beta^2\bar{u} + \mathrm{i}\omega\bar{u}\,, \tag{5.11b}$$

$$(U\bar{v} + V\bar{u})_x + 2(V\bar{v})_y + \beta V\bar{w} + 2\text{Gö}^2 U\bar{u} + \bar{p}_y = \bar{v}_{yy} - \beta^2\bar{v} + \mathrm{i}\omega\bar{v}\,, \tag{5.11c}$$

$$(U\bar{w})_x + (V\bar{w})_y - \beta\bar{p} = \bar{w}_{yy} - \beta^2\bar{w} + \mathrm{i}\omega\bar{w}\,. \tag{5.11d}$$

We will call these equations as linear *nonparallel stability equations* (NSEs).

When Gö $\gtrsim \mathcal{O}(1)$ and $\beta \gtrsim \mathcal{O}(1)$, the normal-mode approach is acceptable and the problem is reducible to a set of ordinary differential equations. Assuming the factorization

$$\bar{u}(x,y) = \hat{u}(y)\mathrm{e}^{\mathrm{i}\alpha x},\ \bar{v}(x,y) = \hat{v}(y)\mathrm{e}^{\mathrm{i}\alpha x},\ \bar{w}(x,y) = \hat{w}(y)\mathrm{e}^{\mathrm{i}\alpha x},\ \bar{p}(x,y) = \hat{p}(y)\mathrm{e}^{\mathrm{i}\alpha x}$$

is admissible, where α is a local complex streamwise wavenumber, yields a quasi-parallel flow approximation

$$\mathrm{i}\alpha\hat{u} + \hat{v}_y + \beta\hat{w} = 0\,, \tag{5.12a}$$

$$U_x\hat{u} + \mathrm{i}\alpha U\hat{u} + V\hat{u}_y + U_y\hat{v} = \hat{u}_{yy} - \beta^2\hat{u} + \mathrm{i}\omega\hat{u}\,, \tag{5.12b}$$

$$\mathrm{i}\alpha U\hat{v}+V_x\hat{u}+V\hat{v}_y+V_y\hat{v}+2\mathrm{G\ddot{o}}^2U\hat{u}+\hat{p}_y=\hat{v}_{yy}-\beta^2\hat{v}+\mathrm{i}\omega\hat{v}\,, \tag{5.12c}$$

$$\mathrm{i}\alpha U\hat{w}+V\hat{w}_y-\beta\hat{p}=\hat{w}_{yy}-\beta^2\hat{w}+\mathrm{i}\omega\hat{w}\,. \tag{5.12d}$$

Note that, when $\omega=0$, α is purely imaginary. If β, ω, Gö, $U(x,y)$ and $V(x,y)$ are given, the equations constitute the eigenvalue problem in α. We denote these equations as linear *locally-nonparallel stability equations* (LNSEs).

The equations can be reduced even more by neglecting downstream variations of the basic flow (i.e., assuming local parallelism of the flow streamlines), yielding

$$\mathrm{i}\alpha\hat{u}+\hat{v}_y+\beta\hat{w}=0\,, \tag{5.13a}$$

$$\mathrm{i}\alpha U\hat{u}+U_y\hat{v}=\hat{u}_{yy}-\beta^2\hat{u}+\mathrm{i}\omega\hat{u}\,, \tag{5.13b}$$

$$\mathrm{i}\alpha U\hat{v}+2\mathrm{G\ddot{o}}^2U\hat{u}+\hat{p}_y=\hat{v}_{yy}-\beta^2\hat{v}+\mathrm{i}\omega\hat{v}\,, \tag{5.13c}$$

$$\mathrm{i}\alpha U\hat{w}-\beta\hat{p}=\hat{w}_{yy}-\beta^2\hat{w}+\mathrm{i}\omega\hat{w}\,. \tag{5.13d}$$

This approximation is close in essence to that used by Görtler (1941) with the exception that he considered the temporal formulation of instability to stationary (zero-frequency) disturbances and, consequently, assumed ω to be purely imaginary and put $\alpha=0$. This set of equations will be denoted as linear *locally-parallel stability equations* (LPSEs). It was shown using asymptotic methods by Hall (1982, 1983) that the neutral stability curve in the limit of $\beta\gg 1$ and $\omega=0$ is described by the formula

$$\mathrm{G\ddot{o}}^2=2.95\beta^4\left[1+\frac{0.96}{\beta}+\mathcal{O}\left(\frac{1}{\beta^2}\right)\right].$$

According to experimental observations, the dimensional wavelength of the vortices λ does not change downstream. That is, their transverse wavenumbers normalized by the characteristic boundary-layer thickness grow downstream and sooner or later β exceeds $\mathcal{O}(1)$. In practice, however, it happens at supercritical Görtler numbers. Hence, in the framework of the normal-mode approach, solutions inside the instability region rather than the neutral stability curve are of primary interest. Similarly to the use of the frequency parameter F independent of x in describing the downstream behavior of the Tollmien–Schlichting waves, it is reasonable to consider the wavelength parameter independent of the boundary-layer thickness

$$\Lambda=\frac{\lambda U_\infty}{\nu}\sqrt{\frac{\lambda}{R}}=\left(\frac{2\pi}{\beta}\right)^{3/2}\mathrm{G\ddot{o}}\,.$$

The numerical technique developed for solving the Orr–Sommerfeld equation in Chap. 3 can be easily adapted to (5.12). Corresponding growth rates $-\alpha_{\mathrm{i}}$ of stationary vortices are given in Fig. 5.7. The most short-wavelength neutrally stable stationary vortex has $\Lambda_{\mathrm{crit}}=44.29$, while the most unstable (at large Görtler numbers) has $\Lambda\approx 210$. For comparison, results of some experimental studies are also given in Fig. 5.7.

The problem of solving (5.11) is different. Let us write this system of equations compactly in an operator form as

$$(\mathscr{M}\mathbf{q})_x=\mathscr{L}\mathbf{q}\,,$$

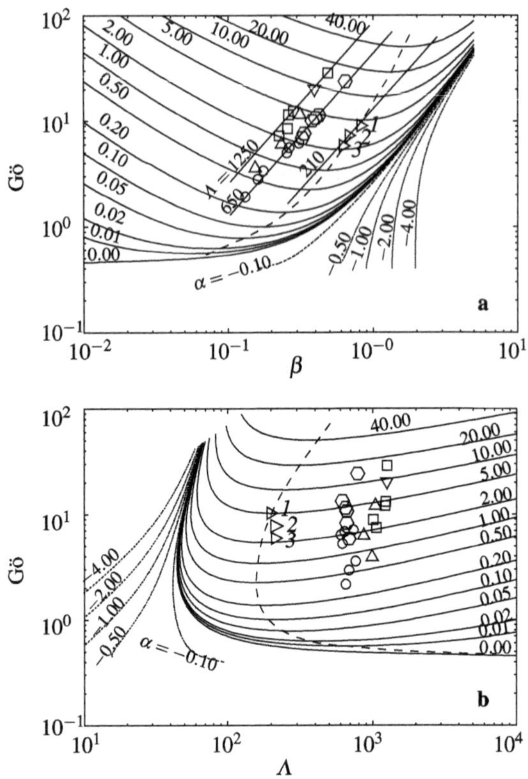

Fig. 5.7 Isolines of the increments (growth rates) $-\alpha_i$ (*solid lines*) and decrements (*dotted lines*), as well as the line of the maximum growth (*dashed lines*) for the Blasius boundary-layer velocity profile as functions of Gö and β (**a**) and Gö and Λ (**b**), as well as experimental data on the growing stationary disturbances Tani and Sakagami (1962) ($\circ, \square, \triangle, \triangledown, \diamond$), Bippes (1972) ($\triangleright$); after (Floryan and Saric, 1982)

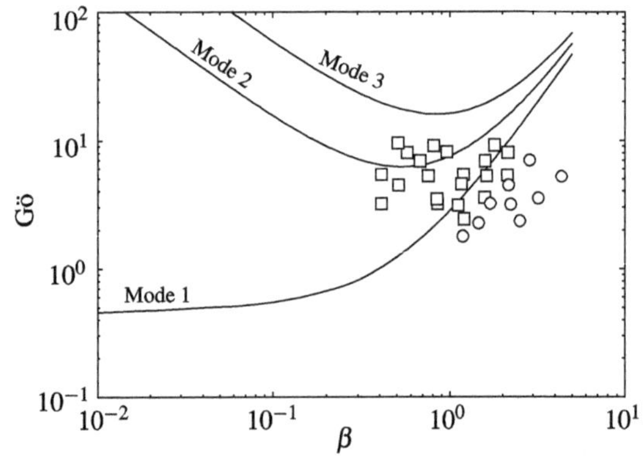

Fig. 5.8 Neutral stability curves for three first Görtler-instability normal modes in the Blasius boundary-layer and experimental data on growing ($\square$) and decaying($\circ$) disturbances (Bippes, 1972); after (Floryan and Saric, 1984)

where $\mathbf{q} = (u, v, w, p)^T$; $\mathscr{M}$ and $\mathscr{L}$ are operators depending only on y. Such an equation can be approximated in the streamwise direction by a finite-difference scheme, e.g.,

$$\mathscr{M}^{(i+1)}\mathbf{q}^{(i+1)} - \mathscr{M}^{(i)}\mathbf{q}^{(i)} = \Delta x \mathscr{L}^{(i+1)}\mathbf{q}^{(i+1)}, \tag{5.14}$$

where $i = 1, 2, \ldots, n-1$, $n\Delta x = x_f - x_0$: x_0 and x_f are the initial and final streamwise positions. We can approximate the operators $\mathscr{M}^{(i)}$ and $\mathscr{L}^{(i)}$ at each step i in x by corresponding collocation matrices $\mathbf{M}^{(i)}$ and $\mathbf{L}^{(i)}$ using the Chebyshev polynomials and solve the resultant one-dimensional boundary-value problems:

$$\mathbf{q}^{(i+1)} = \left[\left(\mathbf{M}^{(i+1)} - \Delta x \mathbf{L}^{(i+1)}\right)^{-1} \mathbf{M}^{(i)}\right] \mathbf{q}_i .$$

See function `GoertlerPS` implementing the algorithm of the solution for stationary vortices.

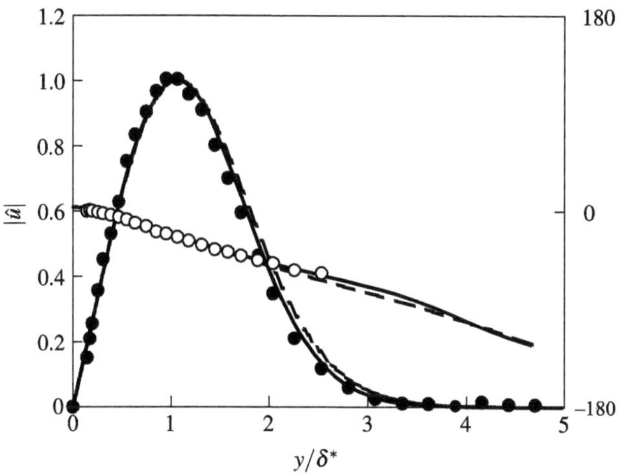

Fig. 5.9 Experimental amplitude (•) and phase (∘) profiles of the boundary-layer streamwise disturbance velocity in comparison with those calculated by NSEs (*solid lines*) and LPSEs (*dashed lines*) for the first Görtler mode at Gö = 17.2, $\Lambda = 149$, $F = 5.67 \times 10^{-6}$ (Boiko et al., 2010)

A specific feature of the Görtler instability in the normal-mode approach is the appearance of several unstable modes at large Gö. Corresponding neutral stability curves for the first three stationary modes are given in Fig. 5.8. An example of the first-mode velocity profile is given in Fig. 5.9. According to these and numerous other experimental data, the streamwise velocity of unstable vortices in the initial (linear) stage of the laminar–turbulent transition has only one maximum (at least at $F \lesssim 20 \times 10^{-6}$) that is a characteristic of the first mode. Numerically obtained growth rates of the first mode also are dominant at least to Gö = 100. Hence, the junior modes are hardly of practical importance in the most interesting range of parameters.

An example of experimentally and numerically determined increments and phase velocities of the non-stationary Görtler vortex with $\Lambda = 149$ as functions of the frequency parameter F are shown in Fig. 5.10. Generally, both theoretical approaches agree very well with the measurements at the given values of Λ and Gö: as seen, the increments decrease with frequency, the flow becoming neutrally stable

```
function [uz,vz,wz]=...
         GoertlerPS(x,U,Uy,V,Vy,b,Go,D,u0,v0,w0)
% GOERTLERPS Goertler instability calculation using
%                  parabolic equations.
% x          equidistant ascending x coordinates
% U,V,Uy,Vy basic velocities and their derivatives in
%            Gauss-Lobatto points from x(1) to x(end)
% b          transverse wavenumber 2*pi*L/lambda
% Go         Goertler number
% D          the matrix collection returned by CHDER1
% u0,v0,w0   initial disturbance velocities
% uz,vz,wz   resulting disturbance velocities

% useful constants
nodes=length(x); N=length(U{1}); dx=x(2)-x(1);
[Z,I]=deal(zeros(N),eye(N)); Db2=D{2}-b^2*I;
%
% useful indices
LL=2:N-1; N1=1:N-2; N2=N1+N-2; N3=N2+N-2;
LR=[LL LL+N LL+2*N (1:N)+3*N];

% placeholders
[uz{1:nodes},vz{1:nodes},wz{1:nodes}]=deal(zeros(N,1));
uz{1}=u0; vz{1}=v0; wz{1}=w0;

% initial conditions
Mf=[u0;U{1}.*u0;U{1}.*v0+V{1}.*u0;U{1}.*w0];
Mf=Mf(LR);

for k=2:nodes,
   VD=diag(V{k})*D{1}; % useful local terms
   VDVy=VD+diag(Vy{k});

   M=[I          Z         Z         Z
     diag(U{k})  Z         Z         Z
     diag(V{k}) diag(U{k}) Z         Z
       Z          Z       diag(U{k}) Z];
   L=[Z              -D{1}           -b*I          Z
      Db2-VD          diag(-Uy{k}) Z              Z
    diag(-2*Go^2*U{k}) Db2-2*VDVy  diag(-b*V{k}) -D{1}
       Z              Z             Db2-VDVy      b*I];

   f=(M(LR,LR)-L(LR,LR)*dx)\Mf;
   Mf=M(LR,LR)*f;
   uz{k}(LL)=f(N1); vz{k}(LL)=f(N2); wz{k}(LL)=f(N3);
end
```

at $F \approx 22 \times 10^{-6}$–$23 \times 10^{-6}$. In contrast, the phase velocities have values between 0.60 and 0.65 and depend slightly on frequency.

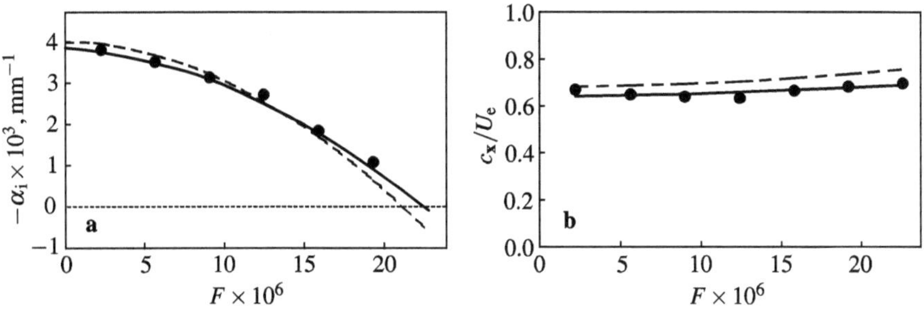

Fig. 5.10 Experimental (•), NSEs (*solid lines*) and LPSEs (*dashed lines*) increments (**a**) and phase velocities (**b**) of the Görtler vortices with $\Lambda = 149$ averaged in the disturbance-source far-field (around $\text{Gö}_{\delta^*} \approx 15$) vs. frequency parameter (Boiko et al., 2010)

Despite a quite strong frequency dependence, the variations of the increments at frequencies $F \lesssim 2.27 \times 10^{-6}$ corresponding to quasi-stationary vortices are extremely weak as seen in Fig. 5.10. The Görtler instability diagrams obtained for stationary and low-frequency disturbances also indicate that the stability diagrams for $F = 0$ and $F = 0.57 \times 10^{-6}$ are the same at least at $\text{Gö}_{\delta^*} < 100$ and $\Lambda < 10^4$. This is also true for $F = 2.27 \times 10^{-6}$ excluding a 'fringe region' at $\Lambda \gtrsim 2000$ and $\text{Gö}_{\delta^*} \gtrsim 40$. Thus, the quasi-stationary Görtler vortices can play an important role in the laminar–turbulent transition.

The stability diagrams obtained for essentially non-stationary (high-frequency) modes in Blasius boundary layer on a curved plate are presented in Fig. 5.11. It turned out that another type of instability can be present in the boundary layer at high frequencies. This are the oblique Tollmien–Schlichting modes, which are clearly observed at $\Lambda > 2000$ around the Görtler number of several tens (corresponding in the experiments to Re_{δ^*} of about 2000-3000). However, the Tollmien–Schlichting instability of the viscous type depends strongly on the Reynolds number. This is in contrast to the inviscid Görtler instability, for which the surface curvature, rather than the Reynolds number, is important. Less oblique Tollmien–Schlichting waves in a flat-plate boundary layer are usually considered to have a streamwise scale substantially different from that of Görtler vortices, being of the order of the boundary-layer thickness δ. This implies fast oscillatory streamwise variations of disturbances, i.e., $\alpha_r \sim 1/\delta$. As far as $\omega = \alpha_r c_x$, the assumption that the phase velocity c_x is of the order of the free-stream velocity U_0 yields the Tollmien–Schlichting wave angular frequency $\omega \sim U_0/\delta$.

As the disturbances under study are oscillatory, the question arises on whether one can attribute them purely to the Görtler instability rather than to the Tollmien–Schlichting instability? To be able to formulate a common set of the locally-nonparallel stability equations for the Görtler and Tollmien–Schlichting instabilities, one has to preserve the streamwise pressure gradient term and the terms with the squared streamwise wavenumber in (5.12). This yields

$$\mathrm{i}\alpha\hat{u} + \hat{v}_y + \beta\hat{w} = 0\,, \tag{5.15a}$$

$$U_x\hat{u}+\mathrm{i}\alpha U\hat{u}+V\hat{u}_y+U_y\hat{v}+\frac{\mathrm{i}\alpha}{\mathrm{Re}^2}\hat{p}=\hat{u}_{yy}-\frac{\alpha^2}{\mathrm{Re}^2}\hat{u}-\beta^2\hat{u}+\mathrm{i}\omega\hat{u}\,,\qquad(5.15\mathrm{b})$$

$$\mathrm{i}\alpha U\hat{v}+V_x\hat{u}+V\hat{v}_y+V_y\hat{v}+2\mathrm{G\ddot{o}}^2U\hat{u}+\hat{p}_y=\hat{v}_{yy}-\frac{\alpha^2}{\mathrm{Re}^2}\hat{v}-\beta^2\hat{v}+\mathrm{i}\omega\hat{v}\,,\qquad(5.15\mathrm{c})$$

$$\mathrm{i}\alpha U\hat{w}+V\hat{w}_y-\beta\hat{p}=\hat{w}_{yy}-\frac{\alpha^2}{\mathrm{Re}^2}\hat{w}-\beta^2\hat{w}+\mathrm{i}\omega\hat{w}\,,\qquad(5.15\mathrm{d})$$

where it is assumed that $\mathrm{Re}=U_0\delta/\nu$ and $\delta=x/\sqrt{\mathrm{Re}_\mathbf{x}}$. Note that this set of equations depends both on Re and Gö. This means that it is impossible generally to draw a universal neutral stability curve on the Gö–Λ-plane at a fixed frequency parameter F; instead, one has to plot a three-dimensional surface in coordinates $(\mathrm{G\ddot{o}},\Lambda,\mathrm{Re})$ or $(\mathrm{G\ddot{o}},\Lambda,R)$, where R is the dimensional radius of curvature.

Figure 5.11 shows some cuts of the surfaces of constant amplification rates corresponding to experimental parameters. Such a representation allows retaining both the Görtler and the Reynolds number axes on one graph. The diagrams were obtained on solving (5.15) by transforming the nonlinear eigenvalue problem to a standard form with the companion matrix method and applying the QZ-algorithm to find α. For the plot, the data are renormalized using the Blasius boundary layer scale δ^* for an easier comparison with the experiment. As seen, at low frequencies, high Reynolds numbers and large transverse wavelengths the boundary between the Görtler and Tollmien–Schlichting instabilities is smeared, i.e., the regions of both instabilities can overlap in a specific range of F, Λ, Gö, and Re.

The solid curves in Fig. 5.11 display the results obtained for the most amplified mode. The regions of the Tollmien–Schlichting instability calculated with the Orr–Sommerfeld equation are shown by dark-gray areas. The regions of the second mode instability are shown in light-gray. The experimentally observed growing and attenuating disturbances are presented with open and filled symbols respectively, while the experimental neutral-stability points are indicated with asterisks.

As frequency grows, the stability diagram undergos significant changes. At $F=5.67\times10^{-6}$, the flow becomes more stable to disturbances with $\Lambda\gtrsim1000$ and $\mathrm{G\ddot{o}}_{\delta^*}\gtrsim40$ (or $\mathrm{Re}_{\delta^*}\gtrsim2000$). At higher frequencies, the instability region continues to shrink and, by $F=19.29\times10^{-6}$, it splits into three domains. One of them disappears at $F=22.70\times10^{-6}$, while two others remain. As shown above, the domain located at relatively high Görtler numbers and large transverse wavelengths corresponds to the Tollmien–Schlichting instability. The domain located in the center of the plots (at moderate values of both Gö and Λ) corresponds to the first (leading) discrete mode of the Görtler instability.

Starting with $F=19.29\times10^{-6}$, the domain corresponding to the amplifying first Görtler mode located in the center of the plots becomes isolated and occupies a restricted area on the stability diagram. For instance, at $F=22.70\times10^{-6}$, the flow is unstable only at $1.9\lesssim\mathrm{G\ddot{o}}_{\delta^*}\lesssim16$. This means, in particular, that an *increase* of the Görtler number is able to *stabilize* the flow with respect to the first-mode Görtler vortices with all transverse wavenumbers and given fixed frequencies. This result is somewhat paradoxical. Indeed, the Görtler number is proportional to the square root of the surface curvature, i.e., $\mathrm{G\ddot{o}}_{\delta^*}\sim(1/R)^{1/2}$. Therefore, its growth can be

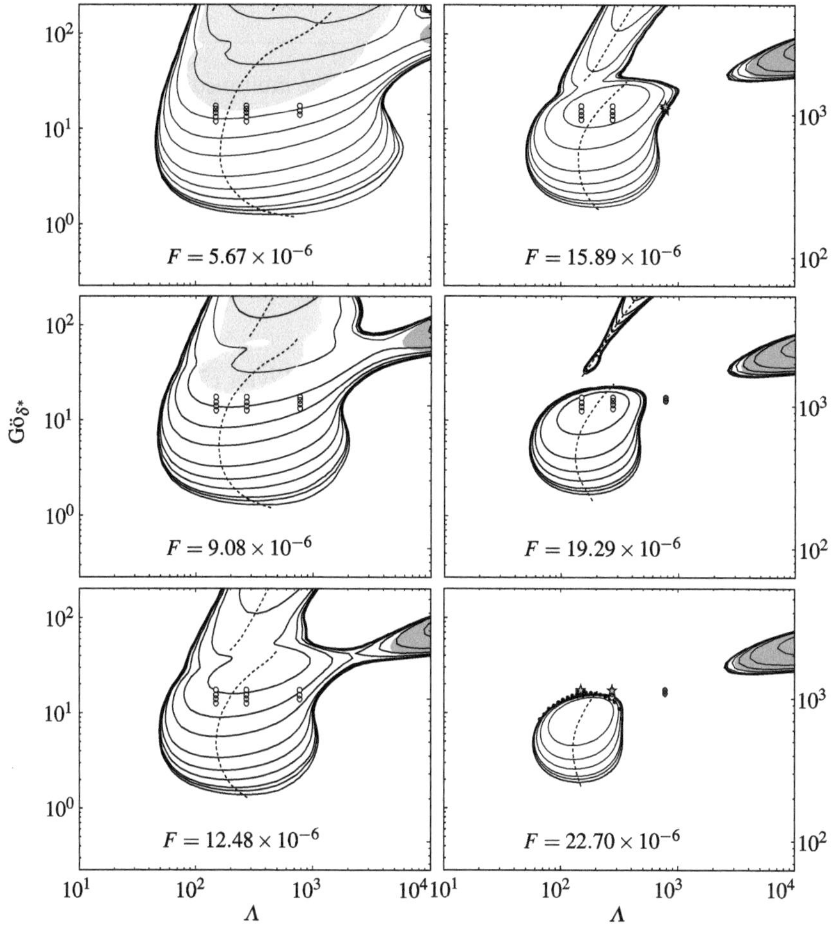

Fig. 5.11 Frequency dependence of the Görtler instability diagram for moderate and high frequencies: instability regions for the second Görtler mode (*light gray areas*); the Tollmien–Schlichting-wave instability domain obtained by solving the Orr–Sommerfeld equation (*dark gray areas*); the line of maximum growth (*dotted line*); a part of the neutral stability curve obtained by NSE-calculations (.); amplified (○) and attenuating (•) disturbances in the experiment; experimentally detected neutral-stability points (★). The isolines $-\alpha_i = 0, 0.01, 0.02, 0.05, 0.1, 0.2, 0.5, 1, 2, 5, 10, 20$, and 40 are shown (Boiko et al., 2010)

interpreted as a relative increase of the wall curvature and, hence, an increase of centrifugal forces, which represent the main cause of the Görtler instability. This growth, however, is shown to be able to lead to flow stabilization. The only obvious cause of this phenomenon might be associated with viscosity, which provides stabilization for non-stationary Görtler vortices, at least in the region of frequencies and Reynolds numbers characteristic for the Tollmien–Schlichting instability, stabilization being enhanced with F and Re.

Additional material for this chapter is given in the multimedia file No. 5, which supplements the book, titled 'Görtler Instability.'

Exercises

5.1. Write programs for the numerical solution of the Taylor (5.7) and Dean (5.9) problems assuming the narrow-gap approximation. Try to obtain neutral stability curves for different regimes of rotation.

5.2. Write a program for the numerical solution of the Görtler equations (5.12) for the Blasius boundary layer assuming $\omega = 0$. Reproduce isolines of amplification growth given in Figs. 5.7.

5.3. Extend the program of Exercise 5.2 for solving (5.11) to treat time-dependent disturbances. Reproduce Fig. 5.11.

5.4. Using the approach developed in Sect. 2.6, describe the structure of the continuous spectrum of Eqs. 5.13. How many branches does it have? Try to find analytical expressions for the position of points on the ω-plane in which the branch(es) appear(s). Compare your conclusions with the results of computations in Exercise 5.3.

5.5. Write a MATLAB function to calculate the eigenfunctions for modes of the continuous spectrum of Eqs. 5.13. Compare them with the eigenfunctions for modes of the Orr–Sommerfeld and Squire equations.

References

Bippes H (1972) Experimentelle Untersuchung des laminar-turbulenten Umschlags an einer parallel angeströmten konkaven Wand. In: Sitzungberichte der Heidelberger Akademie der Wissenschaften Mathematischnaturwissenschaftliche Klasse 3, Springer, Berlin, pp 103–180, translated as NASA TM 72243, 1978.

Boiko AV, Ivanov AV, Kachanov YS, Mischenko DA (2010) Steady and unsteady Görtler boundary-layer instability on concave wall. Eur J Mech B/Fluids 29(2):61–83

Clauser M, Clauser F (1937) The effect of curvature on the transition from laminar to turbulent boundary layer. TN 613, NACA

Dean WR (1928) Fluid motion in a curved channel. Proc R Soc Lond A 121:402–420

Floryan JM, Saric WS (1982) Stability of Görtler vortices in boundary layers. AIAA J 20(3):316–324

Floryan JM, Saric WS (1984) Wavelength selection and growth of Görtler vortices. AIAA J 22(11):1529–1538

Görtler H (1941) Instabilität laminaren Grenzschichten an konkaven Wänden gegenüber gewissen dreidimensionalen Störungen. Z Angew Math Mech 21:250–252

Gregory N, Walker WS (1956) The effect on transition of isolated surface excrescences in the boundary-layer. R&M 2779, ARC

Hall P (1982) Taylor–Görtler vortices in fully developed or boundary-layer flows: Linear theory. J Fluid Mech 124:475–494

Hall P (1983) The linear development of Görtler vortices in growing boundary layers. J Fluid Mech 130:41–58

Ito A (1987) Visualization of boundary layer transition along a concave wall. In: Proc. 4th Int. Symp. Flow Visualization, Hemisphere, Washington, pp 339–344

Liepmann HW (1943) Investigations on laminar boundary layer stability and transition on curved boundaries. Wartime Report W-107, NACA

Lord Rayleigh (1916) On the dynamics of revolving fluids. Scientific papers 6:447–453

Schultz MP, Volino RJ (2003) Effects of concave curvature on boundary layer transition under high free-stream turbulence conditions. ASME J Fluids Eng 125:18–27

Synge JL (1933) The stability of heterogeneous fluids. Trans R Soc, Canada 27(3):1–18

Tani I (1962) Production of longitudinal vortices in the boundary-layer along a curved wall. J Geophys Res 67:3075–3080

Tani I, Sakagami J (1962) Boundary-layer instability at subsonic speeds. In: Proc. Int. Council Aerosp. Sci. 3rd Congress, Spartan, Washington D.C., pp 391–403

Taylor GI (1923) Stability of a viscous liquid contained between two rotating cylinders. Proc R Soc Lond A 223:289–343

Further Reading

Bassom AP, Hall P (1991) Concerning the interaction of non-stationary crossflow vortices in a three-dimensional boundary layer. Q J Mech Appl Math 44:147–172

Bottaro A, Luchini P (1999) Görtler vortices: Are they amenable to local eigenvalue analysis? Eur J Mech B/Fluids 18(1):47–65

Cebeci T (1999) An engineering approach to the calculation of aerodynmic flows. Springer–Verlag, Berlin

Chomaz JM, Perrier M (1991) Nature of the Görtler instability in a forced experiment. In: The Geometry of Turbulence, Plenum, New York, pp 23–32

Denier JP, Hall P, Seddougui SO (1991) On the receptivity problem for Görtler vortices: Vortex motions induced by roughness. Philos Trans R Soc Lond A 335:51–85

Drazin PG, Reid WH (1981) Hydrodynamic stability. Cambridge University Press, Cambridge

Floryan JM (1991) On the Görtler instability of boundary layers. J Aerosp Sci 28:235–271

Gaponenko VR, Ivanov AV, Kachanov YS, Crouch JD (2002) Swept-wing boundary-layer receptivity to surface non-uniformities. J Fluid Mech 461:93–126

Hall P (1988) The nonlinear development of Görtler vortices in growing boundary layers. J Fluid Mech 193:243–266

Hall P (1990) Görtler vortices in growing boundary layers: The leading edge receptivity problem, linear growth and the nonlinear breakdown stage. Mathematika 37:151–189

Hämmerlin G (1955) Über das Eigenwertproblem der dreidimensionalen Instabilität laminarer Grenzschichten an konkaven Wänden. J Rat Mech Anal 4(2):279–321

von Kàrmán T (1934) Some aspects of the turbulence problem. In: Proc. 4th Int. Congr. Appl. Mech., Cambridge University Press, Cambridge, England, pp 54–91

Luchini P, Bottaro A (1998) Görtler vortices: A backward in time approach to the receptivity problem. J Fluid Mech 363:1–23

Murphy JS (1965) Extensions of the Falkner–Skan similar solutions to flows with surface curvature. AIAA J 3(11):2043–2049

Saric WS (1994) Görtler vortices. Ann Rev Fluid Mech 26:379–409

Smith AMO (1955) On the growth of Taylor–Görtler vortices along highly concave walls. Quart Appl Math 13:223

Chapter 6
Some other basic factors of shear-layer stability

This chapter describes the results of theoretical, numerical and experimental studies to show how different isolated factors affect the linear stability of parallel and quasi-parallel flows. The palette of these factors includes surface geometry, volume forces, temperature effects, presence of particles in the fluid, wall permeability and compliance. Certainly, this set is not exhaustive. In particular, the stability of magnetohydrodynamic, unsteady flows, etc. is beyond the present scope. However, the set is diverse enough to provide a general view of the basic aspects of the stability analysis as applied to some problems related to engineering applications.

6.1 Axial flow symmetry

The simplest types of axisymmetric flows are the Hagen–Poiseuille flow in a round pipe and the boundary layer over a round cylinder at a zero angle of attack. In both cases, a cylindrical coordinate system is preferable.

The momentum equations for the axisymmetric boundary layer are invariable if the local boundary-layer thickness is much smaller than the local radius of curvature $R(x)$. In this case, only the continuity equation is modified as:

$$\frac{\partial(RU)}{\partial x} + \frac{\partial(RV)}{\partial y} = 0 \,.$$

As seen, if R is constant, the continuity equation also reduces to the form of the equation in the Cartesian coordinates. The case of $\delta/R = \mathcal{O}(1)$ is described by the boundary-layer theory of a high order and can hardly be described by a self-similar solution. Except for this difficulty, the solution of the stability problem for an axisymmetric boundary layer (with the quasi-parallel flow approximation) is simpler than that for the pipe, because in the last case one has to apply special boundary conditions at $r = 0$. Hence, we consider the pipe flow in more detail. Note that, with some additional effort, the analysis can be extended to elliptic channels as well (see Kerswell and Davey, 1996).

A.V. Boiko et al., *Physics of Transitional Shear Flows*,
Fluid Mechanics and Its Applications 98, DOI 10.1007/978-94-007-2498-3_6,

The velocity distribution **U** in a developed pipe flow is not changed downstream and in the azimuthal direction, i.e., the flow streamlines are parallel to each other. The corresponding velocity profile is an exact solution of the Navier–Stokes equations and has the form $W(r) = 1 - r^2$, so that $\mathbf{U} = (U = 0, V = 0, W(r))^T$. Let u, v, and w be the disturbance velocities in the azimuthal (θ), radial (r) and axial (z) directions, respectively, and p be the pressure disturbance. Then, the linearized Navier–Stokes equations and the continuity equation take the form

$$\begin{aligned}
\frac{\partial u}{\partial t} + W\frac{\partial u}{\partial z} \quad &= -\frac{1}{r}\frac{\partial p}{\partial \theta} + \frac{1}{\mathrm{Re}}\left[\nabla^2 u - \frac{u}{r^2} + \frac{2}{r^2}\frac{\partial u}{\partial \theta}\right],\\
\frac{\partial v}{\partial t} + W\frac{\partial v}{\partial z} \quad &= -\frac{\partial p}{\partial r} + \frac{1}{\mathrm{Re}}\left[\nabla^2 v - \frac{v}{r^2} - \frac{2}{r^2}\frac{\partial u}{\partial \theta}\right],\\
\frac{\partial w}{\partial t} + W\frac{\partial w}{\partial z} + W_r v &= -\frac{\partial p}{\partial z} + \frac{1}{\mathrm{Re}}\nabla^2 w,\\
\frac{1}{r}\frac{\partial u}{\partial \theta} + \frac{1}{r}\frac{\partial (rv)}{\partial r} + \frac{\partial w}{\partial z} &= 0\,.
\end{aligned}$$

The subscript 'r' denotes here the first derivative with respect to r, the Reynolds number is defined as $\mathrm{Re} = W_\mathrm{a}R/\nu$, where W_a is the basic flow velocity on the centerline, and the Laplace operator

$$\nabla^2 = \frac{1}{r^2}\frac{\partial^2}{\partial \theta^2} + \frac{1}{r}\frac{\partial}{\partial r}\left(r\frac{\partial}{\partial r}\right) + \frac{\partial^2}{\partial z^2}\,.$$

As the flow is homogeneous in azimuth, assuming disturbance periodicity in z, the disturbances can be expressed as

$$\begin{aligned}
u(\theta,r,z,t) &= \hat{u}(r)\mathrm{e}^{\mathrm{i}\alpha z + \mathrm{i}n\theta - \mathrm{i}\omega t}, \quad & v(\theta,r,z,t) &= \hat{v}(r)\mathrm{e}^{\mathrm{i}\alpha z + \mathrm{i}n\theta - \mathrm{i}\omega t},\\
w(\theta,r,z,t) &= \hat{w}(r)\mathrm{e}^{\mathrm{i}\alpha z + \mathrm{i}n\theta - \mathrm{i}\omega t}, \quad & p(\theta,r,z,t) &= \hat{p}(r)\mathrm{e}^{\mathrm{i}\alpha z + \mathrm{i}n\theta - \mathrm{i}\omega t}\,.
\end{aligned}$$

where α is the streamwise wavenumber, and the integer n is the discrete azimuthal wavenumber. The presence of two homogeneous directions makes its possible to reduce the stability problem to consideration of only two ordinary differential equations for radial velocity $\hat{v}$ and radial vorticity $\hat{\eta}$ using a procedure similar to that described in Chap. 2 for plane flows:

$$\left[(\mathrm{i}\alpha W - \mathrm{i}\omega)\mathscr{T} - \frac{\mathrm{i}\alpha}{r}\frac{\mathrm{d}}{\mathrm{d}r}\left(\frac{W_r}{\varkappa^2 r}\right) - \frac{1}{\mathrm{Re}}\mathscr{T}\left(\varkappa^2 r^2 \mathscr{T}\right)\right]\hat{\Phi} = \frac{2\alpha n^2}{\mathrm{Re}}\mathscr{T}\hat{\Omega}\,, \tag{6.1a}$$

$$\left[(\mathrm{i}\alpha W - \mathrm{i}\omega)\varkappa^2 r^2 - \frac{1}{\mathrm{Re}}\mathscr{R}\right]\hat{\Omega} = \left[\frac{2\alpha}{\mathrm{Re}}\mathscr{T} - \frac{W_r}{r}\right]\hat{\Phi}\,, \tag{6.1b}$$

where

$$\varkappa^2 = \alpha^2 + \frac{n^2}{r^2}\,,$$

$$\mathscr{T} = \frac{1}{r^2} - \frac{1}{r}\frac{\mathrm{d}}{\mathrm{d}r}\left(\frac{1}{\varkappa^2 r}\frac{\mathrm{d}}{\mathrm{d}r}\right), \quad \mathscr{R} = k^4 r^2 - \frac{1}{r}\frac{\mathrm{d}}{\mathrm{d}r}\left(\varkappa^2 r^3 \frac{\mathrm{d}}{\mathrm{d}r}\right),$$

$$\hat{\Phi} \equiv -\mathrm{i}r\hat{v}\,, \quad \hat{\Omega} \equiv \frac{\alpha r\hat{u} - n\hat{w}}{n\varkappa^2 r^2} = \frac{\hat{\eta}}{\mathrm{i}n\varkappa^2 r}\,.$$

On solving these equations, the streamwise and azimuthal velocities can be found as

$$\hat{u} = -\frac{n}{\varkappa^2 r^2}\frac{\mathrm{d}\hat{\Phi}}{\mathrm{d}r} + \alpha n r \hat{\Omega}\,,$$
$$\hat{w} = -\frac{\alpha}{\varkappa^2 r}\frac{\mathrm{d}\hat{\Phi}}{\mathrm{d}r} - n^2\hat{\Omega}\,.$$

With given α, n, and Re, Eqs. (6.1) constitute an eigenvalue problem whose solution provides a set of eigenvalues ω. To solve it, we must apply boundary conditions for $\hat{\Phi}$ and $\hat{\Omega}$. The usual no-slip boundary condition is satisfied at the wall:

$$\hat{\Phi}(r=1) = \hat{\Phi}_r(r=1) = \hat{\Omega}(r=1) = 0\,.$$

Note, because of the symmetry of the basic flow velocity profile, the disturbances can be categorized according to their parity, similarly to the case of the plane Poiseuille flow in Sect. 3.3. Thus, it is desirable to consider only one half of the radial domain, namely, only the region $0 < r < 1$. However, due to the presence of terms $\sim 1/r$, it is necessary in such a case to apply nontrivial conditions at the pipe axis to provide smoothness and boundedness of the eigenfunctions $\hat{\Phi}$ and $\hat{\Omega}$ to be found. These conditions can be obtained by requiring azimuthal independence of the solution:

$$\begin{aligned}\lim_{r\to 0}\frac{\partial \mathbf{v}}{\partial\theta} &= \frac{\partial}{\partial\theta}(u\mathbf{e}_\theta + v\mathbf{e}_\mathbf{r} + w\mathbf{e}_\mathbf{z})\\ &= \frac{\partial u}{\partial\theta}\mathbf{e}_\theta + u\frac{\partial\mathbf{e}_\theta}{\partial\theta} + \frac{\partial v}{\partial\theta}\mathbf{e}_\mathbf{r} + v\frac{\partial\mathbf{e}_\mathbf{r}}{\partial\theta} + \frac{\partial w}{\partial\theta}\mathbf{e}_\mathbf{z} + w\frac{\partial\mathbf{e}_\mathbf{z}}{\partial\theta} = 0\,,\end{aligned}$$

where $\mathbf{e}_\mathbf{z}$, $\mathbf{e}_\mathbf{r}$, and $\mathbf{e}_\theta$ are the unit vectors in the axial, radial, and azimuthal directions, respectively. Taking into account that $\partial\mathbf{e}_\mathbf{z}/\partial\theta = 0$, $\partial\mathbf{e}_\mathbf{r}/\partial\theta = \mathbf{e}_\theta$, and $\partial\mathbf{e}_\theta/\partial\theta = -\mathbf{e}_\mathbf{r}$ yields

$$(\hat{u} + in\hat{v})\mathbf{e}_\theta + (in\hat{u} - \hat{v})\mathbf{e}_\mathbf{r} + in\hat{w}\mathbf{e}_\mathbf{z} = 0\,.$$

As the azimuthal dependence should be absent for each velocity component individually, the last condition is equivalent to the following set:

$$\begin{aligned}\hat{u} + in\hat{v} &= 0\,,\\ in\hat{u} - \hat{v} &= 0\,,\\ in\hat{w} &= 0\,.\end{aligned}$$

After simple algebraic manipulations it is found that they are equivalent to:

$$\begin{aligned}\hat{\Phi}(0) = \hat{\Phi}_r(0) &= 0 \quad \text{at} \quad n = 0\,,\\ \|\hat{\Phi}_r(0)\| < \infty \quad \text{and} \quad \hat{\Phi}(0) = \hat{\Omega}(0) &= 0 \quad \text{at} \quad |n| = 1\,,\\ \hat{\Phi}(0) = \hat{\Phi}_r(0) = \hat{\Omega}(0) &= 0 \quad \text{at} \quad |n| \geq 2\,.\end{aligned}$$

Equations (6.1) can be written more compactly as

$$\mathscr{L}\hat{\mathbf{q}} = \mathrm{i}\omega\mathscr{M}\hat{\mathbf{q}}\,,$$

where

$$\mathscr{M} = \begin{pmatrix} \mathscr{T} & 0 \\ 0 & \varkappa^2 r^2 \end{pmatrix}, \; \mathscr{L} = \begin{pmatrix} \mathscr{L}_{\mathrm{OS}} & 2\alpha n^2 \mathscr{T}/\mathrm{Re} \\ -\mathrm{i}W_r/r - 2\alpha\mathscr{T}/\mathrm{Re} & \mathscr{L}_{\mathrm{Sq}} \end{pmatrix}, \; \hat{\mathbf{q}} = \begin{pmatrix} \hat{\Phi} \\ \hat{\Omega} \end{pmatrix}$$

and

$$\mathscr{L}_{\mathrm{OS}} = -\mathrm{i}\alpha W \mathscr{T} + \frac{\mathrm{i}\alpha}{r}\frac{\mathrm{d}}{\mathrm{d}r}\left(\frac{W_r}{\varkappa^2 r}\right) + \frac{1}{\mathrm{Re}}\mathscr{T}(\varkappa^2 r^2 \mathscr{T}),$$

$$\mathscr{L}_{\mathrm{Sq}} = -\mathrm{i}\varkappa^2 r^2 \alpha W + \frac{1}{\mathrm{Re}}\mathscr{R}.$$

It is seen that, in contrast to the plane case, non-diagonal elements depending on Re appear in $\mathscr{L}$; at a finite Reynolds number they vanish simultaneously only if $\alpha = 0$. However, one of these non-diagonal element in the first row of $\mathscr{L}$ vanishes in the case of axisymmetric disturbances, i.e., when $n = 0$. Then, the structure of the operator becomes equivalent to that of its plane counterpart. Hence, in these two particular cases the flow stability can be established by considering only (6.1a). It is also possible to show analytically that, when $n = 0$, modes specific to the second equation are always stable.

Moreover, numerical simulations indicate that the flow in a round pipe is linearly stable to all other disturbances as well. Hence, the cylindrical geometry stabilizes the flow. Similarly, in boundary layers, the axial flow symmetry makes the critical Reynolds number Re_{L} larger, but not infinite.

6.2 Two-dimensional geometry

As a prototypical problem, we consider below the stability of a flow in a rectangular duct in which the basic flow depends on two Cartesian coordinates. For the first time the flow stability in the rectangular duct was considered theoretically and numerically by Tatsumi and Yoshimura (1996). As the flow in a round pipe, such a flow has parallel streamlines. However, the equations describing its stability are not reducible to ordinary differential equations, as the separation of variables in this case can be done only partly. Hence, the instability, if exists, is an example of the so-called global one.

To be specific, let us deal with the Poiseuille flow of a viscous incompressible fluid in a duct with a constant cross section $(-d < x < d) \times (-1 < y < 1) \times (-\infty < z < \infty)$, where x, y, and z are the dimensionless Cartesian coordinates. The dimensionless velocity distribution $W = W(x,y)$ in the streamwise direction z can be obtained by solving the Poisson equation

$$\frac{\partial^2 W}{\partial x^2} + \frac{\partial^2 W}{\partial y^2} = -2 \tag{6.2}$$

with homogeneous Dirichlet boundary conditions. The right-hand-side value provides a particular scaling of the velocity. Note that, if $d \to \infty$, this scaling gives the velocity in the channel center equal to one. However, the results can be anytime rescaled according to any other normalization condition.

Denoting the dimensionless disturbance velocities along x, y, z and pressure by u, v, w, p, respectively, and the Reynolds number by Re, we can write the linearized viscous incompressible flow equations for the disturbance as

$$\frac{\partial u}{\partial t} + W\frac{\partial u}{\partial z} + \frac{\partial p}{\partial x} = \frac{\nabla^2}{\mathrm{Re}} u\,, \tag{6.3a}$$

$$\frac{\partial v}{\partial t} + W\frac{\partial v}{\partial z} + \frac{\partial p}{\partial y} = \frac{\nabla^2}{\mathrm{Re}} v\,, \tag{6.3b}$$

$$\frac{\partial w}{\partial t} + \frac{\partial W}{\partial x}u + \frac{\partial W}{\partial y}v + W\frac{\partial w}{\partial z} + \frac{\partial p}{\partial z} = \frac{\nabla^2}{\mathrm{Re}} w\,, \tag{6.3c}$$

$$\frac{\partial u}{\partial x} + \frac{\partial v}{\partial y} + \frac{\partial w}{\partial z} = 0\,, \tag{6.3d}$$

with homogeneous boundary conditions for u, v, w, and the Laplace operator

$$\nabla^2 = \frac{\partial^2}{\partial x^2} + \frac{\partial^2}{\partial y^2} + \frac{\partial^2}{\partial z^2}\,.$$

The problem of linear stability in time of the flow under consideration consists of investigating the solutions that are harmonic in the streamwise direction but arbitrary in the transverse directions:

$$u = \tilde{u}(x,y,t)\mathrm{e}^{\mathrm{i}\alpha z},\ v = \tilde{v}(x,y,t)\mathrm{e}^{\mathrm{i}\alpha z},\ w = \tilde{w}(x,y,t)\mathrm{e}^{\mathrm{i}\alpha z},\ p = \tilde{p}(x,y,t)\mathrm{e}^{\mathrm{i}\alpha z}, \tag{6.4}$$

where $\tilde{u}$, $\tilde{v}$, $\tilde{w}$, and $\tilde{p}$ are functions of x, y, and t but not z, and α is a real wavenumber. If $\alpha = 0$, then the solution (6.4) is real. Otherwise, it is complex, but its real part, possessing a physical meaning, satisfies (6.3) as well.

Upon inserting (6.4) into (6.3), we obtain the following system for $\tilde{\mathbf{v}} = (\tilde{u}, \tilde{v}, \tilde{w})^T$ and $\tilde{p}$:

$$\frac{\partial \tilde{\mathbf{u}}}{\partial t} = \mathscr{J}\tilde{\mathbf{u}} + \mathscr{G}\tilde{p}\,, \quad \mathscr{F}\cdot\tilde{\mathbf{v}} = 0\,, \tag{6.5}$$

where

$$\mathscr{J} = \begin{pmatrix} \mathscr{L} & 0 & 0 \\ 0 & \mathscr{L} & 0 \\ -W_x & -W_y & \mathscr{L} \end{pmatrix}, \quad \mathscr{G} = \begin{pmatrix} -\partial/\partial x \\ -\partial/\partial y \\ -\mathrm{i}\alpha \end{pmatrix},$$

$$\mathscr{F} = (\partial/\partial x, \partial/\partial y, \mathrm{i}\alpha)\,,$$

$$\mathscr{L} = -\mathrm{i}\alpha W + \frac{1}{\mathrm{Re}}\left(\frac{\partial^2}{\partial x^2} + \frac{\partial^2}{\partial y^2} - \alpha^2\right).$$

Considering elementary solutions

$$\tilde{u} = \hat{u}(x,y)\mathrm{e}^{-\mathrm{i}\omega t}, \quad \tilde{v} = \hat{v}(x,y)\mathrm{e}^{-\mathrm{i}\omega t}, \quad \tilde{w} = \hat{w}(x,y)\mathrm{e}^{-\mathrm{i}\omega t}, \quad \tilde{p} = \hat{p}(x,y)\mathrm{e}^{-\mathrm{i}\omega t}\,,$$

and eliminating the pressure $\hat{p}$ and the streamwise velocity $\hat{w}$ from the resulting equations, we can reduce the system to a pair of coupled partial differential equations, which can be written compactly as

$$\mathscr{E}(x,y)\hat{v} + \mathscr{O}(x,y)\hat{u} = 0\,, \tag{6.6a}$$

$$\mathscr{E}(y,x)\hat{u} + \mathscr{O}(y,x)\hat{v} = 0\,, \tag{6.6b}$$

where

$$\mathscr{E}(x,y) = \left[(\alpha W - \omega) + \frac{\mathrm{i}}{\mathrm{Re}}\left(\frac{\partial^2}{\partial x^2} + \frac{\partial^2}{\partial y^2}\right)\right]\left(\frac{\partial^2}{\partial x^2} - \alpha^2\right) - \alpha W_{xx} ,$$

$$\mathscr{O}(x,y) = \left[(\alpha W - \omega) + \frac{\mathrm{i}}{\mathrm{Re}}\left(\frac{\partial^2}{\partial x^2} + \frac{\partial^2}{\partial y^2}\right)\right]\frac{\partial^2}{\partial x \partial y} + \alpha\left(W_x \frac{\partial}{\partial y} - W_y \frac{\partial}{\partial x} - W_{xy}\right) ,$$

with the following boundary conditions:

$$\hat{u} = \hat{v} = \hat{u}_x = 0 \quad \text{at} \quad x = \pm d ,$$

$$\hat{u} = \hat{v} = \hat{v}_y = 0 \quad \text{at} \quad y = \pm 1 .$$

This system of equations with the boundary conditions constitute an eigenvalue problem.

Table 6.1 Symmetries of normal modes in the rectangular duct

	$\hat{u}$		$\hat{v}$	
	x	y	x	y
I	o	o	e	e
II	o	e	e	o
III	e	o	o	e
IV	e	e	o	o

As the basic flow velocity is an even function of x and y, the system has modes of the solution of the following four classes with different parity as shown in Table 6.1, where, e.g., $\hat{u}$ of the second class corresponds to a mode that is odd and even in the x and y directions, respectively. For a channel with a square cross section, i.e., when $d = 1$, the eigenvalues of classes *I* and *IV* are identical, while classes *II* and *III* can be further subdivided to subclasses according to their parity with respect to diagonals. In contrast, when $d \to \infty$, classes *I* and *II* approach symmetric $\hat{v}$-disturbances, and classes *III* and *IV* approach antisymmetric ones. It is known (see Chap. 3) that symmetric disturbances in the plane Poiseuille flow can be unstable at high Reynolds numbers. Hence, we may expect that the same happens with modes of classes *I* and *II*, while this cannot be foreseen for the other classes. It appears that the flow is unstable at some finite Reynolds numbers at $d \gtrsim 3$, the modes of the first class becoming unstable first. Corresponding critical Reynolds numbers are given in Table 6.2.

Table 6.2 Re_L and corresponding α for the class *I* modes (Tatsumi and Yoshimura, 1996)

d	3.5	4	5	6	8	∞
R_L	36,600	18,400	10,400	8,200	6,800	5,772
α	0.71	0.80	0.91	0.94	0.98	1.02

The solution for the Poisson equation (6.2) can be found analytically in a series form as

$$W(x,y)=1-y^2-4\left(\frac{2}{\pi}\right)^3\sum_{n=0}^{\infty}\frac{(-1)^n}{(2n+1)^3}\frac{\cosh[(2n+1)\pi x/2]\cos[(2n+1)\pi y/2]}{\cosh[(2n+1)\pi d/2]}\,.$$

However, as the stability characteristics can be found only numerically, it makes sense to find the basic flow also numerically on the same grid.

The first step in the numerical solution of the multidimensional problem (6.5) is its reduction to a one-dimensional problem using the Kronecker tensor product. To explain the procedure, let us consider, e.g., a *square* region with a grid of 5×5 nodes enumerated sequentially as shown in Fig. 6.1 (for simplicity, we plotted the nodes spaced similarly in both directions). We enumerate the values of a function $h(x,y)$, defined on these nodes in the same way, so that a discrete vector can be composed:

$$\mathbf{h}=\begin{pmatrix} h_1 \\ h_2 \\ \vdots \\ h_{25} \end{pmatrix}\,.$$

Then, the derivatives in the x and y directions for the vector $\mathbf{h}$ can be written as

$$\mathbf{D}_{55}^{(x)}=\mathbf{I}_5\otimes\mathbf{D}_5\,,$$
$$\mathbf{D}_{55}^{(y)}=\mathbf{D}_5\otimes\mathbf{I}_5\,,$$

where $\otimes$ denotes the Kronecker tensor multiplication and $\mathbf{D}_5$ is any discrete matrix analog of the 'one-dimensional' first derivative on the grid of five nodes, for example, the collocation derivative introduced in Sect. 3.3. For an arbitrary number of collocation points $m,n\in\mathbb{N}$ for the x and y axes, respectively, this can be generalized as

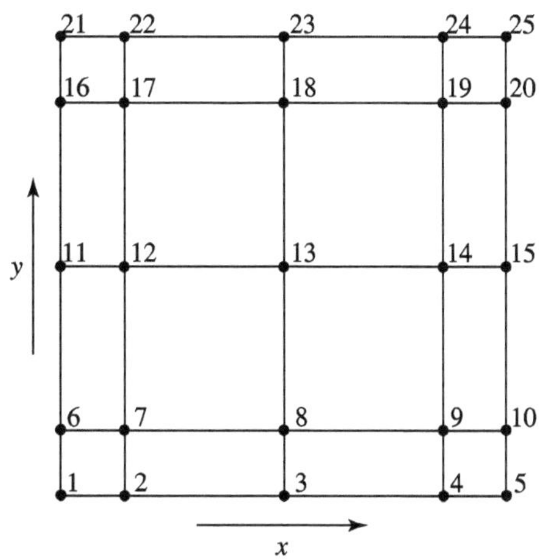

Fig. 6.1 Square grid with the lexicographic order of nodes

$$\mathbf{D}_{mn}^{(x)} = \mathbf{I}_m \otimes \mathbf{D}_n \,,$$
$$\mathbf{D}_{mn}^{(y)} = \mathbf{D}_m \otimes \mathbf{I}_n \,.$$

Higher directional and mixed two-dimensional differentiation matrices can be obtained by the Kronecker tensor multiplication of corresponding one-dimensional differentiation matrices, e.g.,

$$\mathbf{D}_{mn}^{(xx)} = \mathbf{I}_m \otimes \mathbf{D}_n^2 \,,$$
$$\mathbf{D}_{mn}^{(xy)} = \mathbf{D}_m \otimes \mathbf{D}_n \,,$$
$$\mathbf{D}_{mn}^{(yy)} = \mathbf{D}_m^2 \otimes \mathbf{I}_n \,.$$

In what follows, we will need also separate derivative matrices for pressure. Note, that in contrast to the velocities, there is no evolution equation for the pressure in the incompressible Navier–Stokes equations (in full or linearized, as above, form). It adjusts itself instantaneously in a time-dependent flow to satisfy the divergence-free condition (the continuity equation) and requires no initial or boundary conditions. Mathematically such a problem is well-posed. However, if the same Gauss–Lobatto points are used for the interpolation nodes both for the velocities and pressure, a spurious nonzero pressure modes can appear whose gradient is zero at all nodes, but which can have arbitrary value that deteriorate results of the stability computations.

One of the ways to overcome the problem is to restrict the expansion of the pressure to a polynomial of a lower degree. This removes the spurious modes but violates the divergence-free condition. Hence, the continuity equation is satisfied in this case only approximately, but the error is usually negligible when the number of nodes is large enough. Using definitions of Sect. 3.3, the one-dimensional derivative matrix for pressure, which requires no boundary conditions at $\varsigma \pm 1$, is defined as

$$\frac{\partial \hat{p}}{\partial \varsigma}(\varsigma_i) = (\mathbf{D}_n^{(p)} \hat{p})_i \,, \quad i = 1, \ldots, n-2 \,,$$

This is achieved if

$$(\mathbf{D}_n^{(p)})_{ij} = \frac{1-\varsigma_j^2}{1-\varsigma_i^2} (\mathbf{D}_n)_{ij} + \frac{2\varsigma_i}{1-\varsigma_i^2} \delta_{ij} \,, \quad i,j = 1, \ldots, n-2 \,.$$

An implementation of the matrices discussed above is given in function `chder2`.

To extend this procedure to rectangular regions, we need only to account for the x-coordinate stretching. Hence, the solution of the Poisson equation (6.2) reduces to the discrete problem

$$\left(\frac{\mathbf{D}_{mn}^{(xx)}}{d^2} + \mathbf{D}_{mn}^{(yy)} \right) \mathbf{W}_{mn} = -2\mathbf{I}_{mn} \,,$$

where $\mathbf{I}_{mn}$ is the $m \times n$ column vector of unities and $\mathbf{W}_{mn}$ is the $m \times n$ column vector of the discrete velocities. For further details (e.g., for application of the boundary conditions), see function `DuctBV` implementing the algorithm of the solution.

```
function [Dx,Dy,Dpx,Dpy,Dxx,Dyy,Dxy,x,y]=chder2(m,n)
% CHDER2 caculates 2D derivative matrices
% for a square region [-1;1]x[-1;1]
% m   - number of nodes in x direction
% n   - number of nodes in y direction
% x,y - corresponding nodes
% Dx,Dy,Dxx,Dyy,Dxy - resulting matrices

Dm=chder1(m);
Dn=chder1(n);
Im=eye(m);
In=eye(n);

Dpm=(1-Dm{5}.^2)*(1./(1-Dm{5}.^2))'.*Dm{1}+...
       diag(2*Dm{5}./(1-Dm{5}.^2));

Dpn=(1-Dn{5}.^2)*(1./(1-Dn{5}.^2))'.*Dn{1}+...
       diag(2*Dn{5}./(1-Dn{5}.^2));

Dx =kron(Dm{1},In);
Dy =kron(Im    ,Dn{1});
Dxx=kron(Dm{2},In);
Dyy=kron(Im    ,Dn{2});
Dxy=kron(Dm{1},Dn{1});
Dpx=kron(Dpm   ,In);
Dpy=kron(Im    ,Dpn);

x=Dm{5};
y=Dn{5};
```

Instead of solving numerically system (6.6), let us apply an algebraic reduction to the initial system (6.5). The transformation leads to a discrete eigenvalue problem of the same size as the size of the discrete counterpart of (6.6), but it is easier to program and high derivatives, as a rule deteriorating the numerical accuracy, do not appear.

After approximating (6.5) with respect to the transverse variables x, y and applying the homogeneous Dirichlet boundary conditions, we obtain an algebraic differential system of the form

$$\frac{\mathrm{d}\tilde{\mathbf{u}}}{\mathrm{d}t} = \mathbf{J}\tilde{\mathbf{u}} + \mathbf{G}\tilde{p}\,, \quad \mathbf{F}\tilde{\mathbf{u}} = 0\,, \tag{6.7}$$

where $\tilde{\mathbf{u}} \in \mathbb{C}^{n_v}$ and $\tilde{p} \in \mathbb{C}^{n_p}$ are the vectors of the discrete velocities and pressure, respectively ($n_p < n_v$, $n_p = (m-2)(n-2)$, $n_v = 3n_p$); $\mathbf{J}$ is a square matrix of order n_v; $\mathbf{G}$ and $\mathbf{F}$ are rectangular matrices of sizes $n_v \times n_p$ and $n_p \times n_v$, respectively.

We assume that (6.7) inherits from (6.5) the possibility of eliminating the pressure from the momentum equations with the help of the continuity equation. It implies that the matrix $\mathbf{FG}$ is nonsingular or that (6.7) can be reduced to a system of

```
function [W,x,y]=DuctBV(a,m,n)
% DuctBV calculates the basic velocity distribution
% in a rectangular channel [-a;a]x[-1;1]
% m   - number of nodes in x direction
% n   - number of nodes in y direction
% a   - relative width of the duct
% x,y - corresponding nodes
% W   - the velocity

[Dx,Dy,Dpx,Dpy,Dxx,Dyy,Dxy,x,y]=chder2(m,n);

I=true(n,m);

% boundaries are marked by zeros
I([1,n],:)=0;
I(:,[1,m])=0;

% solving
DD=Dxx/a+Dyy;
W0=DD(I,I)\(-2*ones((m-2)*(n-2),1));

% restoring boundaries
W=zeros(n,m);
W(2:n-1,2:m-1)=reshape(W0,[n-2,m-2]);

x=x*a;
```

the same form and with the matrix possessing this property by eliminating some columns of $\mathbf{G}$ and rows of $\mathbf{F}$.

Let

$$\mathbf{G} = \mathbf{X}\begin{pmatrix} \mathbf{R}_\mathrm{G} \\ 0 \end{pmatrix}, \quad \mathbf{F}^H = \mathbf{Y}\begin{pmatrix} \mathbf{R}_\mathrm{F} \\ 0 \end{pmatrix}$$

be the QR-decompositions, where the superscript 'H' denotes the Hermitian transpose (complex conjugate and transpose) of a matrix, $\mathbf{X}$ and $\mathbf{Y}$ are unitary matrices of order n_v and $\mathbf{R}_\mathrm{G}$ and $\mathbf{R}_\mathrm{F}$ are upper triangular ones of order n_p. Partition the matrices $\mathbf{X}$ and $\mathbf{Y}$ as follows: $\mathbf{X} = (\mathbf{X}_1, \mathbf{X}_2)$, $\mathbf{Y} = (\mathbf{Y}_1, \mathbf{Y}_2)$, where $\mathbf{X}_j$ and $\mathbf{Y}_j$ are matrices of size $n_v \times n_j$ with $n_1 = n_p$, $n_2 = n_v - n_p$, and $\mathbf{J}_{ij} = \mathbf{X}_i^H \mathbf{J} \mathbf{Y}_j$, $\mathbf{C}_{ij} = \mathbf{X}_i^* \mathbf{Y}_j$.

As shown by Boiko and Nechepurenko (2008), if $\mathbf{FG}$ is nonsingular, then $\mathbf{C}_{22}$ is also nonsingular, and the solution $\tilde{\mathbf{u}}(t;\tilde{\mathbf{u}}_0)$ of (6.7), with any $\tilde{\mathbf{u}}_0 \in \mathbb{C}^{n_v}$ such that $F\tilde{\mathbf{u}}_0 = 0$, is uniquely defined by

$$\tilde{\mathbf{u}}(t;\tilde{\mathbf{u}}_0) = \mathbf{Y}_2 \exp(t\mathbf{H})\, \mathbf{Y}_2^H \tilde{\mathbf{u}}_0\,,$$

where $\mathbf{H} = \mathbf{C}_{22}^{-1}\mathbf{J}_{22}$. Hence, the solution of the corresponding eigenvalue problem reduces to the solution of the equivalent eigenvalue problem

$$\mathrm{i}\mathbf{H}\hat{\mathbf{u}} = \omega \mathbf{I}\hat{\mathbf{u}}\,.$$

```
function [e,W,x,y]=Duct(m,n,a,alpha,Re)
% DUCT solves the linearized NS-equations in temporal
% formulation for a flow in a rectangular duct
% [-a;a]x[-1;1].
% m     - number of nodes in x direction
% n     - number of nodes in y direction
% a     - relative width of the duct
% alpha - streamwise wavenumber
% W     - the basic velocity
% x,y   - corresponding nodes
% e     - vector of calculated eigenvalues

 % find derivatives and basic velocity
[Dx,Dy,Dpx,Dpy,Dxx,Dyy]=chder2(m,n);
Dx =Dx/a; Dpx=Dpx/a; Dxx=Dxx/a^2;
[W,x,y]=DuctBV(a,m,n);
Wx=diag(Dx*W(:));
Wy=diag(Dy*W(:));

% useful constants and expressions
np=(m-2)*(n-2);
I=eye(m*n); Z=zeros(m*n);
L =(Dxx+Dyy-alpha^2*I)/Re-1i*alpha*diag(W(:));

% form the matrices for NS-equations
J=[ L   Z   Z
    Z   L   Z
  -Wx -Wy   L];
G=[-Dpx;-Dpy;-1i*alpha*I];
F=[ Dx , Dy , 1i*alpha*I];

% apply homogeneous boundary conditions for u,v, and w
I=true(n,m); I([1,n],:)=false; I(:,[1,m])=false;
K=[I(:);I(:);I(:)];
J=J(K,K); G=G(K,I); F=F(I,K);

% reduce the formed matrices
[X,~]=qr(G ); X2=X(:,np+1:end);
[Y,~]=qr(F'); Y2=Y(:,np+1:end);

% calculate eigenvalues
e=1i*eig((X2'*Y2)\(X2'*J*Y2));
```

The described above approach is implemented in function `Duct`.

6.3 Transverse flow periodicity

Transverse flow periodicity occurs over a corrugated surface or a surface with some technological elements, e.g., streamwise riblets or grooves. Research data on the 'natural riblets' on shark skin in the mid-1960s resulted in an idea to apply them for turbulent drag reduction. As for the linear stability application, wind-tunnel data and numerical results show that the wall grooves reduce the critical Reynolds number $\mathrm{Re_L}$ in channel flows and boundary-layers. Transverse-periodic flow appears also at a plane surface when the flow is modulated by stationary streamwise vortices of different nature. If the amplitude of the vortices changes slowly downstream, the flow can be considered as quasi-parallel and can experience the so-called secondary instability, see Chap. 11.

Below the linear stability analysis of the channel flow with grooved walls is outlined. The approach is based on application of the theory of ordinary differential equations with periodic coefficients by Floquet and its extension to partial differential equations by Bloch.

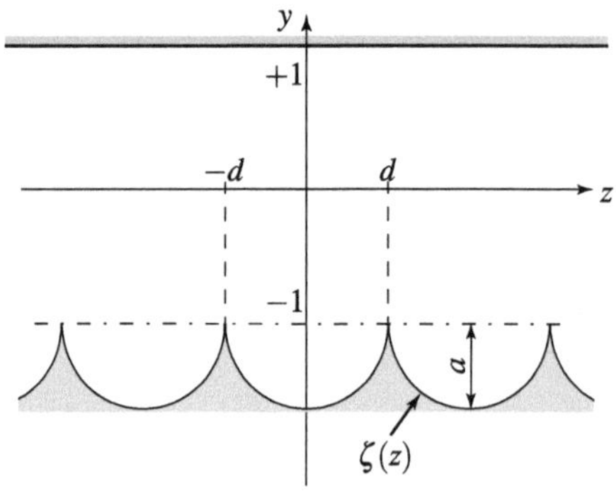

Fig. 6.2 Grooved channel

Let us consider a channel with a grooved bottom surface, see Fig. 6.2, whose shape is described by a periodic function $\zeta(z)$. A half of the height between the upper wall and the groove tips, h, is used to scale the lengths. Then, $d = l/h$ is the dimensionless half of the spacing between the tips (l is the dimensional half distance between the tips). The profile of basic flow velocity $U(y,z)$ in the region $z \in [-d;+d]$ and $y \in [-1+\zeta(z);+1]$ is described by the Poisson equation

$$\frac{\partial^2 U}{\partial y^2} + \frac{\partial^2 U}{\partial z^2} = -2\,.$$

The boundary conditions are the periodicity conditions of U, i.e., $U(y,-d) = U(y,+d)$, and $\partial U/\partial z(y,-d) = \partial U/\partial z(y,+d)$ as well as the no-slip conditions at the walls.

In contrast to the rectangular duct flow, an analytic solution to the present problem is unknown. To solve the equation by the collocation technique, first of all, we

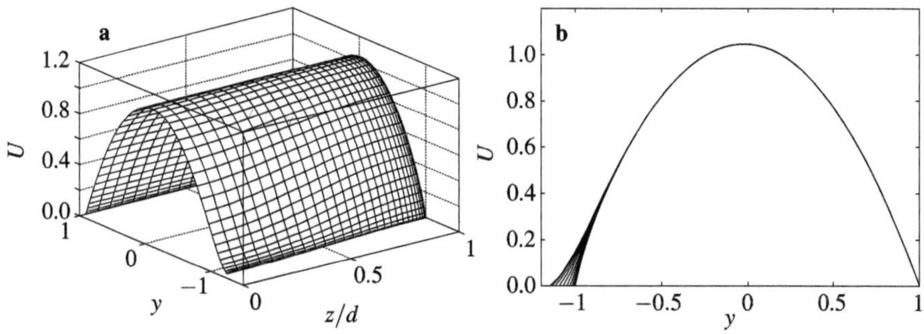

Fig. 6.3 Surface plot of one period of the velocity distribution in a grooved channel (**a**) and 16 equidistant velocity profiles (**b**) at $d = 0.25$; $a = 0.15$

perform a transformation of the physical domain to a square computational domain $\bar{r} \in [-1;+1]$ $\bar{s} \in [-1;+1]$, so that

$$\bar{r} = \frac{z}{d}\,,$$
$$\bar{s} = (1-\bar{\gamma})y + \bar{\gamma}\,, \quad \bar{\gamma} = \frac{\zeta(z)}{\zeta(z)-2}$$

or, inversely,

$$z = f(\bar{r},\bar{s}) = rl\,,$$
$$y = g(\bar{r},\bar{s}) = \frac{\bar{s}-\bar{\gamma}}{1-\bar{\gamma}}\,.$$

This leads to the following expression for the physical collocation derivatives:

$$\frac{\partial}{\partial z} = \frac{1}{J}\left(\frac{\partial g}{\partial \bar{s}}\frac{\partial}{\partial \bar{r}} - \frac{\partial g}{\partial \bar{r}}\frac{\partial}{\partial \bar{s}}\right)\,,$$
$$\frac{\partial}{\partial y} = \frac{1}{J}\left(-\frac{\partial f}{\partial \bar{s}}\frac{\partial}{\partial \bar{r}} + \frac{\partial f}{\partial \bar{r}}\frac{\partial}{\partial \bar{s}}\right)\,,$$
$$J = \frac{\partial f}{\partial \bar{r}}\frac{\partial g}{\partial \bar{s}} - \frac{\partial f}{\partial \bar{s}}\frac{\partial g}{\partial \bar{r}}\,.$$

In matrix notation these transformations take the form

$$\mathbf{D}_{mn}^{(z)} = \mathbf{J}_{mn}^{-1}(\ \ \mathbf{G}_m \otimes \mathbf{D}_n - \mathbf{D}_m \otimes \mathbf{G}_n)\,,$$
$$\mathbf{D}_{mn}^{(y)} = \mathbf{J}_{mn}^{-1}(-\mathbf{F}_m \otimes \mathbf{D}_n + \mathbf{D}_m \otimes \mathbf{F}_n)\,,$$

where $\mathbf{G}_m$, $\mathbf{G}_n$, $\mathbf{F}_m$, and $\mathbf{F}_n$ are some diagonal matrices and $\mathbf{J}_{mn}$ is the Jacobian of the transformation.

Similarly, one can obtain higher derivatives (see, e.g., Boyd, 1999, Appendix E). Hence, the problem of finding the discrete velocity distribution $\mathbf{U}_{mn}$ in the channel reduces to the numerical solution of the equation

$$\left(\mathbf{D}_{mn}^{(yy)} + \mathbf{D}_{mn}^{(zz)}\right)\mathbf{U}_{mn} = -2\mathbf{I}_{mn}$$

with homogeneous boundary conditions at solid boundaries and periodic boundary conditions at $z = \pm d$. An example of the solution for the groove shape described by the equation $\zeta(z) = -a\cos^2(z\pi/2d)$ is shown in Fig. 6.3. Appearance of inflections in some local velocity profiles is seen in Fig. 6.3**b**. For a velocity distribution depending only on one coordinate, it would be a clear indication of inviscid flow instability, as described in Sect. 2.1.2. However, in the present case, the theory is not applicable and we have to rely only on experimental data and numerical results.

The periodic boundary conditions can be applied, e.g., by adding the last column of the one-dimensional first derivative matrix to the first column and, then, deleting the last column and the last row. Higher-order derivatives can be obtained by matrix multiplications. Alternatively, one can use interpolation by trigonometric rather than Chebyshev polynomials in $\bar{r}$, as the periodicity is embedded in their definition and application of explicit periodic conditions is redundant. In this case, the nodes are equidistant on the interval $[-1;+1]$:

$$\bar{r}_i = \frac{2}{m} i , \quad i = -m/2, \ldots, m/2 - 1 ,$$

where m is even. The first two collocation derivatives, coordinates of the Fourier nodes and integration metric coefficients (weights) can be obtained, e.g., by MATLAB function `foder1`.

```
function D=foder1(m)
% FODER1 returns 1st 2 derivative matrices m x m (m is even)
% in D{1:2}, coordinates of the Fourier nodes in
% D{3} and integration metric w to find
% Int[f(y)dy]=Sum[f(y)*w] and scalar products in D{4}.

h=2*pi/m;

% 1st derivative
col=[0 .5*(-1).^(1:m-1).*cot((1:m-1)*h/2)]';
D{1}=toeplitz(col,col([1 m:-1:2]))*pi;

% 2nd derivative
col=[-pi^2/(3*h^2)-1/6 ...
    -.5*(-1).^(1:m-1)./(sin((1:m-1)*h/2)).^2]';
D{2}=toeplitz(col,col([1 m:-1:2]))*pi^2;

% nodes
D{3}=2/m*(-m/2:m/2-1)';

% weight
D{4}=2/m*ones(1,m);
```

The flow is homogeneous in the streamwise direction; hence, the disturbance velocities and pressure in the framework of the modal analysis can be expressed as

$$u(x,y,z,t) = \bar{u}(y,z)e^{i(\alpha x-\omega t)}, \quad v(x,y,z,t) = \bar{v}(y,z)e^{i(\alpha x-\omega t)},$$
$$w(x,y,z,t) = \bar{w}(y,z)e^{i(\alpha x-\omega t)}, \quad p(x,y,z,t) = \bar{p}(y,z)e^{i(\alpha x-\omega t)}.$$

Using the procedure developed in Sect. 2.1, the system of linearized governing equations for disturbances is obtained:

$$i\alpha\bar{u} + \bar{v}_y + \bar{w}_z = 0\,, \tag{6.8a}$$

$$-i\omega\bar{u} + i\alpha U\bar{u} + U_y\bar{v} + U_z\bar{w} = -i\alpha\bar{p} + \frac{\bar{u}_{yy} + \bar{u}_{zz} - \alpha^2\bar{u}}{\mathrm{Re}}\,, \tag{6.8b}$$

$$-i\omega\bar{v} + i\alpha U\bar{v} = -\bar{p}_y + \frac{\bar{v}_{yy} + \bar{v}_{zz} - \alpha^2\bar{v}}{\mathrm{Re}}\,, \tag{6.8c}$$

$$-i\omega\bar{w} + i\alpha U\bar{w} = -\bar{p}_z + \frac{\bar{w}_{yy} + \bar{w}_{zz} - \alpha^2\bar{w}}{\mathrm{Re}}\,. \tag{6.8d}$$

Let us assume that the disturbances $\bar{u}(y,z)$, $\bar{v}(y,z)$, $\bar{w}(y,z)$, and $\bar{p}(y,z)$ are periodic in z, with periods greater than or equal to $2d$. In accordance with the Floquet theory, they can be expressed as

$$\bar{u}(y,z) = \hat{u}(y,z)e^{i\beta z},\ \bar{v}(y,z) = \hat{v}(y,z)e^{i\beta z},\ \bar{w}(y,z) = \hat{w}(y,z)e^{i\beta z},\ \bar{p}(y,z) = \hat{p}(y,z)e^{i\beta z},$$

where $\hat{u}(y,z)$, $\hat{v}(y,z)$, $\hat{w}(y,z)$, and $\hat{p}(y,z)$ are functions with a period $2d$ and β is the so-called Floquet parameter. Introducing the main transverse wavenumber $\beta_0 = \pi/d$, it is readily shown that, for the stability analysis, it is sufficient to consider

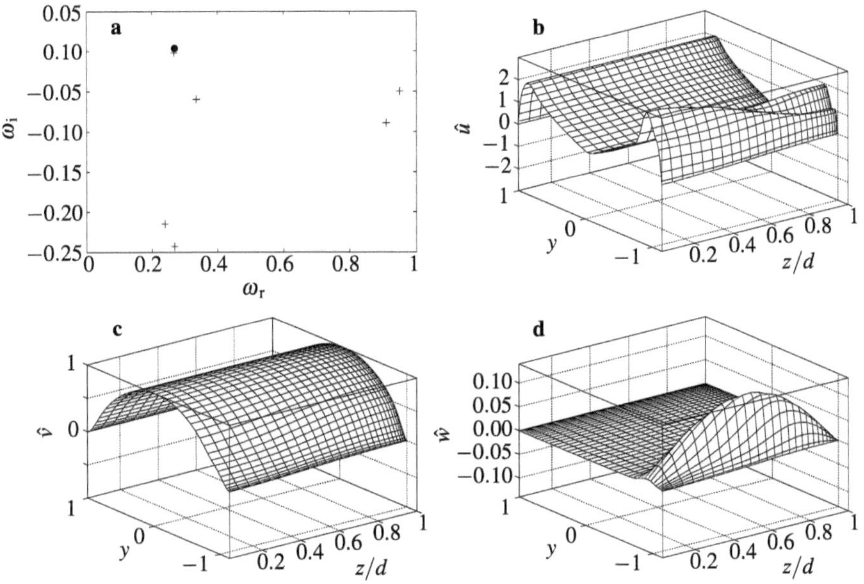

Fig. 6.4 Leading eigenvalues in plane (×) and grooved (•) channels (**a**), as well as amplitude distributions of eigenfunctions $|\hat{u}|$ (**b**), $|\hat{v}|$ (**c**), and $|\hat{w}|$ (**d**) in a grooved channel at $d = 0.25$, $a = 0.15$, $\mathrm{Re}_h = 5000$

Table 6.3 Symmetries of normal modes in the periodic channel with symmetric basic flow velocity

	$\hat{u}$	$\hat{v}$	$\hat{w}$	$\hat{p}$
varicose	e	e	o	e
sinuous	o	o	e	o

only the range $\beta \in [0;\beta_0/2]$. It is customary to denote the mode with $\beta = 0$ as fundamental, with $\beta/2$ as subharmonic, and the other modes as detuned.

After discretization, the problem reduces to the solution of the eigenvalue problem defined by the following system of matrix equations:

$$\begin{aligned}
\mathrm{i}\alpha\mathbf{I}_{mn}\hat{u} + \mathbf{D}^{(y)}_{mn}\hat{v} + (\mathbf{D}^{(z)}_{mn} + \mathrm{i}\beta\mathbf{I}_{mn})\hat{w} &= 0\,,\\
\mathrm{i}(\mathrm{diag}(\alpha U) - \omega\mathbf{I}_{mn})\hat{u} + \mathrm{diag}(U_y)\hat{v} + \mathrm{diag}(U_z)\hat{w} &= -\mathrm{i}\alpha\mathbf{I}_{mn}\hat{p} + \frac{\mathbf{L}_{mn}}{\mathrm{Re}}\hat{u}\,,\\
\mathrm{i}(\mathrm{diag}(\alpha U) - \omega\mathbf{I}_{mn})\hat{v} &= -\mathbf{D}^{(y)}_{mn}\hat{p} + \frac{\mathbf{L}_{mn}}{\mathrm{Re}}\hat{v}\,,\\
\mathrm{i}(\mathrm{diag}(\alpha U) - \omega\mathbf{I}_{mn})\hat{w} &= -(\mathbf{D}^{(z)}_{mn} + \mathrm{i}\beta)\mathbf{I}_{mn}\hat{p} + \frac{\mathbf{L}_{mn}}{\mathrm{Re}}\hat{w}\,,
\end{aligned}$$

where $\mathbf{L}_{mn} = \mathbf{D}^{(yy)}_{mn} + (\mathbf{D}^{(zz)}_{mn} + 2\mathrm{i}\beta\mathbf{D}^{(z)}_{mn} - \beta^2\mathbf{I}_{mn}) - \alpha^2\mathbf{I}_{mn}$, with the boundary conditions $\hat{u} = \hat{v} = \hat{w} = 0$ at the walls and the condition of periodicity of $\hat{u}$, $\hat{v}$, $\hat{w}$ and $\hat{p}$ in z with a period $2d$. The structure of the equation is such that the algebraic reduction introduced in Sect. 6.2 can be applied here eliminating, in fact, two of the four equations and, hence, halving the size of the matrices. If ζ is symmetric with respect to $z = 0$, then the basic flow velocity is also symmetric with respect to $z = 0$ and the modes can be separated into two classes according to their parity as shown in Table 6.3, where, e.g., $\hat{u}$ of a varicose mode corresponds to the mode with velocity $\hat{w}$ odd, and the other velocities and pressure even in the z direction. This can halve the size of the equations one more time. An example of the solution is presented in Fig. 6.4 by the leading (Tollmien–Schlichting) mode shown by the bullet and the amplitude distribution for its eigenfunction. For comparison, several first leading eigenvalues for a plane channel are also shown.

6.4 Pressure gradients

Pressure gradients change substantially the stability of near-wall flows and, as a consequence, affect the location of the laminar–turbulent transition and skin friction. In accordance with the boundary-layer theory, the pressure gradient normal to the wall is absent, while the linear stability characteristics are affected by the streamwise and transverse pressure gradients. These two cases are considered in the next subsections.

6.4.1 Streamwise pressure gradient

The influence of the streamwise pressure gradient on the basic flow velocity profile of a two-dimensional boundary layer can be evaluated using the Prandtl boundary-layer equations (Schlichting and Gersten, 2000):

$$U\frac{\partial U}{\partial x}+V\frac{\partial U}{\partial y} = -\frac{1}{\rho}\frac{\mathrm{d}P}{\mathrm{d}x}+\nu\frac{\partial^2 U}{\partial y^2}\,,$$
$$\frac{\partial U}{\partial x}+\frac{\partial V}{\partial y} = 0\,.$$

Taking in account the no-slip boundary conditions, the first equation is reduced at the wall to

$$\frac{1}{\rho}\frac{\mathrm{d}P}{\mathrm{d}x} = \nu\frac{\partial^2 U}{\partial y^2}\,,$$

from which it follows that the curvature of the profile at the wall is completely determined by the pressure gradient. Taking into account that the profile is convex at a large distance from the wall, i.e., $\partial^2 U/\partial y^2 < 0$, it follows that the positive pressure gradient inevitably leads to appearance of an inflection point inside the boundary layer. Then the Rayleigh inflection point criterion (see Sect. 2.1.2) indicates that such a flow should be unstable in the inviscid limit. This is in contrast to the viscous boundary-layer instability, which occurs only at relatively high Reynolds numbers, when $\mathrm{d}P/\mathrm{d}x = 0$ (as discussed in Chap. 4) and when $\mathrm{d}P/\mathrm{d}x < 0$. This is the reason to denote the positive (adverse) pressure gradient as the *unfavorable* and the pressure drop as the *favorable* pressure gradient.

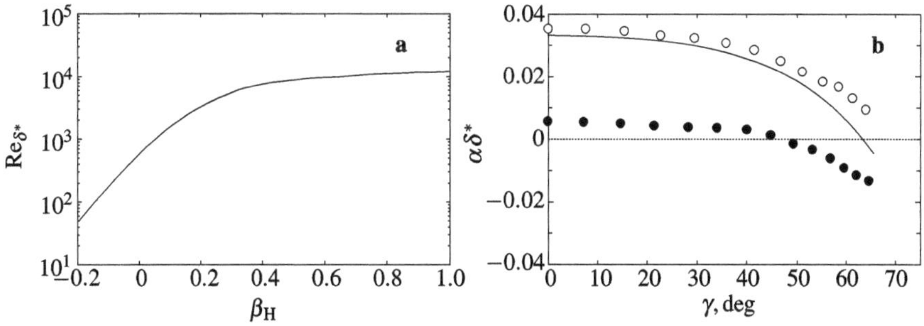

Fig. 6.5 Influence of the streamwise pressure gradient on the critical Reynolds number for the Falkner–Skan velocity profiles (Wazzan et al, 1970) (**a**) and experimental data for $\beta_H = 0$, $F = 94.1\times10^{-6}$ (•) and $\beta_H = -0.115$, $F = 106.8\times10^{-6}$ (○) as well as theoretical (*solid line*) rates of spatial growth of normal instability modes vs. the angle of wavevector inclination to the flow at $Re_{\delta^*} = 1052$ (Kachanov et al, 2000) (**b**)

Destabilization of boundary layer with respect to the Tollmien–Schlichting waves under the influence of an adverse pressure gradient had already been detected in the first experiments on boundary-layer stability of Schubauer and Skramstad (1948).

Currently, the upstream shift of the laminar–turbulent transition in the region of flow deceleration and its delay in the region of flow acceleration are well known from observations of the boundary-layer behavior on airfoils and numerical calculations, see Fig. 6.5**a**. Experimental data correlate well with the results of calculations on stability of the Falkner–Skan profiles as shown in Fig. 6.5**b**, except for large angles of wavevector inclination, where nonparallel effects become pronounced.

6.4.2 Transverse pressure gradient

Intensive studies of the flow stability in boundary layers with transverse pressure gradients are traced to the experiments of Gray (1952), who observed a growth of stationary vortices oriented approximately along an inviscid streamline at low Reynolds numbers, when visualized a swept-wing flow. Later, similar vortices were also found in flows on a rotating disc (Gregory et al, 1955), rotating cone (Kobayashi et al, 1983), and swept cylinder (Poll, 1985).

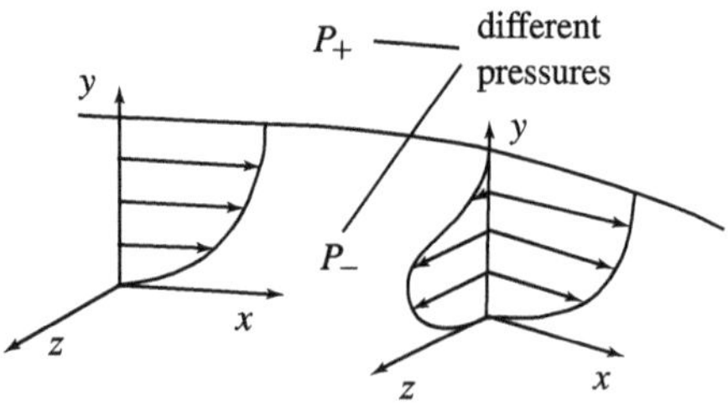

Fig. 6.6 Onset of crossflow in the presence of transverse pressure gradients

Before discussing the stability of such flows, let us consider peculiarities of the flow over an infinite swept wing. Two commonly accepted Cartesian coordinate systems are usually associated with the flow. The system $(\tilde{x}, y, \tilde{z})$ is fitted to the model, i.e., the coordinate $\tilde{x}$ is directed along the chord over the model contour, coordinate $\tilde{z}$ is parallel to the leading edge, coordinate y is normal to the wall, and the coordinate origin is located on the leading edge of the wing. This coordinate system is usually used to calculate the basic flow characteristics, because the basic flow parameters are independent of $\tilde{z}$. It is more convenient, however, to calculate the stability characteristics in a local coordinate system (x, y, z), where x is directed along the local external-flow streamline and z is normal to the x–y-plane.

In the three-dimensional flow, the vector of external-flow velocity does not coincide with the pressure gradient vector. As a result, projections of the external-flow streamlines onto the wing surface are curved, which leads to formation of a spanwise velocity inside the boundary layer. Let us consider initially a two-dimensional boundary layer and assume that a transverse pressure gradient is applied at some position, as shown in Fig. 6.6. This pressure gradient distorts the external-flow streamline, resulting in its curvature to compensate by the centrifugal force for the pressure gradient:

$$\frac{\partial P}{\partial z} = -\frac{\rho U_e^2}{R} ,$$

where R is the streamline curvature.

Closer to the wall with the viscous stresses being neglected, we obtain

$$\frac{\partial P}{\partial z} = -\frac{\rho U^2}{r} ,$$

where r is the local radius of curvature of the streamline in the boundary layer. As the static pressure is constant across the boundary layer, then $\partial P/\partial z = \text{const}$ and $r < R$ at $U < U_e$. Hence, the uncompensated pressure gradient creates a crossflow inside the boundary layer, which is directed to the center of curvature of the external-flow streamline. At the wall, the no-slip boundary conditions provide the absence of the crossflow, and the pressure gradient in the vicinity of the wall is compensated by viscosity.

The theory of a boundary layer at an infinitely long yawing cylinder usually used to describe a non-separated flow over a swept wing is an extension of the Prandtl two-dimensional theory. When the free-stream velocity components behave as $U_e = Q_1 \overline{x}^{\beta_H/(2-\beta_H)}$ and $W_e = Q_2$, where Q_1 and Q_2 are constants, it corresponds to a flow over an infinite swept wedge, and the boundary-layer equations admit self-similar solutions, which can be obtained by solving a set of ordinary differential Falkner–Skan–Cooke equations written in the coordinate system fitted to the model:

$$\Psi_{\eta\eta\eta} + \Psi_{\eta\eta}\Psi + \beta_H \left[1 - (\Psi_\eta)^2\right] = 0 ,$$
$$\Phi_{\eta\eta} + \Phi_\eta \Psi = 0 ,$$

with the boundary conditions

$$\eta = 0 : \Psi = \Psi_\eta = \Phi = 0 ,$$
$$\eta \to \infty : \Psi_\eta = \Phi = 1 ,$$

where the symbols for the first equation and the scaling being the same as in Sect. 4.2. As in the two-dimensional case, the flow is characterized by the Hartree pressure gradient parameter β_H, as well as by the angle between the direction of the external-flow streamline and the wedge chord $\phi = \arctan(W_e/U_e)$. As seen the first equation given above is the Falkner–Skan equation for Ψ and, hence, the streamwise flow described by Ψ is independent of the crossflow and can be computed straight away. Then the velocity profile $W(\eta)$ is obtained by solving the second equation. In the coordinate system aligned with the external-flow streamline

$$U = \Psi_\eta \cos^2\phi + \Phi \sin^2\phi ,$$
$$W = (\Phi - \Psi_\eta)\cos\phi \sin\phi .$$

It appears that the linear stability theory in the parallel-flow approximation is quite applicable to analyze the development of small-amplitude disturbances in a

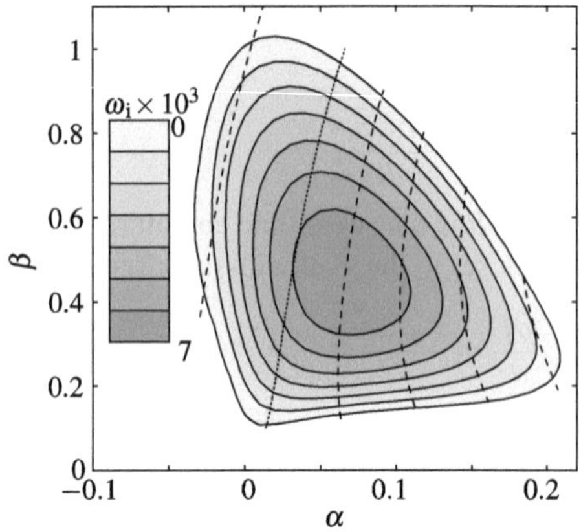

Fig. 6.7 Diagram of stability in time for the Falkner–Skan–Cooke boundary layer at $\beta_H = 0.5$, $\phi_0 = 45°$, $\text{Re}_{\delta^*} = 800$, where δ^* is calculated along the free-stream streamline: isolines ω_i (*solid lines*), isolines ω_r with step 0.02 (*dashed lines*), $\omega_r = 0$ (*dotted lines*) (calculated by A. V. Boiko)

swept-wing boundary layer. The linearized governing equations in this case take the form

$$
\begin{aligned}
\frac{\partial u}{\partial t} + U\frac{\partial u}{\partial x} + W\frac{\partial u}{\partial z} + v\frac{\partial U}{\partial y} &= -\frac{\partial p}{\partial x} + \frac{\nabla^2 u}{\text{Re}}\,,\\
\frac{\partial v}{\partial t} + U\frac{\partial v}{\partial x} + W\frac{\partial v}{\partial z} &= -\frac{\partial p}{\partial y} + \frac{\nabla^2 v}{\text{Re}}\,,\\
\frac{\partial w}{\partial t} + U\frac{\partial w}{\partial x} + W\frac{\partial w}{\partial z} + v\frac{\partial W}{\partial y} &= -\frac{\partial p}{\partial z} + \frac{\nabla^2 w}{\text{Re}}\,,\\
\frac{\partial u}{\partial x} + \frac{\partial v}{\partial y} + \frac{\partial w}{\partial z} &= 0\,,
\end{aligned}
$$

which can be further reduced for the modal analysis to

$$
\begin{aligned}
\mathrm{i}(\alpha U + \beta W - \omega)\,\hat{u} + U_y\hat{v} &= -\mathrm{i}\alpha\hat{p} + \frac{1}{\text{Re}}\left(\mathcal{D}^2 - \varkappa^2\right)\hat{u}\,,\\
\mathrm{i}(\alpha U + \beta W - \omega)\,\hat{v} &= -\mathcal{D}\hat{p} + \frac{1}{\text{Re}}\left(\mathcal{D}^2 - \varkappa^2\right)\hat{v}\,,\\
\mathrm{i}(\alpha U + \beta W - \omega)\,\hat{w} + W_y\hat{v} &= -\mathrm{i}\beta\hat{p} + \frac{1}{\text{Re}}\left(\mathcal{D}^2 - \varkappa^2\right)\hat{w}\,,\\
\mathrm{i}(\alpha\hat{u} + \beta\hat{w}) + \mathcal{D}\hat{v} &= 0\,.
\end{aligned}
$$

Elimination of the disturbance pressure and streamwise velocity results in the following variants of the Orr–Sommerfeld and Squire equations:

$$
\left[\mathrm{i}\left(\alpha\overline{U} - \omega\right)\left(\mathcal{D}^2 - \varkappa^2\right) - \mathrm{i}\alpha\overline{U}_{yy} - \frac{1}{\text{Re}}\left(\mathcal{D}^2 - \varkappa^2\right)^2\right]\hat{v} = 0\,,
$$

$$
\left[\mathrm{i}(\alpha\overline{U} - \omega) - \frac{1}{\text{Re}}(\mathcal{D}^2 - \varkappa^2)\right]\overline{w} = -\mathrm{i}\beta\hat{v}\overline{W}_y\,.
$$

Here $\overline{U} = U + W\beta/\alpha$, $\overline{u} = (\alpha\hat{u} + \beta\hat{w})/\varkappa$ and $\overline{W} = U - W\alpha/\beta$, $\overline{w} = (\beta\hat{u} - \alpha\hat{w})/\varkappa$ are the effective basic and disturbance flow velocities normal and parallel to the wavefront, respectively. In respect to the disturbances, this transformation is equivalent to rotation of the system of coordinates in the plane parallel to the wall, so that the continuity equation becomes

$$\mathrm{i}\varkappa\bar{u} + \hat{v}_y = 0\,.$$

Such a transformation is called the Squire transformation. As after the coordinate transform the equivalent flow is two-dimensional, the inflection point criterion as described in Sect. 2.1.2 can be applied to it. Then, as $W(\eta)$ always has an inflection, directions of the wavevectors are expected along which the effective basic flow velocity $\overline{U}(\eta)$ also has an inflection. Hence, the boundary-layer flow in the presence of the sweep can be treated as inviscidly unstable, i.e., the crossflow instability (CFI) may be found at low Reynolds numbers. Next, it appears that there is a particular direction close to the crossflow direction for which the condition $\overline{U}(\eta_s) = 0$ holds at the inflection point η_s. Hence, one may expect that if there are neutral disturbances for which the critical layer and the inflection point coincide, at higher Reynolds numbers stationary unstable disturbances appear.

Typical stability characteristics for a Falkner–Skan–Cooke profile are shown in Fig. 6.7. It can be seen that the velocity profile gives rise to flow instability with respect to both the stationary vortices ($\omega_r = 0$) and the traveling waves ($\omega_r \neq 0$). The characteristic frequencies for the traveling disturbances are usually much lower than those for the Tollmien–Schlichting waves. The ratio of the wavenumbers β/α for the most unstable traveling waves is quite high. This means that they propagate approximately in the crossflow direction. Note that stationary vortices are not the most amplifying disturbances, and their predominance in the experiments of Gray (1952) and in other studies is stipulated by initial conditions in the form of natural surface roughness of microscopic size being, nevertheless, large enough compared to the boundary-layer thickness at low Reynolds numbers. Otherwise, the traveling modes are observed when the free-stream disturbance level is relatively high and the surface roughness is small.

The stability analyses of Lingwood (1995), Taylor and Peake (1998), and Ryzhov and Terent'ev (1998) show that the cross-flow instability is convective in a typical range of parameters characteristic for laboratory or flight conditions. In contrast to the flat-plate boundary layer, for which the flow independence in the transverse direction presumes only real values of the transverse wavenumber β, the dispersion relation for a general swept-wing flow reads $\alpha = \alpha(\omega, \beta_r, \beta_i)$, where α is also complex. Thus, it is necessary to deal with an additional parameter β_i. This is usually done by getting rid of it with the help of the coordinate system rotation, aligning the x axis with the direction of growth of the disturbances. In numerical calculations, this direction is selected with some conjectures concerning the nature of perturbations and their group velocity (see, e.g., Mack, 1979, for details).

Theoretical results on both stationary and traveling CFI modes are found in good agreement with experimental data (see, e.g., Gaponenko et al, 1995, Kachanov and Tararykin, 1990); an illustration is given in Fig. 6.8.

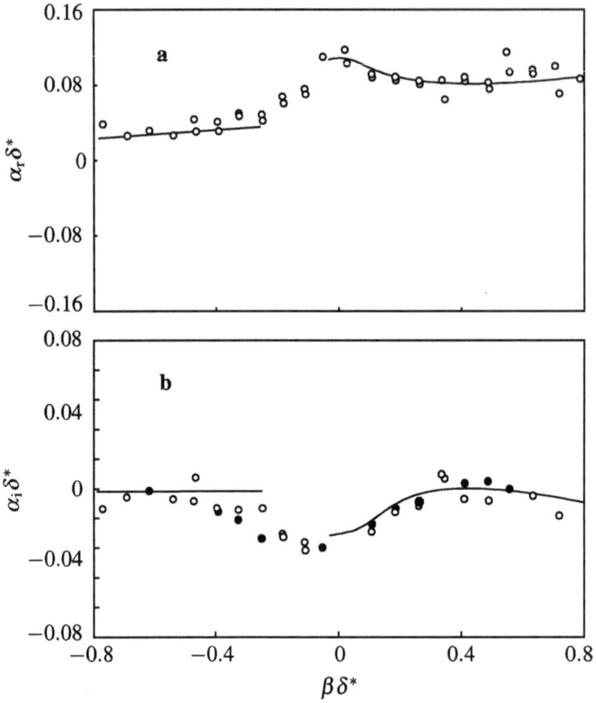

Fig. 6.8 Streamwise wavenumbers (**a**) and amplification rates (**b**) of crossflow-instability normal modes as functions of the spanwise wavenumber in a local coordinate system, $F = 48.8 \times 10^{-6}$: experimental results (○, •) and corresponding theoretical curves (*solid lines*) (Gaponenko et al, 1995)

6.5 Heat transfer

The stability of laminar shear flows is strongly influenced by the temperature difference between the flow and the wall. The resulting heat transfer affects the stability through the dependence of viscosity on temperature. This effect can be qualitatively demonstrated by considering the streamwise momentum equation for a two-dimensional boundary layer with variable viscosity:

$$\rho\left(U\frac{\partial U}{\partial x}+V\frac{\partial U}{\partial y}\right)=U\frac{\mathrm{d}p}{\mathrm{d}y}+\frac{\partial}{\partial y}\left(\mu\frac{\partial U}{\partial y}\right),$$

where μ is the dynamic viscosity. In the absence of the pressure gradient and under no-slip boundary conditions, this equation reduces at the wall ($y = 0$) to

$$\left(\frac{\partial^2 U}{\partial y^2}\right)_{y=0}=-\frac{1}{\mu_{y=0}}\left(\frac{\partial\mu}{\partial y}\right)_{y=0}\left(\frac{\partial U}{\partial y}\right)_{y=0}.$$

The dynamic viscosity μ is a function of temperature: as the temperature grows, it decreases in liquids and increases in gases. Hence, if the wall is heated, i.e., $T_w > T$, where T_w is the wall temperature and T is the fluid temperature, then $d\mu/dT < 0$, $(dT/dy)_{y=0} < 0$ for a liquid and

$$\left(\frac{\partial \mu}{\partial y}\right)_{y=0} = \frac{d\mu}{dT}\left(\frac{dT}{dy}\right)_{y=0} > 0\,.$$

Alternatively, the last inequality is reversed if the wall is cooled. In other words, $U_{yy} < 0$ in the vicinity of a heated boundary, and $U_{yy} > 0$ at the cooled boundary. Comparing this conclusion with the effect of the streamwise pressure gradient on the curvature of the velocity distribution inside the boundary layer discussed in Sect. 6.4.1, we see that cooling leads to appearance of an inflection point $U_{yy} = 0$ in the velocity profile. Hence, the boundary layer becomes unstable in inviscid limit. Similarly, inviscid instability of a gas flow is found when the surface is heated.

Such an influence can be accurately taken into account either by adding a complete description of the temperature field and flow parameters to the linearized Navier–Stokes equations or by admitting only the dynamic viscosity variation in the Orr–Sommerfeld equations. Both approaches give similar results for the flat-plate boundary layer, that is, the connection between the temperature and hydrodynamic disturbances is quite weak. Hence, assuming the second approach, the modified Orr–Sommerfeld equation for two dimensional disturbances takes the form

$$\left[(\alpha U - \omega)\left(\mathcal{D}^2 - \alpha^2\right) - \alpha U_{yy} + \frac{i}{\mathrm{Re}}\tilde{\mu}\left(\mathcal{D}^2 - \alpha^2\right)^2\right]\hat{v} \\ + \left[2\tilde{\mu}_y\left(\mathcal{D}^3 - \alpha^2\mathcal{D}\right) + \tilde{\mu}_{yy}\left(\mathcal{D}^2 + \alpha^2\right)\right]\hat{v} = 0\,,$$

where $\tilde{\mu} = \mu(y)/\mu_e$ and $\mu_e(T_e)$ is the dynamic viscosity at $y \to \infty$, which is also used in determining of the kinematic viscosity for the Reynolds number Re. Corresponding numerical calculations and experiments confirm the substantial boundary-layer stabilization in liquids at moderate surface heating.

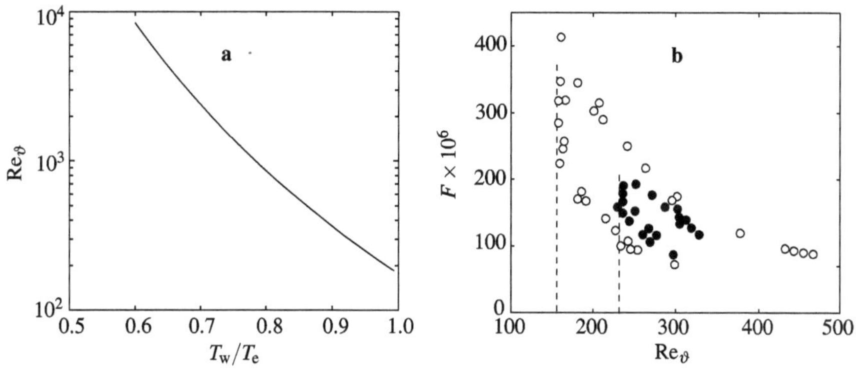

Fig. 6.9 Influence of wall cooling on linear stability of a flat-plate boundary layer in air: theoretical results on the critical Reynolds number of Gaponov and Maslov (1971) (**a**) and experimentally obtained neutral stability curves for isothermal (∘) and wall-cooled flow $T_w/T_e = 0.945$ (•) of Kachanov et al (1974) (**b**)

Theoretical data on the critical Reynolds numbers for a flat-plate boundary layer in air are given in Fig. 6.9**a**. Wind-tunnel results are shown in Fig. 6.9**b**, where the neutral points at the cooled and heat-insulated walls are compared. As seen, cooling moves the neutral stability curve to the region of high Reynolds numbers and reduces the range of amplifying disturbances. The critical Reynolds number Re_ϑ, defined using the boundary-layer momentum thickness

$$\vartheta = \int_0^\infty \frac{U}{U_e}\left(1 - \frac{U}{U_e}\right) \mathrm{d}y \,,$$

shifts from 155 to 235 (i.e., becomes 1.5 times larger).

Results of linear stability calculations and experimental data indicate that, under certain conditions, heating can be used for boundary-layer stabilization in gases. This becomes possible with a nonuniform surface temperature distribution. In a limited flow section, the heat flux is directed from the surface to the gas so that a situation similar—but not identical—to wall cooling occurs downstream of the heating area where the boundary layer develops over the surface that has a relatively low temperature. At a certain regime of heating, despite some destabilization in the initial flow region, subsequent reduction of the growth rates of disturbances produces a net positive effect.

6.6 Fluid suction

Suction produces a two-fold positive effect on the boundary layer. First, it makes the boundary-layer thinner and, hence, the Reynolds number at the same downstream position lower. Second, it results in basic flow velocity profiles with higher critical Reynolds numbers both for the Tollmien–Schlichting waves and crossflow vortices. For two-dimensional boundary layer this effect can be highlighted taking into account the streamwise momentum boundary-layer equation, similarly to that in Sect. 6.4.1 and 6.5. Let us consider a laminar flat-plate boundary layer with uniform suction through, e.g., a porous wall, Fig. 6.10. Then, the boundary conditions take the form

$$U = 0, \quad V = V_w \quad \text{at} \quad y = 0 \,,$$
$$U \to U_e \quad \text{as} \quad y \to \infty \,.$$

The boundary-layer thickness being assumed constant, the streamwise momentum equation reduces to

$$V_w \frac{\partial U}{\partial y} = \nu \frac{\partial^2 U}{\partial y^2} \,. \tag{6.9}$$

As seen, fluid blowing ($V_w > 0$) promotes an inflection point in the basic flow velocity profile; oppositely, suction stabilizes the flow.

In the case of fluid suction through a permeable flat plate, a self-similar solution of the Prandtl boundary-layer equations can be found (see, e.g., Schlichting and

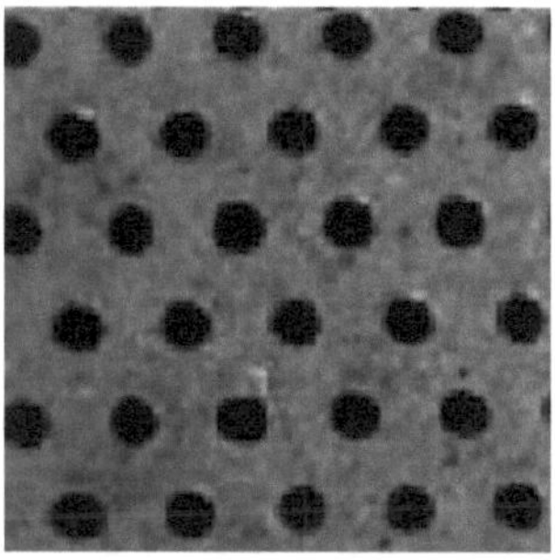

Fig. 6.10 Microscopic view of a modern smooth structured porous surface: porosity 17.1%; orifice diameter 0.175 mm (Boiko and Kornilov, 2010)

Gersten, 2000). If the free-stream velocity U_e is constant along the plate the self-similarity is provided at a constant suction velocity V_w.

The velocity profile

$$\frac{U}{U_e} = 1 - \exp\left(\frac{V_w y}{\nu}\right),$$

which is a solution of (6.9), is called the asymptotic suction profile. It provides, according to calculation results, the critical Reynolds number $Re_{\delta^*} \approx 70,000$ (note that it is about 520 for the Blasius boundary layer).

6.7 Compliant boundaries

Verification of drag reduction by compliant coatings in liquids is a long-standing problem. After the report of Kramer (1957) on significant (up to 60%) drag reduction, no experiment was able to repeat the record. However, in experiments with compliant coatings, certain success in laminar–turbulent transition delay (Gaster, 1988), rather than turbulent drag reduction, substantiated by numerical results (Lucey and Carpenter, 1995) was achieved.

An example of the test coating is shown in Fig. 6.11. The modification of the flow linear stability results from a nontrivial coupling of the flow and coating at the interface between them. The propagation of small disturbances in a compliant material is described by the Navier equations, which in the Cartesian coordinates take the form

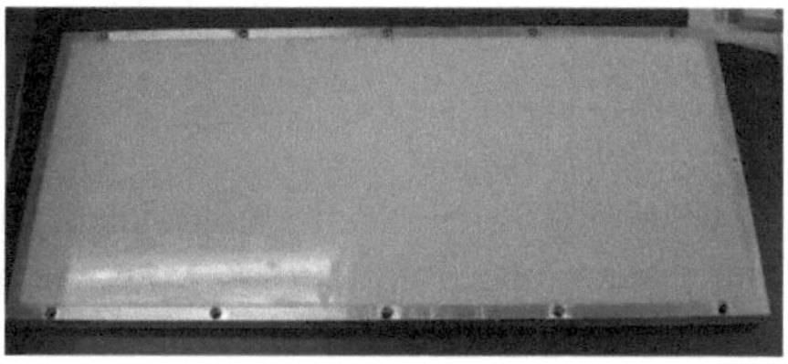

Fig. 6.11 Plane compliant coating manufactured from soft rubber (Lee et al, 2009)

$$\frac{\partial^2 \xi}{\partial t^2} = c_{\mathrm{T}}^2 \left(\frac{\partial^2 \xi}{\partial x^2} + \frac{\partial^2 \xi}{\partial y^2} + \frac{\partial^2 \xi}{\partial z^2} \right) + (c_{\mathrm{T}}^2 - c_{\mathrm{L}}^2) \frac{\partial \hat{\Delta}}{\partial x} ,$$
$$\frac{\partial^2 \eta}{\partial t^2} = c_{\mathrm{T}}^2 \left(\frac{\partial^2 \eta}{\partial x^2} + \frac{\partial^2 \eta}{\partial y^2} + \frac{\partial^2 \eta}{\partial z^2} \right) + (c_{\mathrm{T}}^2 - c_{\mathrm{L}}^2) \frac{\partial \hat{\Delta}}{\partial y} ,$$
$$\frac{\partial^2 \zeta}{\partial t^2} = c_{\mathrm{T}}^2 \left(\frac{\partial^2 \zeta}{\partial x^2} + \frac{\partial^2 \zeta}{\partial y^2} + \frac{\partial^2 \zeta}{\partial z^2} \right) + (c_{\mathrm{T}}^2 - c_{\mathrm{L}}^2) \frac{\partial \hat{\Delta}}{\partial z} ,$$

where

$$\hat{\Delta} = \frac{\partial \xi}{\partial x} + \frac{\partial \eta}{\partial y} + \frac{\partial \zeta}{\partial z} ,$$

$(\xi, \eta, \zeta)^T$ is the displacement vector field, c_{L} and c_{T} are the velocities of dilatation (compression) and equivoluminal (shear) waves in the viscoelastic material. We assume that the coordinates and the displacements have already been scaled by l, the velocities c_{T} and c_{L} by U_{ch}, and t by l/U_{ch}, where l and U_{ch} are fluid characteristic scales.

With regard to the direction of energy transfer, the velocities c_{L} and c_{T} are also called longitudinal and transverse speeds of sound. The velocities can be related to the bulk K and shear G modules as

$$c_{\mathrm{T}} = \sqrt{\frac{G}{\rho_{\mathrm{coat}}}} , \quad c_{\mathrm{L}} = \sqrt{\frac{3K + 4G}{3\rho_{\mathrm{coat}}}} ,$$

where ρ_{coat} is the density of the compliant material. The propagation of the waves is accompanied by their damping, which depends on frequency.

It is known that sound is transmitted through fluids as longitudinal waves, whose absorption is related to what is called the second viscosity, which is frequently neglected. The first (ordinary) viscosity is related to propagation of shear vortical, e.g., Tollmien–Schlichting, waves.

A somewhat similar situation takes place in viscoelastic solids. In this case, the bulk modulus is frequently assumed real (elastic) and the shear modulus is assumed complex (viscous). The dependence of the modules on frequency can be established either in an experiment or rely on a theoretical viscoelastic model. One of the simplest models of this kind is the Voigt model, in which case it is assumed that K is independent of frequency and $G = G_{\mathrm{r}} - \mathrm{i}\omega\check{\mu}$, where G_{r} is real and also independent of frequency, and $\check{\mu}$ is the damping coefficient (similar to the first fluid viscosity) of the solid material. Note that we always have $|c_{\mathrm{L}}| > (4/3)^{1/2}|c_{\mathrm{T}}|$. For typical compliant coating consisting of rubber, K is very large relative to $|G|$, leading to $|c_{\mathrm{L}}| \gg |c_{\mathrm{T}}|$. This means also that Poisson's ratio

$$\frac{3K - 2G}{2(3K + G)}$$

is close to 0.5, i.e., such a material is almost incompressible. Naturally, for absolutely incompressible solid, K would be truly real.

Assuming the normal-mode analysis, the displacements in the viscoelastic material can be considered in the form

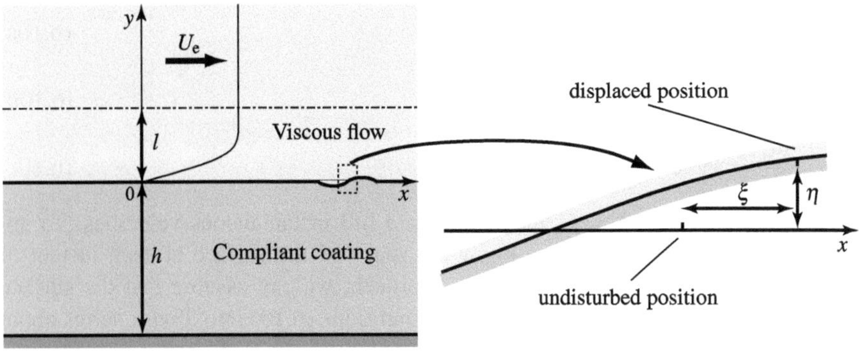

Fig. 6.12 Flow-wall interface

$$\xi = \hat{\xi}(y)e^{i(\alpha x+\beta z-\omega t)}, \quad \eta = \hat{y}(y)e^{i(\alpha x+\beta z-\omega t)}, \quad \zeta = \hat{\zeta}(y)e^{i(\alpha x+\beta z-\omega t)},$$

whose substitution to the Navier equations yields

$$\omega^2\hat{\xi} + c_T^2\left(\mathcal{D}^2 - \varkappa^2\right)\hat{\xi} + \left(c_T^2 - c_L^2\right)\mathrm{i}\alpha\left(\mathrm{i}\alpha\hat{\xi} + \mathcal{D}\hat{\eta} + \mathrm{i}\beta\hat{\zeta}\right) = 0,$$
$$\omega^2\hat{\eta} + c_T^2\left(\mathcal{D}^2 - \varkappa^2\right)\hat{\eta} + \left(c_T^2 - c_L^2\right)\frac{\mathrm{d}}{\mathrm{d}y}\left(\mathrm{i}\alpha\hat{\xi} + \mathcal{D}\hat{\eta} + \mathrm{i}\beta\hat{\zeta}\right) = 0,$$
$$\omega^2\hat{\zeta} + c_T^2\left(\mathcal{D}^2 - \varkappa^2\right)\hat{\zeta} + \left(c_T^2 - c_L^2\right)\mathrm{i}\beta\left(\mathrm{i}\alpha\hat{\xi} + \mathcal{D}\hat{\eta} + \mathrm{i}\beta\hat{\zeta}\right) = 0.$$

The behavior of fluid disturbances is governed by the linearized Navier–Stokes equations. For a flow in a plane channel or in a flat-plate boundary layer, they take the form

$$\begin{aligned}
\mathrm{i}(\alpha U - \omega)\hat{u} + U_y\hat{v} &= -\mathrm{i}\alpha\hat{p} + \frac{1}{\mathrm{Re}}\left(\mathcal{D}^2 - \varkappa^2\right)\hat{u},\\
\mathrm{i}(\alpha U - \omega)\hat{v} &= -\mathcal{D}\hat{p} + \frac{1}{\mathrm{Re}}\left(\mathcal{D}^2 - \varkappa^2\right)\hat{v},\\
\mathrm{i}(\alpha U - \omega)\hat{w} &= -\mathrm{i}\beta\hat{p} + \frac{1}{\mathrm{Re}}\left(\mathcal{D}^2 - \varkappa^2\right)\hat{w},\\
\mathrm{i}(\alpha\hat{u} + \beta\hat{w}) + \mathcal{D}\hat{v} &= 0.
\end{aligned}$$

We consider the flow above a homogeneous isotropic compliant wall of height h bonded firmly to a rigid surface beneath it, as shown for the flat-plate case in Fig. 6.12. Let us examine the kinematic boundary and interface conditions:

1. The kinematic no-slip boundary conditions between the coating and the rigid surface: $\xi = \eta = \zeta = 0$ at $y = -h$.
2. The kinematic no-slip boundary conditions at the other rigid channel wall $\hat{u} = \hat{v} = \hat{w} = 0$ or boundedness of disturbances in the free stream: $\hat{u} < +\infty$, $\hat{v} < +\infty$, $\hat{p} < +\infty$.
3. The kinematic no-slip boundary conditions at the interface between the fluid and the coating:

$$\frac{\partial \xi}{\partial t} = \mathcal{U}(x+\xi, y+\eta, z+\zeta, t) , \tag{6.10a}$$

$$\frac{\partial \eta}{\partial t} = \mathcal{V}(x+\xi, y+\eta, z+\zeta, t) , \tag{6.10b}$$

$$\frac{\partial \zeta}{\partial t} = \mathcal{W}(x+\xi, y+\eta, z+\zeta, t) , \tag{6.10c}$$

where $\mathcal{U} = U + u$, $\mathcal{V} = v$, and $\mathcal{W} = w$ are full instantaneous velocities. To apply them exactly, one has to know the shape of the surface at each instant of time. However, owing to the linear approach, we can assume that the surface deformations are small. Expansion of conditions (6.10) into Taylor series about $(\xi, \eta, \zeta) = (0, 0, 0)$ gives

$$\frac{\partial \xi}{\partial t} = \mathcal{U}(x,0,z,t) + \xi \frac{\partial \mathcal{U}}{\partial x} + \eta \frac{\partial \mathcal{U}}{\partial y} + \zeta \frac{\partial \mathcal{U}}{\partial z} + \dots ,$$

$$\frac{\partial \eta}{\partial t} = \mathcal{V}(x,0,z,t) + \xi \frac{\partial \mathcal{V}}{\partial x} + \eta \frac{\partial \mathcal{V}}{\partial y} + \zeta \frac{\partial \mathcal{V}}{\partial z} + \dots ,$$

$$\frac{\partial \zeta}{\partial t} = \mathcal{W}(x,0,z,t) + \xi \frac{\partial \mathcal{W}}{\partial x} + \eta \frac{\partial \mathcal{W}}{\partial y} + \zeta \frac{\partial \mathcal{W}}{\partial z} + \dots .$$

Separating the basic flow conditions and dropping all nonlinear terms, we finally obtain for the modal disturbances

$$-\mathrm{i}\omega\hat{\xi} = \hat{u} + U_y\hat{\eta} ,$$

$$-\mathrm{i}\omega\hat{\eta} = \hat{v} ,$$

$$-\mathrm{i}\omega\hat{\zeta} = \hat{w} .$$

Let us now examine the dynamic interface conditions. They consist in requirements of continuity of normal (τ_{yy}) and tangential (τ_{xy}, τ_{zy},) stresses between the fluid and the wall (assuming that the interface does not change its position):

$$\tau_{yy}^{\text{coat}} = \tau_{yy}^{\text{fluid}} , \tag{6.11a}$$

$$\tau_{xy}^{\text{coat}} = \tau_{yy}^{\text{fluid}} , \tag{6.11b}$$

$$\tau_{zy}^{\text{coat}} = \tau_{zy}^{\text{fluid}} , \tag{6.11c}$$

where

$$\tau_{yy}^{\text{coat}} = 2\tilde{\rho} c_{\mathrm{T}}^2 \left[\frac{\partial \eta}{\partial y} + \left(\frac{c_{\mathrm{L}}^2}{2c_{\mathrm{T}}^2} - 1 \right) \Delta \right] ,$$

$$\tau_{xy}^{\text{coat}} = \tilde{\rho} c_{\mathrm{T}}^2 \left(\frac{\partial \xi}{\partial y} + \frac{\partial \eta}{\partial x} \right) ,$$

$$\tau_{zy}^{\text{coat}} = \tilde{\rho} c_{\mathrm{T}}^2 \left(\frac{\partial \zeta}{\partial y} + \frac{\partial \eta}{\partial z} \right) ,$$

and $\tilde{\rho} = \rho_{\text{coat}}/\rho$ is the dimensionless density of the coating (the ratio of coating and fluid physical densities). For a Newtonian isotropic fluid, the stresses in dimensionless form are

$$\tau_{ij}^{\text{fluid}} = \frac{1}{\text{Re}}\left(\frac{\partial \mathcal{U}_i}{\partial x_j} + \frac{\partial \mathcal{U}_j}{\partial x_i}\right) - p\delta_{ij}, \quad i,\ j = 1,2,3,$$

where $x_1 \equiv x, x_2 \equiv y, x_3 \equiv z, \mathcal{U}_1 \equiv \mathcal{U}, \mathcal{U}_2 \equiv \mathcal{V}, U_3 \equiv \mathcal{W}$ and δ_{ij} is the Kronecker delta. Extracting the conditions for the disturbances from (6.11) gives

$$2\tilde{\rho}c_{\mathrm{T}}^2\left[\frac{c_{\mathrm{L}}^2}{2c_{\mathrm{T}}^2}\mathcal{D}\hat{\eta} - \mathrm{i}\left(\frac{c_{\mathrm{L}}^2}{2c_{\mathrm{T}}^2} - 1\right)(\alpha\hat{\xi} + \beta\hat{\zeta})\right] = \frac{2}{\text{Re}}\mathcal{D}\hat{v} - \hat{p}\,,$$

$$\tilde{\rho}c_{\mathrm{T}}^2(\hat{\xi} + \mathrm{i}\alpha\hat{\eta}) = \frac{1}{\text{Re}}(\mathcal{D}\hat{u} + \mathrm{i}\alpha\hat{v} + U_{yy}\hat{\eta})\,,$$

$$\tilde{\rho}c_{\mathrm{T}}^2(\hat{\zeta} + \mathrm{i}\beta\hat{\eta}) = \frac{1}{\text{Re}}(\mathcal{D}\hat{w} + \mathrm{i}\beta\hat{v})\,.$$

If $\text{Re} \gg 1$, the first terms in the right-hand side of these equations can be dropped, which indicates that the main coupling between the fluid and the coating is achieved through the disturbance pressure. Moreover, if the coating is stiff enough, the terms depending on the tangential surface displacements ξ and ζ can be also omitted, so that the first equation takes the form

$$\frac{\partial \hat{\eta}}{\partial y} \approx -\frac{1}{c_{\mathrm{L}}^2}\frac{\hat{p}}{\tilde{\rho}}\,.$$

Note that the right-hand-side term in this equation and, hence, the coating response, increases if $\tilde{\rho}$ and c_{L} decrease. For a given material, the values of $\tilde{\rho}$ and c_{L} are determined by the fluid density ρ and the characteristic fluid velocity U_{ch}, respectively. In accordance with these considerations, experiments and numerical simulations show that the coatings are most effective in liquids and can have only a marginal influence on the linear stability in high-speed gas flows. As the right-hand side contains the complex-value factor c_{L}^2, the response of the coating to the impact occurs with some delay, which, in a favorable case, leads to a delay of the laminar–turbulent transition.

As it was shown by Yeo (1992), the Squire (1933) theorem is not applicable in this case. Squire's approach leads to the need of rescaling the parameters of the wall so that it becomes effectively more stiff with respect to the three-dimensional normal modes.

The derived set of equations with corresponding boundary and interface conditions can be solved only numerically. Note, that in the boundary-layer case one has also to take into account that the dimensionless thickness of the coating changes in the downstream direction. An example of numerical calculation of the stability in space for a flat-plate boundary layer is shown in Fig. 6.13. The displacement thickness and external-flow velocity were used as scaling factors to normalize the data, i.e., $l = \delta^*$ and $U_{\text{ch}} = U_{\text{e}}$. The calculations were performed with a spectral collocation procedure as described in Chaps. 3 and 4. One hundred nodes were taken for discretizing the linearized Navier–Stokes equations and the same amount for the Navier equations. The wall Reynolds number $\text{Re}_h = U_{\text{e}}h/\nu$ was used to rescale the coating thickness at different flow Reynolds numbers Re. The results coincide essentially with the data presented by Yeo (1988).

As seen, three different regions of instability are distinguished. The region A corresponds to the usual Tollmien–Schlichting instability or TSI. The instabilities in

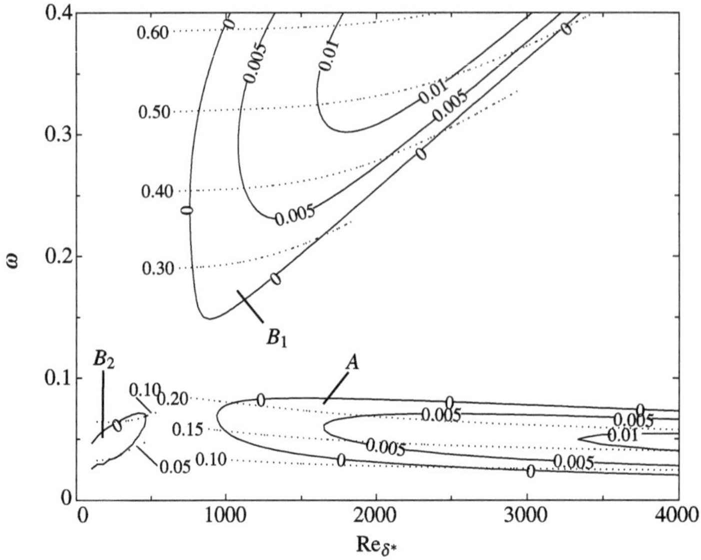

Fig. 6.13 Stability diagram for a Blasius boundary layer over a compliant wall with $G_r = 0.7$, $\tilde{\rho} = 1$, $\breve{\mu} = 0.0049$, and $K = 500$, $Re_h = 20 \times 10^3$: values $-\alpha_i$ (*solid lines*) and α_r (*dotted lines*). After Yeo (1988)

regions B_1 and B_2 are due to the viscoelasticity of the wall. They are frequently denoted as the compliance-induced flow instabilities or CIFIs. The stabilizing effect of the wall compliance on TSI can be easily traced by comparing the data with those for a rigid wall, Fig. 4.6. First, the critical Reynolds number shifts from about 520 to 935. Second, the unstable frequency band shrinks. Finally, at any given Reynolds number inside the instability region, the growth rates become smaller. The drawback in this particular case is the appearance of the regions B_1 and B_2 at high frequencies and low Reynolds numbers, respectively. At the low Reynolds numbers the boundary-layer and parallel-flow approximations are invalid, and, hence, the region B_2 requires an additional elaboration. In contrast, the region B_1 can be treated with the quasi-parallel approximation quite accurately. The growth rates of disturbances in it are larger than in the TSI region at the same Reynolds numbers, while the critical Reynolds number $Re_{L,\delta^*} \approx 758$ is smaller. Thus, in this example, the CIFI rather than TSI modes can trigger the laminar–turbulent transition.

6.8 Dusty flow

Addition of small solid particles to a fluid flow can radically change its characteristics. For the first time, the hydrodynamic stability problem for dispersed parallel flows was formulated by Saffman (1962). He showed that the problem of linear stability for a fluid–particle mixture with a uniform distribution of suspended particles is reduced to the solution of the Orr–Sommerfeld equation with an effective (generally complex-value) velocity profile.

Traditionally, the hydrodynamic description of a two-phase mixture is constructed on the basis of multi-fluid hydrodynamic equations. Let us consider an incompressible dilute fluid-particle mixture. The solid particles are small enough and spherical with the radius equal to R. The Brownian motion of dispersed particles is not taken into account. It is assumed that the particles are so rarefied that we can neglect interparticle interactions. That is, the total volume occupied by the particles is much smaller than the volume of the fluid. The net effect of the dispersed phase on the carrier fluid is modeled by an extra force, which, taken per unit volume, is linearly proportional to the relative velocity of these phases:

$$\mathbf{F}_\mathrm{p} = bn_\mathrm{p}(\mathscr{U}_\mathrm{p} - \mathscr{U}) ,$$

where b is the drag constant depending on the geometry of the particles, n_p is the number density of the particles, and $\mathscr{U}_\mathrm{p}(x,y,z,t)$ and $\mathscr{U}(x,y,z,t)$ are the velocities of the particles and fluid, respectively. As the particles are spherical and their relative motion through the viscous fluid is slow, we can use the Stokes expression for the drag constant

$$b = 6\pi R\rho\nu ,$$

where ρ and ν are the density and kinematic viscosity of the unpolluted fluid, respectively. The opposite force $\mathbf{F} = -\mathbf{F}_\mathrm{p}$ is exerted by the fluid on the particles. Based on these assumptions, the continuity and momentum equations of the carrier fluid and suspended particles, if we introduce the dimensionless variables using the ordinary scheme, take the form

$$\nabla\cdot\mathbf{U} = 0 , \tag{6.12a}$$

$$\frac{\partial\rho}{\partial t} + \nabla\cdot(\rho\mathbf{U}_\mathrm{p}) = 0 , \tag{6.12b}$$

$$\left[\frac{\partial\mathbf{U}}{\partial t} + (\mathbf{U}\cdot\nabla)\mathbf{U}\right] = -\nabla P + \frac{1}{\mathrm{Re}}\left[\nabla^2\mathbf{U} + \frac{\rho}{S}(\mathbf{U}_\mathrm{p} - \mathbf{U})\right] , \tag{6.12c}$$

$$\left[\frac{\partial\mathbf{U}_\mathrm{p}}{\partial t} + (\mathbf{U}_\mathrm{p}\cdot\nabla)\mathbf{U}_\mathrm{p}\right] = -\frac{1}{S\mathrm{Re}}(\mathbf{U}_\mathrm{p} - \mathbf{U}) , \tag{6.12d}$$

where $\mathbf{U} = \mathscr{U}/U_\mathrm{ch}$, $P = p/(\rho U_\mathrm{ch}^2)$, $\tilde{\rho} = \rho_\mathrm{p}/\rho$ (ρ_p is the density of the particle material), $S = 2/9(R/l)^2\tilde{\rho}$, $\mathrm{Re} = U_\mathrm{ch}l/\nu$, and, finally, U_ch and l are the characteristic velocity and length scales of the basic flow, respectively. Note that here we neglected the effects of buoyancy and gravity. Model (6.12) is the simplest one for two-phase dispersed media. The parameter $\tilde{\rho}$ determines how many particles are present in the flow and R/l is the relative particle size.

Let the two-phase flow along the Cartesian x-axis be steady, plane, and parallel. Thus, it is determined by the velocity profiles $\mathbf{U} = (U(y),0,0)^T$ and pressure $P(y)$. In accordance with the normal-mode approach, we study the flow stability to small disturbances in the form of plane waves

$$u = \hat{u}\mathrm{e}^{\mathrm{i}(\alpha x+\beta z-\omega t)} ,\ v = \hat{v}\mathrm{e}^{\mathrm{i}(\alpha x+\beta z-\omega t)} ,\ w = \hat{w}\mathrm{e}^{\mathrm{i}(\alpha x+\beta z-\omega t)} ,\ p = \hat{p}\mathrm{e}^{\mathrm{i}(\alpha x+\beta z-\omega t)} ,$$

for fluid disturbances and

$$u_{\mathrm{p}} = \hat{u}_{\mathrm{p}} e^{\mathrm{i}(\alpha x + \beta z - \omega t)}, \; v_{\mathrm{p}} = \hat{v}_{\mathrm{p}} e^{\mathrm{i}(\alpha x + \beta z - \omega t)}, \; w_{\mathrm{p}} = \hat{w}_{\mathrm{p}} e^{\mathrm{i}(\alpha x + \beta z - \omega t)} .$$

for particle disturbances. Saffman (1962) using the arguments of Squire (1933) showed that the Squire theorem holds for the uniform distribution of particles. Hence, below we put $\hat{w}_{\mathrm{p}} = \hat{w} = 0$ and $\beta = 0$.

Substituting the expressions for disturbances in (6.12) and neglecting nonlinear terms, we obtain a system of linear hydrodynamic stability equations. This system has eight scalar equations from which we can split off a closed subsystem of the following five equations

$$\begin{aligned}
\mathrm{i}\alpha\hat{u} + \mathcal{D}\hat{v} &= 0 , \\
\mathrm{i}(\alpha U - \omega)\hat{u} + U_y\hat{v} &= -\mathrm{i}\alpha\hat{p} + \frac{1}{\mathrm{Re}}\left(\mathcal{D}^2 - \alpha^2\right)\hat{u} + \frac{\rho}{S\mathrm{Re}}(\hat{u} - \hat{u}_{\mathrm{p}}) , \\
\mathrm{i}(\alpha U - \omega)\hat{v} &= -\mathcal{D}\hat{p} + \frac{1}{\mathrm{Re}}\left(\mathcal{D}^2 - \alpha^2\right)\hat{v} + \frac{\rho}{S\mathrm{Re}}(\hat{v} - \hat{v}_{\mathrm{p}}) , \\
\mathrm{i}(\alpha U - \omega)\hat{u}_{\mathrm{p}} + U_y\hat{v}_{\mathrm{p}} &= -\frac{1}{S\mathrm{Re}}(\hat{u} - \hat{u}_{\mathrm{p}}) , \\
\mathrm{i}(\alpha U - \omega)\hat{v}_{\mathrm{p}} &= -\frac{1}{S\mathrm{Re}}(\hat{v} - \hat{v}_{\mathrm{p}}) ,
\end{aligned}$$

and homogeneous boundary conditions are assumed. Note that the last two equations are algebraic. Solving them with respect to $\hat{u}_{\mathrm{p}}$ and $\hat{v}_{\mathrm{p}}$ yields

$$\begin{aligned}
\hat{u}_{\mathrm{p}} &= \frac{1}{1 + \mathrm{i}S\mathrm{Re}(\alpha U - \omega)}\left(\hat{u} - \frac{S\mathrm{Re}}{1 + \mathrm{i}S\mathrm{Re}(\alpha U - \omega)} U_y \hat{v}\right) , \\
\hat{v}_{\mathrm{p}} &= \frac{1}{1 + \mathrm{i}S\mathrm{Re}(\alpha U - \omega)}\hat{v} .
\end{aligned}$$

On eliminating the pressure and streamwise velocity disturbances, the consideration of the flow stability reduces to the eigenvalue problem for the single equation resembling the Orr–Sommerfeld equation

$$\mathrm{i}\left(\alpha\tilde{U} - \omega\right)\left(\mathcal{D}^2 - \alpha^2\right)\hat{v} - \mathrm{i}\alpha\left[\tilde{U}_{yy}\hat{v} - \frac{\mathrm{d}^2\left(\hat{v}J\rho_y\right)}{\mathrm{d}y^2}\right] = \frac{1}{\mathrm{Re}}\left(\mathcal{D}^2 - \alpha^2\right)^2\hat{v} , \tag{6.13}$$

where

$$\tilde{U} = U + \tilde{\rho}J , \quad J = \frac{\alpha U - \omega}{\alpha[1 + \mathrm{i}S\mathrm{Re}(\alpha U - \omega)]} .$$

However, it has a more complex structure than the Orr–Sommerfeld equation for a one-phase fluid. In particular, an additional term appears, which depends on the mass concentration distribution $\tilde{\rho}(y)$, while the velocity profile $U(y)$ is replaced by the new complex effective velocity $\tilde{U}(y)$. If the solid particles are uniformly distributed ($\tilde{\rho} = \mathrm{const}$), then the equation reduces to the Orr–Sommerfeld equation in the form

$$\left[\mathrm{i}(\alpha\tilde{U} - \omega)\left(\mathcal{D}^2 - \alpha^2\right) - \mathrm{i}\alpha\tilde{U}_{yy} - \frac{1}{\mathrm{Re}}\left(\mathcal{D}^2 - \alpha^2\right)^2\right]\hat{v} = 0 . \tag{6.14}$$

For limit cases of coarse and fine particles, the problem can be further simplified.

If both the particle size and the flow velocity are small, then $S\mathrm{Re} \ll 1$ and

$$\alpha\tilde{U} - \omega = \alpha U + \tilde{\rho}\frac{\alpha U - \omega}{1 + \mathrm{i}S\mathrm{Re}(\alpha U - \omega)} - \omega \approx (1 + \tilde{\rho})(\alpha U - \omega)\,.$$

Hence, (6.13) becomes the classical Orr–Sommerfeld equation for a one-phase fluid with the effective Reynolds number

$$\mathrm{Re}_{\mathrm{eff}} = \mathrm{Re}(1 + \tilde{\rho})\,.$$

This means that the two-phase flow behaves as a homogenous fluid with a smaller kinematic viscosity, i.e., one can expect a destabilizing effect of the particles. The physical reasoning for this effect is quite clear. Small particles simply increase the effective density of the mixture, while the dynamic viscosity is preserved. This directly leads to a decrease of the effective kinematic viscosity.

On the contrary, when the particle sizes or the flow velocity are such that $S\mathrm{Re} \gg 1$, we come to

$$\alpha\tilde{U} - \omega = \approx \alpha U - \left(\omega + \frac{\mathrm{i}\tilde{\rho}}{S\mathrm{Re}}\right)\,.$$

Comparing the last expression with the corresponding term in the Orr–Sommerfeld equation for a one-phase flow, one can see that (6.14) adopts exactly its form after the change of variables

$$\tilde{\omega} = \omega_{\mathrm{r}} + \mathrm{i}[\omega_{\mathrm{i}} + \tilde{\rho}/(S\mathrm{Re})]\,.$$

As $\tilde{\rho}/(S\mathrm{Re}) > 0$, it follows that the flow with coarse particles is more stable, i.e., inertial particles damp fluid disturbances.

Exercises

6.1. Modify functions `DuctBV`, `Duct`, and (if necessary) `chder2` to account for the symmetry of the channel cross section. Compare the spectra of modes of the different classes. Try to obtain some of the critical Reynolds numbers from Table 6.2.

6.2. Write a function implementing the two-dimensional mapping as outlined in Sect. 6.3.

6.3. Using a function from Exercise 6.2 write a program for the numerical solution of the Poisson equation in a grooved channel. Modify it for the case of identical grooves at both walls. Reproduce Fig. 6.3

6.4. Modify the program from Exercise 6.3 to account for the symmetries of the channel cross section. Design grooves with smooth shape (without sharp tips) and use them in computations to minimize computational expenses. Compare the results obtained for the sharp and smooth grooves with `chder1` and `foder1` to approximate derivatives in r direction.

6.5. Using functions from Exercise 3.4 write a program for numerical solution of Eqs. 6.8 at given values of α, β, and Re, which finds separately even and odd modes. Reproduce Fig. 6.4.

6.6. Using function `FS` from Sect. 4.2 as a starting point, write a MATLAB function to solve the Falkner–Skan–Cooke equation.

6.7. Modify the program from Exercise 4.5 to obtain and draw the growth rates of the leading crossflow modes for the boundary layer described by the Falkner–Skan–Cooke equations as shown for temporal formulation in Fig. 6.7. What is the value of Re_L in this case? How does the neutral stability curve look like in the spatial formulation?

References

Arnal D, Michel R (eds) (1990) Laminar–turbulent Transition, IUTAM Symposium. Springer–Verlag, Berlin

Boiko AV, Kornilov VI (2010) Turbulent boundary-layer control by passive and active methods. Progress and problems. In: Fomin VM (ed) Proc. Ninth Internat. Conf. on Methods of Aerophysical Research, Parallel, Novosibirsk, vol 1, pp 49–50

Boiko AV, Nechepurenko YM (2008) Numerical spectral analysis of temporal stability of laminar flows in ducts of constant cross-sections. Comput Maths Math Phys 48:1699–1714

Boyd JP (1999) Chebyshev and Fourier Spectral Methods, 2nd edn. Dover, NY

Gaponenko VR, Ivanov AV, Kachanov YS (1995) Experimental study of a swept-wing boundary-layer stability with respect to unsteady disturbances. Thermophys Aeromech 2(4):287–312

Gaponov SA, Maslov AA (1971) Stability of compressible boundary layer at subsonic flow velocities. Izv Sib Otd Akad Nauk SSSR, Ser Tekhn Nauk 1(3):24–27

Gaster M (1988) Is the dolphin a red herring? In: Liepmann HW, Narasimha R (eds) Turbulent Management and Relaminarization, Springer–Verlag, Berlin, pp 285–304

Gray WE (1952) The effect of wing sweep on laminar flow. TM Aero 255, RAE

Gregory N, Stuart JT, Walker WS (1955) On the stability of three-dimensional boundary layers with application to the flow due to a rotating disk. Philos Trans R Soc Lond A 248:155–199

Kachanov YS, Tararykin OI (1990) The experimental investigation of stability and receptivity of a swept-wing flow. In: Arnal and Michel (1990), pp 499–509

Kachanov YS, Kozlov VV, Levchenko VY (1974) Experimental study on the effect of wall cooling upon laminar boundary-layer stability. Izv Sib Otd Akad Nauk SSSR, Ser Tekhn Nauk 9(8):75–79

Kachanov YS, Koptsev DB, Smorodsky BV (2000) Three-dimensional stability of self-similar boundary layer with a negative Hartree parameter. Thermophys Aeromech 7(3):341–351

Kerswell RR, Davey A (1996) On the linear instability of elliptic pipe flow. J Fluid Mech 316:307–324

Kobayashi R, Kohama Y, Kurosawa M (1983) Boundary-layer transition on a rotating cone in axial flow. J Fluid Mech 127:341–352

Kramer MO (1957) Boundary layer stabilization by distributed damping. J Aerosp Sci 24(6):459

Lee I, Chun HH, Kulik VM, Boiko AV (2009) Experimental investigation of drag reducing mechanism of compliant coating with optimal viscoelastic properties. In: Johansson AV, Sung HJ (eds) Sixth International Symposium on Turbulence and Shear Flow Phenomena. Seoul National University, Seoul, pp 891–895

Lingwood RJ (1995) Absolute instability of the boundary layer on a rotating disk. J Fluid Mech 299:17–33

Lucey AD, Carpenter PW (1995) Boundary layer instability over compliant wall: Comparison between theory and experiment. Phys Fluids 7(10):2355–2363

Mack LM (1979) Three-dimensional effects in boundary layer stability. In: Naval hydrodynamics. National Academy of Sciences, Washington, DC, pp 63–76

Poll DIA (1985) Some observations on the transition process on the windward face of a long yawed cylinder. J Fluid Mech 150:329–356

Ryzhov OS, Terent'ev ED (1998) Streamwise absolute instability of a three-dimensional boundary layer at high Reynolds numbers. J Fluid Mech 373:111–153

Saffman PG (1962) On the stability of a laminar flow to a dusty gas. J Fluid Mech 13:120–128

Schlichting H, Gersten K (2000) Boundary layer theory, 8th edn. Springer–Verlag, Berlin

Schubauer GB, Skramstad HK (1948) Laminar-boundary layer oscillations and transition on a flat plate. TN 909, NACA

Squire HB (1933) On the stability for three-dimensional disturbances of viscous fluid between parallel walls. Proc R Soc Lond A 142:621–628

Tatsumi T, Yoshimura T (1996) Stability of the laminar flow in a rectangular duct. J Fluid Mech 316:307–324

Taylor MJ, Peake N (1998) The long-time behaviour of incompressible swept-wing boundary layer subject to impulsive forcing. J Fluid Mech 355:359–381

Wazzan AR, Okamura TT, Smith AMO (1970) The stability and transition of heated and cooled incompressible laminar boundary layers. In: Grigull V, Hahne E (eds) Fourth Internat. Heat Transfer Conf., Elsevier, Amsterdam, vol 2, paper No. FC-14.

Yeo KS (1988) The stability of boundary-layer flow over single and mulri-layer viscoelastic walls. J Fluid Mech 196:359–408

Yeo KS (1992) The three-dimensional stability of boundary-layer flow over compliant walls. J Fluid Mech 238:537–577

Further Reading

Arnal D, Michel R (eds) (1990) Laminar–Turbulent Transition, IUTAM Symposium, Springer–Verlag, Berlin

Arnal D, Juillen JC, Reneaux J, Gasparian G (1997) Effect of wall suction on leading edge contamination. Aerosp Sci Technol 8:505–517

Bippes H, Wiegel M, Bertolotti F (1999) Experiments on the control of cross-flow instability with the aid of suction through perforated walls. In: Meier GEA, Viswanath PR (eds) Mechanics of Passive and Active Flow Control, Kluwer, Dordrecht, pp 165–170

Carpenter PW (1990) Status of transition delay using compliant walls. In: Bushnell DM, Heffner JN (eds) Viscous Drag Reduction in Boundary Layers, AIAA, New York, pp 79–113

Cooke JC (1950) The boundary layer of a class of infinite yawed cylinders. Proc Camb Philos Soc 46:645–648

Dagenhart JR, Saric WS, Hoos JA, Mousseux MC (1990) Experiments on swept-wing boundary layers. In: Arnal and Michel (1990), pp 369–380

Deyhle H, Bippes H (1996) Disturbance growth in an unstable three-dimensional boundary layer and its dependence on environmental conditions. J Fluid Mech 316:73–113

Dovgal AV, Levchenko VY, Timofeev VA (1990) Boundary layer control by a local heating of the wall. In: Arnal and Michel (1990), pp 113–121

Ehrenstein U (1996) On the linear stability of channel flow over riblets. Phys Fluids 8(11):3194–3196

Grek GR, Kozlov VV, Titarenko SV (1996) An experimental study on the influence of riblets on transition. J Fluid Mech 315:31–49

Kazakov AV, Kogan MN, Kuparev VA (1985) On increase of subsonic boundary-layer stability at surface heating close the leading edge. Dokl Akad Nauk 283(2):333–335

Lebedev YB, Fomichev VM (1987) Stability of the fluid boundary layer with nonuniform surface temperature distribution. J Eng Phys Thermophys 52(6):948–953

Lowell RL, Reshotko E (1974) Numerical study of the stability of a heated water boundary layer. FT AS/TR 73–93, Case Western Reserve University, Cleveland

Luchini P (1995) Asymptotic analysis of laminar boundary-layer flow over finely grooved surfaces. Eur J Mech B/Fluids 14(2):169–195

Luchini P, Manzo M, Pozzi A (1991) Resistance of a grooved surface to parallel flow and cross-flow. J Fluid Mech 228:87–109

Mack LM (1980) On the stabilization of three-dimensional boundary layers by suction and cooling. In: Eppler R, Fasel H (eds) Laminar–Turbulent Transition, Springer–Verlag, Berlin, IUTAM Symposium, pp 223–238

Mack LM (1984) Boundary layer stability theory: Special course on stability and transition of laminar flow. Report 709, AGARD

Malik MR, Orszag SA (1980) Comparison of methods for prediction of transition by stability analysis. AIAA J 18(12):1485–1489

Nitschke-Kowsky P, Bippes H (1988) Instability and transition of a three-dimensional boundary layer on a swept flat plate. Phys Fluids 31(4):786–795
Rosenhead L (1963) Laminar boundary layers. Oxford University Press, Oxford
Struminskiy VV, Lebedev YB, Fomichev VM (1986) Effect of streamwise pressure gradient on laminar boundary-layer extension in gas flow. Dokl Akad Nauk 289(4):813–816
Takagi S, Itoh N (1994) Observation of traveling waves in the three-dimensional boundary layer along a yawed cylinder. Fluid Dyn Res 14:167–189

Chapter 7
Instability of separated boundary layers

In this chapter we discuss flow instability at laminar boundary-layer separation that can happen on airfoils and wings, in diffusors, at sudden surface variations such as steps and bumps, etc. Within this topic, a good deal of research data is available for local separation regions generated when the separated layer reattaches back to the surface (Fig. 7.1). Such flows, also known as 'separation bubbles,' are in focus of the following consideration.

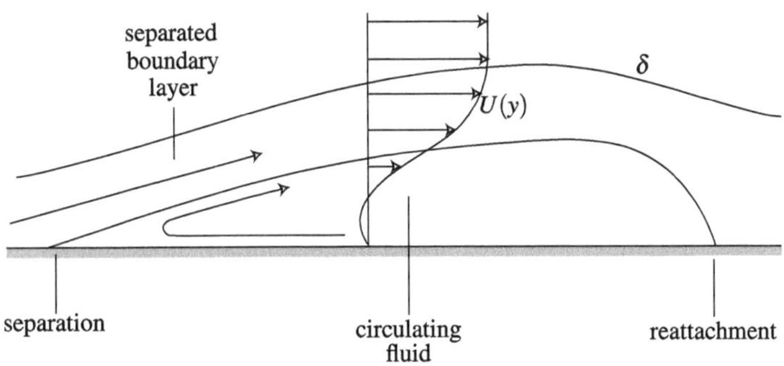

Fig. 7.1 Separation bubble

7.1 Basic features of separated flow instability

Even at rather low Reynolds numbers, boundary-layer separation results in laminar flow destabilization, an increase of flow disturbances, and onset of unsteady fluid motion. The perturbed separated flow pattern comes from an interplay of different phenomena, which can be distinguished by their appearances.

A.V. Boiko et al., *Physics of Transitional Shear Flows*,
Fluid Mechanics and Its Applications 98, DOI 10.1007/978-94-007-2498-3_7,

One of them is the transition to turbulence in the separated shear layer. In early studies, it was revealed that in some cases, e.g., on airfoils, the behavior of separation bubbles strongly depends on flow turbulization. A number of semi-empirical models of such flows have been proposed, the transition location being taken from experimental correlations or predicted through linear stability analysis (van Ingen, 1975). Another phenomenon competing with the transition to small-scale chaotic motion is the formation of large-scale quasi-periodic two- and three-dimensional vortices shedding into the reattached boundary layer. The contribution of coherent vortices to the separated flow pattern depends on the configuration of the basic flow, its mean parameters, and environmental perturbations. *Vortex shedding* is typically found in separation regions induced by quite strong external pressure gradients, behind sharp edges, high steps, etc. Also, separation bubbles are often subject to low-frequency oscillations or the so-called '*flapping*.'

Basically, a wide spectrum of laminar flow disturbances in separation regions originates from both amplification of environmental perturbations and generation of self-sustained oscillations. That is, a good deal of research data testify to the spatial growth of velocity perturbations permanently generated in separation bubbles by external forcing (Gaster, 1967). These disturbances are dominated by local separated flow characteristics similarly to that in other convectively unstable shear layers. On the other hand, there are strong indications on self-excitation of separation regions associated with global unsteadiness of the flow pattern in the form of, e.g., vortex shedding. In what follows, we are mainly interested in such flows as noise amplifiers by focusing on the oscillations of the separated shear layer in terms of the classic stability theory rather than on the global dynamics of separation regions. The latter was examined in a series of recent studies; one can find them among other references on the present topic at the end of this chapter.

One expects that, at least at a low level of background perturbations, the initial growth of the separated-layer disturbances is predicted by a linear stability analysis. If so, the process of laminar flow breakdown in separation regions can be approached in the same way as in other convectively unstable systems by focusing on several aspects (Fig. 7.2). These are the behavior of small-amplitude instability waves, their excitation by environmental disturbances, and nonlinear effects of the high-amplitude oscillations. Some key points of the above aspects of separated flow instability are mentioned in this chapter as well as in Chaps. 10 and 12.

We begin with linear instability at flow separation in two-dimensional plane configurations. Then, the consideration is complemented by the axial symmetry and three-dimensionality of the separating boundary layer.

7.2 Linear instability at flow separation in plane configurations

First of all, we notice that the validity of the linear stability theory for small-amplitude perturbations of separation bubbles is not immediately obvious. The point is that the entire separation region, including its upstream sections where instabil-

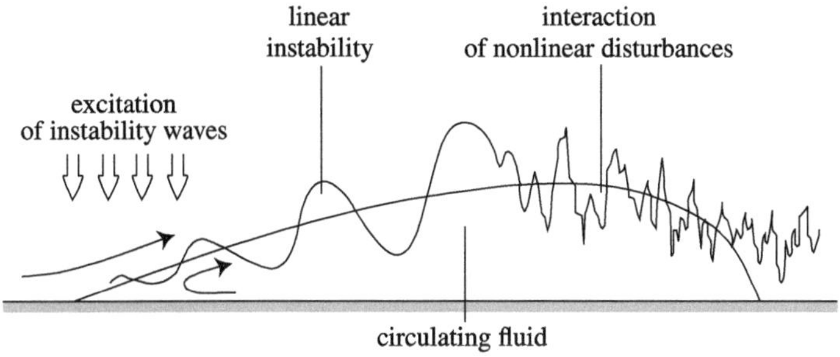

Fig. 7.2 Aspects of laminar flow breakdown in the separation region

ity waves grow, is affected by the disturbed flow in the aft part of the bubble. As a result, the excitation of perturbations causes a basic flow variation, depending on their initial amplitudes (Fig. 7.3). This makes the stability analysis, which deals with oscillations superimposed on a prescribed basic flow and linearity of perturbations, a nontrivial feature of locally separated flows. Nevertheless, when applied to separation bubbles, the linear stability theory appears really meaningful, which is supported by comparisons of theoretical results with experimental, numerical data and solutions of the parabolized stability equations.

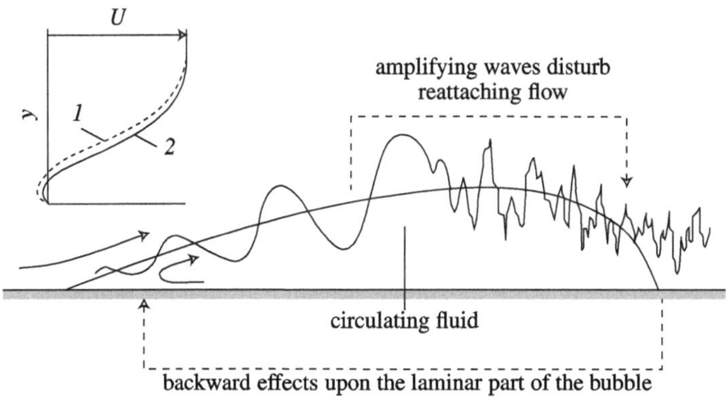

Fig. 7.3 Backward effect of disturbances in a separation bubble: mean velocity profile under natural conditions (*1*) and with excitation of perturbations (*2*)

An explanation of the above-described paradox is that the influence of amplifying disturbances upon the separation region consists, most of all, in variations of its length and depth, while the local basic flow properties in the separated shear layer virtually do not change. Figure 7.4 shows hot-wire data on the mean-velocity profiles in a separation bubble measured under natural experimental conditions and with controlled generation of an instability wave. The excited separation region is diminished so that the distance between the separated boundary layer and the wall

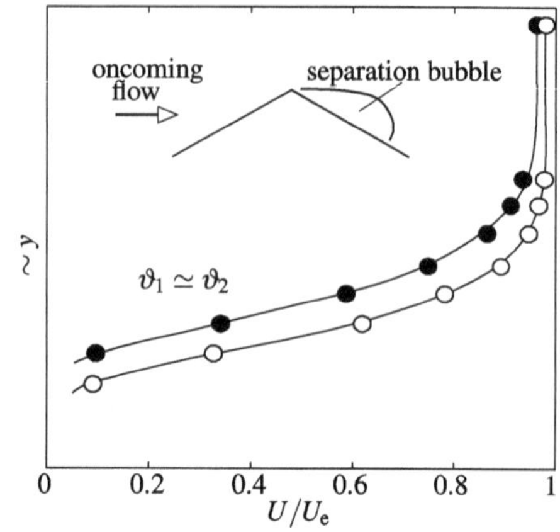

Fig. 7.4 Mean-flow profiles in the laminar part of the separation bubble behind a two-dimensional transverse surface inflection (Dovgal and Kozlov, 1983). Natural conditions (•); under excitation of instability waves (○)

becomes smaller. However, the velocity profiles close to the inflection point near the edge of the circulation region (actually, the momentum thickness ϑ) are almost the same. Thus, at small-enough variations of the velocity profile in the separated shear layer and of the distance between the layer and the wall, the characteristics of instability waves in the upstream part of the separation bubble do not essentially depend on their amplitude.

7.2.1 Waveform

In different conditions of flow separation, fairly universal amplitude and phase velocity distributions of a two-dimensional instability wave across a separation bubble are observed; an illustration is given in Fig. 7.5. In this example, the oscillations are induced in the pre-separated boundary layer upstream of a two-dimensional bump on a flat-plate surface. Neglecting the streamline curvature around the bump, it can be said that the profiles are those of the streamwise velocity of the disturbance.

Penetrating into the separation bubble, the perturbation exhibits the amplitude profile with three maxima near the outer edge of the separated boundary layer, at the position of the largest basic flow velocity gradient and close to the wall in the circulation region. Further downstream at laminar flow reattachment, the disturbance turns back into the instability wave of the attached boundary layer; see the right-hand section in Fig. 7.5. The phase profiles of the perturbation are distinguished by two 'jumps' across the separation region (instead of one in the Tollmien–Schlichting wave of the Blasius boundary layer) with a phase difference close to 180 degrees between the oscillations near the wall and in the separated layer.

Similar profiles of two-dimensional instability waves are also found through wind-tunnel experiments in separation regions induced by an external pressure gradient along a flat plate, behind backward-facing steps, and on airfoils, both in natural and controlled conditions of separated flow instability, (e.g., Cousteix and Pailhas,

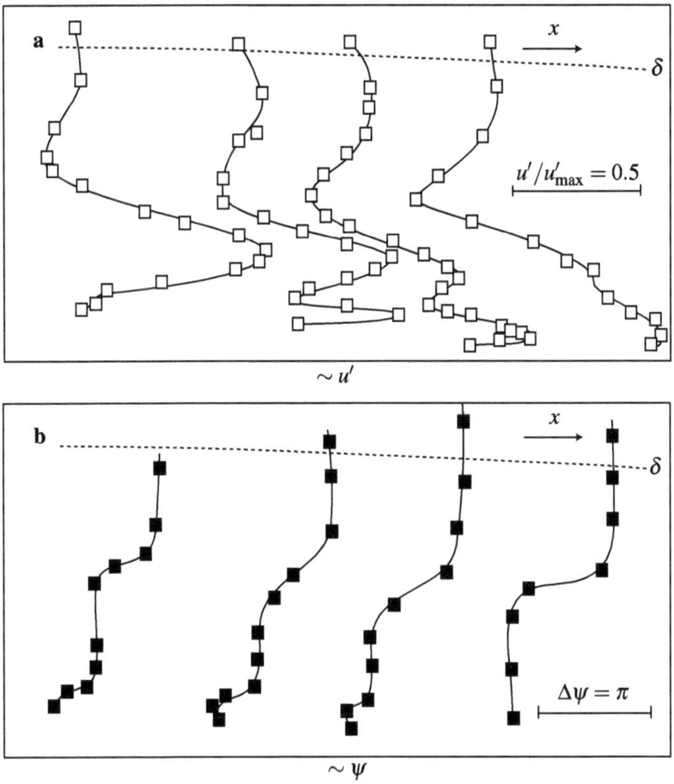

Fig. 7.5 Amplitude (**a**) and phase (**b**) profiles of the instability wave propagating in the separation region downstream of a two-dimensional bump on a flat plate from *left* to *right* at: $\Delta x/h = 6.1$, 24.2, 42.2 and 78.8 from separation; the boundary-layer edge (*dashed lines*). The Reynolds number based on the bump height h is $\mathrm{Re}_h = U_e h/\nu = 730$ (Dovgal and Kozlov, 1990)

1979, Häggmark et al, 2000, Sinha et al, 1981). The above features are well captured by direct numerical simulations and linear stability solutions (see, e.g., Gruber et al, 1987, Maucher et al, 1994, Nayfeh et al, 1988).

7.2.2 Growth rates

In terms of a local stability analysis, the destabilizing influence of separation upon a near-wall laminar flow is explained by the appearance of basic flow velocity profiles with an inflection point. This results in a wide spectrum of disturbances growing much faster than those in the attached boundary layer.

At some distance downstream of the separation point, the disturbances amplify almost exponentially. Thus, in this region, one can derive a constant spatial growth rate $-\alpha_i$. Determined for different frequencies, the growth rates form a linear instability range for the disturbances. Obtained in a series of experiments for nominally two-dimensional separation bubbles, they are shown in Fig. 7.6. The data in the figure are normalized by the external-flow velocity U_e and the mean value of the momentum thickness ϑ at the stage of exponential growth. For each curve, the shape factor $H = \delta^*/\vartheta$ is given, averaged over the region of amplification rate determination. Frequencies and increments of the disturbances are compared with theoretical predictions for the free shear layer with a tanh-like profile and the Blasius boundary layer.

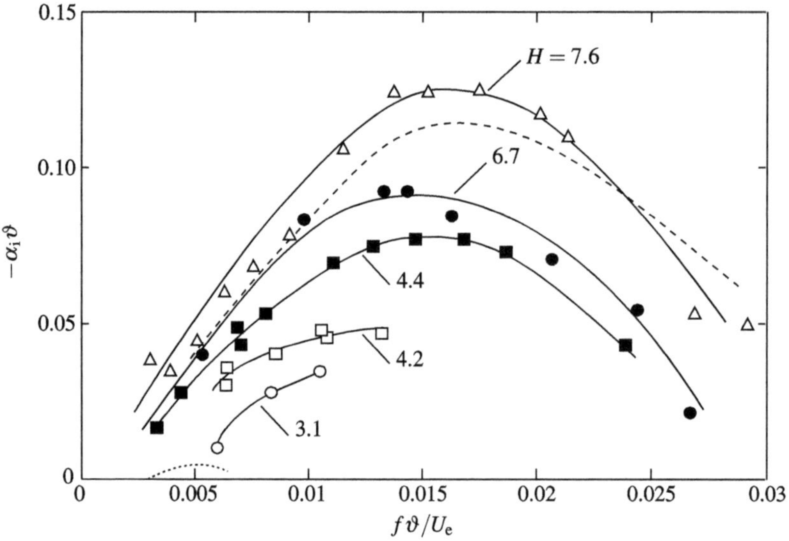

Fig. 7.6 Growth rates of two-dimensional small-amplitude disturbances: experimental data for separation bubbles induced by two-dimensional surface inflections (Dovgal and Kozlov, 1983) (△) and (Dovgal and Kozlov, 1984) (•), a two-dimensional bump on a flat plate (Dovgal and Kozlov, 1990) (■) and backward-facing steps (Boiko et al, 1990) (□, ○); results of stability calculations for the mixing layer $U \sim \tanh y$ (Monkewitz and Huerre, 1982) (*dashed line*), and for the Blasius boundary layer at $Re_{\delta^*} = 1320$ (Levchenko et al, 1975) (*dotted line*)

The experimental data show that the flow in separation bubbles becomes more unstable with their enlargement (accompanied by increase of the shape factor). This observation agrees with theoretical and numerical results, which indicate that the larger the distance between the inflection point in the basic flow velocity profile and the wall, the higher the growth rates of small-amplitude perturbations (Michalke, 1990, Nayfeh et al, 1990, Rist et al, 1996, Taghavi and Wazzan, 1974). The maximum values of experimentally determined amplification rates in Fig. 7.6 are comparable with those calculated in the inviscid limit for a free shear layer. The viscosity effects become more pronounced when a separation bubble gets smaller. The flow behind the small backward-facing step in the depth of the laminar boundary layer ($H = 3.1$) is stable to low-frequency oscillations, but unstable to them in the larger separation regions ($H = 4.2$ to 7.6) as well as in inviscid solutions. The same trend is found through linear stability calculations revealing that, as the inflection point approaches the wall, the effect of a finite Reynolds number increases.

Considering the growth rates of small-amplitude disturbances, we notice that two-dimensional waves have quite high increments in the wavenumber spectrum of oscillations and, thus, make the main contribution to perturbations in the separation region. Furthermore, at laminar boundary-layer separation, two-dimensional waves are not necessarily the most amplified ones in the region of linear instability. It was shown that if the nonparallel flow effects are taken into account, oblique waves may have larger growth rates (Stewart and Smith, 1987).

7.2.3 Propagation velocities

Noteworthy is the relatively small dispersion of propagation velocities of instability waves in separation bubbles. This refers to two-dimensional waves traveling in the external-flow direction with phase velocities that are almost independent of their frequency and to oblique waves of a fixed frequency whose streamwise wavenumber varies slightly with the angle of propagation over a broad range of transverse wavenumbers.

As an example, Fig. 7.7 shows the phase velocities of two-dimensional instability waves obtained in separation bubbles behind a rounded bump on a flat plate and on an airfoil. The results are compared with the dispersion curve in a free shear layer. One can observe that the phase velocity in the separation regions is almost constant over a wide frequency range, which is considerably different from the free shear layer when the wall is removed to infinity. The wind-tunnel data correlate with stability solutions, indicating diminution of phase velocities at low frequencies of oscillations when the distance of the shear layer to the wall is reduced.

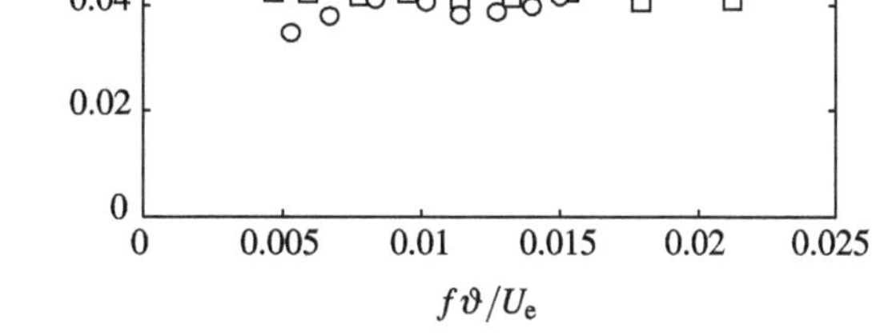

Fig. 7.7 Phase velocities of two-dimensional waves: experimental data for separation bubbles behind a bump on a flat plate (Boiko and Dovgal, 1992) (□) and on an airfoil (Boiko et al, 1989) (○); results of calculations for the mixing layer $U \sim \tanh y$ (Monkewitz and Huerre, 1982) (*solid line*)

Experimental results on dispersion of oblique waves are given in Fig. 7.8. For a fixed frequency of oscillations, each curve in the figure shows a variation with the wave angle γ of the streamwise wavenumber normalized by that of the two-dimensional wave. Similar data for the Blasius boundary layer are plotted for comparison. It turns out that oblique waves amplifying in separation bubbles have a fairly small dependence of the streamwise wavenumber on the angle between the wavevector and the mean flow direction.

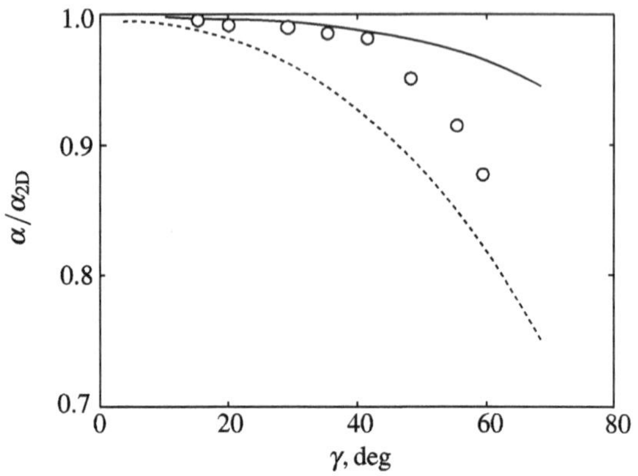

Fig. 7.8 Variation of the streamwise wavenumber of oblique waves with the wave angle: separation bubbles behind a bump on a flat plate (Boiko et al, 1991) (○) and on an airfoil (Gilev et al, 1988) (*solid line*); the Blasius boundary layer (Kachanov, 1985) (*dashed line*)

The above-mentioned characteristics of small-amplitude disturbances are significant at the nonlinear stage of the transition to turbulence in separation regions, following the initial exponential growth. The weak dispersion of two-dimensional and oblique waves means constancy of phase coherence between different components of the frequency and wavenumber spectra during their streamwise evolution, stimulating their subsequent resonant interactions.

7.3 Instability of axisymmetric separation bubbles

When turning to separation of an axisymmetric flow in the stability problem, an additional curvature parameter appears as the ratio of the viscous-layer thickness to the body radius. The axial symmetry makes the flow less unstable as compared to plane separation bubbles. However, its effect on an instability mode may be different depending on other mean-flow parameters (Reynolds number and the amount of back flow near the wall) and the frequency of oscillations. In particular, with increasing the curvature parameter, the growth rates of high-frequency axisymmetric disturbances become lower and the number of amplifying helical modes is reduced. At the same time, a higher separated-flow instability with respect to the first helical mode may occur (Michalke et al, 1995).

Similarly to that in plane configurations, the axisymmetric flow is destabilized as the thickness of the separation bubble grows; that is, the maximum amplification

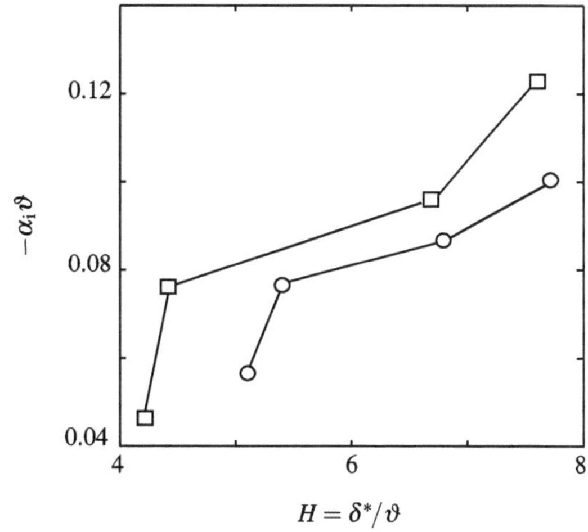

Fig. 7.9 Maximum amplification rates of the axisymmetric mode of oscillations at flow separation on a body of revolution (Dovgal et al, 1995) (○) as compared to similar data from Fig. 7.6 for plane separation bubbles (□)

rate of instability waves gets higher when the inflection point in basic flow velocity profiles moves away from the wall, see Fig. 7.9.

7.4 Flow instability at separation of a swept-wing boundary layer

The research results on two-dimensional separation bubbles support the quite simple notion that their convective instability is associated with amplification of disturbances of the separated shear layer. Generally, the stability problem is more complex to apply to three-dimensional flows, in which different instability mechanisms are expected. A related topic is amplification of velocity perturbations in swept geometries where the basic flow is often taken as a superposition of the main (in the external-flow direction) and the crossflow velocities. A generic flow configuration is boundary-layer separation that can be induced by an adverse pressure gradient in the aft part of a swept wing. In this case, the main flow behind the line of separation is unstable and the crossflow velocity is large enough to trigger the growth of velocity perturbations.

The effect of three-dimensionality on laminar flow disturbances in a separation bubble on a swept-wing is illustrated in Fig. 7.10. In this example, a wave packet of controlled harmonic oscillations is excited in the upstream part of a swept wing by periodic air blowing through a tiny hole in the surface of the test model; the excitation frequency is close to the spectral maximum of naturally amplifying instability waves. The disturbances generated in the accelerated flow initially decay and then grow in the adverse pressure gradient boundary layer and further downstream in the separation region. Obviously, the wave packet spreading over the wing span becomes asymmetric to the external-flow direction as is seen in the right-hand part of Fig. 7.10.

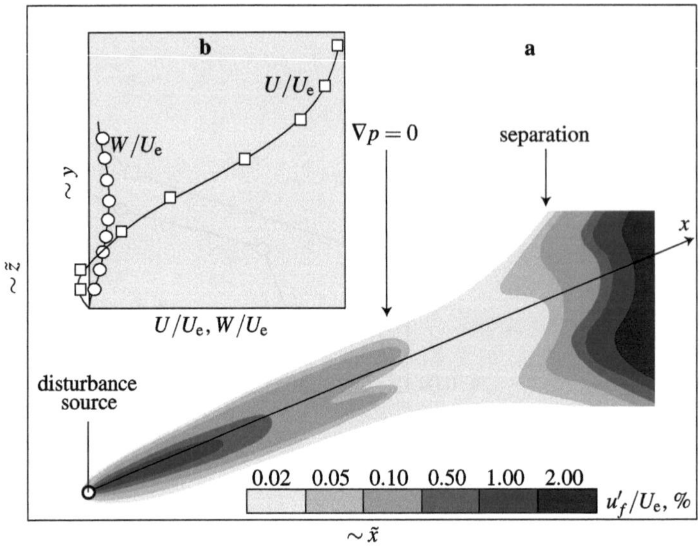

Fig. 7.10 Wave packet in a swept-wing flow: isolines of disturbance amplitude as percentage of the oncoming flow velocity (**a**) and mean-flow profiles in the separation region (**b**); the sweep angle is 30°, the Reynolds number based on the wing chord length $\mathrm{Re}_l = 2.6 \times 10^5$ (Dovgal et al, 1988a)

The dominant instability mechanism at flow separation in such configurations is clarified when comparing the amplification of perturbations in plane and swept separation bubbles. Figure 7.11 shows the growth rates of velocity oscillations measured at close Reynolds numbers behind a two-dimensional surface inflection and that constituting a 30° sweep angle. In both cases, the spectral range of amplifying disturbances is almost the same. This means that flow instability in the region of three-dimensional boundary-layer separation is principally initiated by the separated shear layer. The latter is dominated by the main flow while the maximum of the crossflow is closer to the wall inside the region of circulation. Thus, the effect

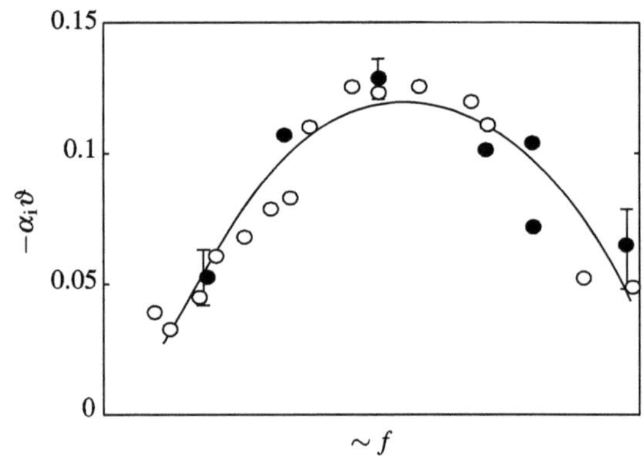

Fig. 7.11 Spatial growth rates of disturbances: two-dimensional separation bubble of Fig. 7.4 (○) and separation region in the same configuration at the 30° sweep angle (●) (Dovgal et al, 1988b)

of the three-dimensional basic flow on the perturbations (see Fig. 7.10) is due to the particular flow configuration rather than the crossflow instability. Further details on instability of swept separation bubbles are available through numerical simulations (Hetsch and Rist, 2009, Kaltenbach and Janke, 2000).

7.5 Outline of approaches for linear stability calculations in separation regions

The main difficulty with respect to numerical investigation of linear stability in local separation regions is related to the accurate calculation of steady laminar basic flow. The Falkner–Skan equation admits self-similar solutions with back flow. However, they are probably of a little practical importance, as due to the back flow in circulation region, the boundary-layer approximation fails and more involved approaches such as viscous-inviscid interaction model or DNS are necessary to calculate adequate velocity distributions in separation bubbles.

Next, as we mentioned in Sect. 7.1, the separation bubbles are often subject to the so-called flapping, i.e., to low-frequency oscillations of the entire separation region even at rather low Reynolds numbers. This physical phenomenon is stipulated by a global bubble dynamics. In numerical calculations, it leads to difficulty in finding a steady solution of the Navier–Stokes equations for the laminar flow. Hence, if the flapping is present, in the framework of the local approach, the linear stability calculations should be performed after averaging the basic flow in time.

Alternatively, one may rely on experimental data with measured mean velocity profiles, which, after approximation by smooth functions, can be used in the stability calculations. Indeed, a shortcoming of the hot-wire anemometry with stationary one-wire probes which is faced at examination of nonparallel two-dimensional and complex three-dimensional flows is that it does not identify the flow direction. However, at least in some cases, inaccuracy of the mean velocity profiles in the back flow region has rather small effect on the results of the linear stability calculations (Dovgal et al, 1995).

An alternative to the hot-wire measurements is the particle image velocimetry (PIV) which is able to recognize virtually instantaneous velocity vectors in a flow region under investigation. An example of amplitude and phase distributions of the streamwise and wall-normal velocities of oscillations excited by sound at fixed frequency is given in Fig. 7.12. The data were obtained by a high-speed PIV in the middle of circulation region behind a backward-facing step. They are compared with results of the local linear stability analysis in which the measured mean streamwise velocity profile (after its approximation by a cubic spline), also shown in Fig. 7.12**a**, was used. For the comparison, amplitude of the disturbance is normalized by its maximum value. As seen, an overall agreement between the wind tunnel data on the perturbation velocity distribution and the calculation results takes place.

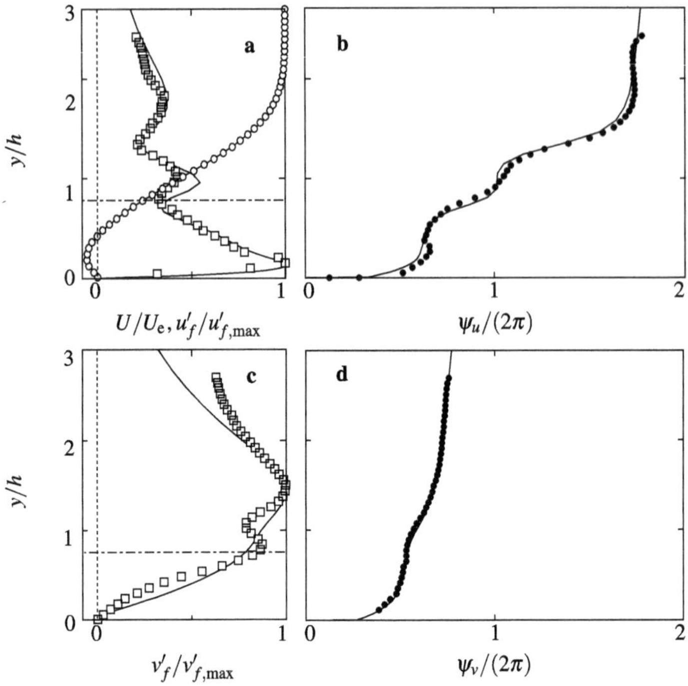

Fig. 7.12 PIV data downstream of a backward-facing step at $x/h = 14.6$: amplitude (□) (**a**) and phase (•) (**b**) distributions of the streamwise component of the forced oscillations and amplitude (□) (**c**) and phase (•) (**d**) distributions of the wall-normal component of the forced oscillations and corresponding profiles calculated through the Orr–Sommerfeld equation (*solid lines*); streamwise mean velocity profile obtained by averaging of the PIV data (○); location of the critical layer (*dash-dotted line*); $\mathrm{Re}_h = 1{,}060$; $fh/U_e = 0.055$ ($f\vartheta/U_e = 0.010$; $u'_{f,\max}/U_e = 2.19\%$ (Boiko et al, 2011)

> Additional material for this chapter is given in the multimedia file No. 6, which supplements the book, titled 'Instability of laminar separating flows.'

Exercises

7.1. Write a MATLAB function to find the Stewartson solution of the Falkner–Skan equation for $-0.199 \lesssim \beta_H < 0$ using the false position method. Hint: $\Psi_2(0) < 0$ for the Stewartson solution. Is it possible to obtain it using a procedure similar to `FS` given in Chap. 4?

7.2. For the Stewartson profile with $\beta_H = -0.198$ try to build the neutral stability curve similar to that shown in Fig. 2.4 for pre-separation Hartree boundary-layer profile with the same value of the pressure gradient parameter. Compare these two neutral stability curves. How low is the critical Reynolds number Re_L in both cases?

References

Boiko A, Dovgal A, Hein S, Henning A (2011) Particle image velocimetry of a low-Reynolds-number separation bubble. Exp Fluids 50(1):13–21

Boiko AV, Dovgal AV (1992) Instability of locally separated flows to small-amplitude disturbances. Sib Fiz-Techn Zh 34(3):19–24, In Russian.

Boiko AV, Dovgal AV, Kozlov VV (1989) Nonlinear interactions between perturbations in transition to turbulence in the zone of laminar boundary-layer separation. Soviet J Appl Phys 3(2):46–52

Boiko AV, Dovgal AV, Kozlov VV, Scherbakov VA (1990) Instability and receptivity of a boundary layer close to 2D surface imperfections. Izv Sib Otd Akad Nauk SSSR, Ser Tekhn Nauk 1:50–56

Boiko AV, Dovgal AV, Simonov OA, Scherbakov VA (1991) Effects of laminar–turbulent transition in separation bubbles. In: Kozlov and Dovgal (1991) pp 565–572

Cousteix J, Pailhas G (1979) Étude exploratoire d'un processus de transition laminaire–turbulent au voisinage du décollement d'une couche laminaire. Rech Aérospat 1979(3):213–218

Dovgal AV, Kozlov VV (1983) Instability of flow separating from a wall with an inflexion. Dokl Akad Nauk 270(6):1356–1358

Dovgal AV, Kozlov VV (1984) Stability of separated flow in a dihedral corner. Fluid Mech – Soviet Res 13(1):137–143

Dovgal AV, Kozlov VV (1990) Hydrodynamic instability and receptivity of small scale separation regions. In: Arnal D, Michel R (eds) Laminar–Turbulent Transition IUTAM Symposium, Springer–Verlag, Berlin, pp 523–531

Dovgal AV, Kozlov VV, Simonov OA (1988a) Development of a spatial wave packet of disturbances in a swept-wing boundary layer. Izv Sib Otd Akad Nauk SSSR, Ser Tekhn Nauk 3(11):43–47

Dovgal AV, Kozlov VV, Simonov OA (1988b) Stability of three-dimensional flow generated upon separation from a wall with an inflexion. Soviet J Appl Phys 2(4):18–24

Dovgal AV, Kozlov VV, Michalke A (1995) Contribution to the instability of laminar separating flows along axisymmetric bodies. Part 2: Experiment and comparison with theory. Eur J Mech B/Fluids 14(3):351–365

Gaster M (1967) The structure and behavior of separation bubbles. R&M 3595, ARC

Gilev VM, Dovgal AV, Kachanov YS, Kozlov VV (1988) Development of three-dimensional disturbances in a boundary layer with pressure gradient. Fluid Dyn 23(3):393–399

Gruber K, Bestek H, Fasel H (1987) Interaction between a Tollmien–Schlichting wave and a laminar separation bubble. AIAA Paper 87–1256

Häggmark CP, Bakchinov AA, Alfredsson PH (2000) Experiments on a two-dimensional laminar separation bubble. Philos Trans R Soc Lond A 358:3193–3205

Hetsch T, Rist U (2009) The influence of sweep on the linear stability of a series of swept laminar separation bubbles. Eur J Mech B/Fluids 28:494–505

van Ingen JL (1975) On the calculation of laminar separation bubbles in two dimensional incompressible flow. In: AGARD-CP-168

Kachanov YS (1985) Development of spatial wave packets in boundary layer. In: Kozlov VV (ed) Laminar–Turbulent Transition, Springer–Verlag, Berlin, IUTAM Symposium, pp 115–123

Kaltenbach HJ, Janke G (2000) Direct numerical simulation of a flow separation behind a swept, rearward-facing step at $Re_H = 3000$. Phys Fluids 12(9):2320–2337

Kozlov VV, Dovgal AV (eds) (1991) Separated flows and Jets, IUTAM Symposium, Springer–Verlag, Berlin

Levchenko VY, Volodin AG, Gaponov SA (1975) Stability characteristics of boundary layers. Nauka. Sib. Otd., Novosibirsk, In Russian.

Maucher U, Rist U, Wagner S (1994) Direct numerical simulation of airfoil separation bubbles. In: Wagner S, Hirschel EH, Périaux J, Piva R (eds) Computational Fluid Dynamics '94, Wiley, New York, pp 471–477

Michalke A (1990) On the inviscid instability of wall-bounded velocity profiles close to separation. Z Flugwiss Weltraumforsch 14:24–31

Michalke A, Kozlov VV, Dovgal AV (1995) Contribution to the instability of laminar separating flows along axisymmetric bodies. Part 1: Theory. Eur J Mech B/Fluids 14(3):333–350

Monkewitz PA, Huerre P (1982) Influence of the velocity ratio on the spatial instability of mixing layers. Phys Fluids 25(7):1137–1143

Nayfeh AH, Ragab SA, Al-Maaitah AA (1988) Effects of bulges on the stability of boundary layers. Phys Fluids 31(4):796–806

Nayfeh AH, Ragab SA, Masad JA (1990) Effect of a bulge on the subharmonic instability of boundary layers. Phys Fluids A 2(6):937–948

Rist U, Maucher U, Wagner S (1996) Direct numerical simulation of some fundamental problems related to transition in laminar separation bubbles. In: Désidéri JA, Hirsch C, Tallec PL, Pandolfi M, Périaux J (eds) Computational Fluid Dynamics '96, Wiley, New York, pp 319–325

Sinha SN, Gupta AK, Oberai MM (1981) Laminar separating flow over backsteps and cavities. Part 1: Backsteps. AIAA J 19(12):1527–1530

Stewart PA, Smith FT (1987) Three-dimensional instabilities in steady and unsteady non-parallel boundary layers, including effects of Tollmien–Schlichting disturbances and cross flow. Proc R Soc Lond A 409:229–248

Taghavi H, Wazzan AR (1974) Spatial stability of some Falkner–Skan profiles with reversed flow. Phys Fluids 17(12):2181–2183

Further Reading

Al-Maaitah AA, Nayfeh AH, Ragab SA (1990a) Effect of cooling on the stability of compressible subsonic flows over smooth humps and backward-facing steps. Phys Fluids A 2(3):381–389

Al-Maaitah AA, Nayfeh AH, Ragab SA (1990b) Effect of suction on the stability of compressible subsonic flows over backward-facing steps. AIAA J 28(11):1916–1924

Alam M, Sandham ND (2000) Direct numerical simulation of 'short' laminar separation bubbles with turbulent reattachment. J Fluid Mech 410:1–28

Alizard F, Cherubini S, Robinet JC (2009) Sensitivity and optimal forcing response in separated boundary layer flows. Phys Fluids 21(6):064108, 1–13

Allen T, Riley N (1995) Absolute and convective instabilities in separation bubbles. Aeronaut J 99:439–448

Arena AV, Mueller TJ (1980) Laminar separation, transition, and turbulent reattachment near the leading edge of airfoils. AIAA J 18(7):747–753

Barkley D, Gomes MGM, Henderson RD (2002) Three-dimensional instability in flow over a backward-facing step. J Fluid Mech 473:167–190

Bestek H, Gruber K, Fasel H (1989) Self-excited unsteadiness of laminar separation bubbles caused by natural transition. In: The Prediction and Exploitation of Separated Flow, Roy. Aeronaut. Soc., London, pp 14.1–14.16

Bestek H, Gruber K, Fasel H (1993) Direct numerical simulation of unsteady separated boundary-layer flows over smooth backward-facing steps. In: Gersten K (ed) Physics of Separated Flows – Numerical, Experimental, and Theoretical Aspects, Notes on Numerical Fluid Mechanics, vol 40, Vieweg, Braunschweig, pp 73–80

Brendel M, Mueller TJ (1988) Boundary-layer measurements on an airfoil at low Reynolds numbers. J Aircraft 25(7):612–617

Cebeci T, Egan DA (1989) Prediction of transition due to isolated roughness. AIAA J 27(7):870–875

Dallmann U, Herberg T, Gebing H, Su WH, Zhang HQ (1995) Flow field diagnostics: Topological flow changes and spatio-temporal flow structure. AIAA Paper 95–0791

van Dam CP, Elli S (1992) Simulation of nonlinear Tollmien–Schlichting wave growth through a laminar separation bubble. In: Hussaini MY, Kumar A, Streett CL (eds) Instability, Transition, and Turbulence, Springer–Verlag, Berlin, pp 311–321

Dini P, Selig MS, Maughmer MD (1992) Simplified linear stability transition prediction method for separated boundary layers. AIAA J 30(8):1953–1961

Dovgal AV, Sorokin AM (2001) Instability of a laminar separation bubble to vortex shedding. Thermophys Aeromech 8(2):189–197

Dovgal AV, Sorokin AM (2002) Experimental modelling of vortex shedding at flow separation behind a backward-facing step. Thermophys Aeromech 9(2):183–190

Dovgal AV, Kozlov VV, Michalke A (1994) Laminar boundary layer separation: Instability and associated phenomena. Progr Aerosp Sci 30:61–94

Drela M, Giles MB (1987) Viscous-inviscid analysis of transonic and low Reynolds number airfoils. AIAA J 25(10):1347–1355

Eaton JK, Johnston JP (1981) A review of research on subsonic turbulent flow reattachment. AIAA J 19(9):1093–1100

Ehrenstein U, Gallaire F (2008) Two-dimensional global low-frequency oscillations in a separating boundary-layer flow. J Fluid Mech 614:315–327

Fasel HF, Saric WS (eds) (2000) Laminar–Turbulent Transition, IUTAM Symposium, Springer–Verlag, Berlin

Gates EM (1980) Observations of transition on some axisymmetric bodies. In: Eppler R, Fasel H (eds) Laminar–Turbulent Transition, Springer–Verlag, Berlin, IUTAM Symposium, pp 351–363

Hammond DA, Redekopp LG (1998) Local and global instability properties of separation bubbles. Eur J Mech B/Fluids 17(2):145–164

Hasan MAZ (1992) The flow over a backward-facing step under controlled perturbation: Laminar separation. J Fluid Mech 238:73–96

Hein S (2000) Linear and nonlinear nonlocal instability analysis for two-dimensional laminar separation bubbles. In: Fasel and Saric (2000), pp 681–686

Hein S, Theofilis V, Dallmann U (1998) Unsteadiness and three-dimensionality of steady two-dimensional laminar separation bubbles as result of linear instability mechanisms. IB 223-98 A 39, DLR, Göttingen

Hetsch T, Rist U (2009) An analysis of the structure of laminar separation bubbles in swept infinite geometries. Eur J Mech B/Fluids 28:486–493

Horton HP (1967) A semi-empirical theory for the growth and bursting of laminar separation bubbles. CP 1073, Aeronaut. Research Council

Hudy LM, Naguib A, Humphreys WM (2007) Stochastic estimation of a separated-flow field using wall-pressure-array measurements. Phys Fluids 19(2):024103, 1–18

van Ingen JL (1977) Transition, pressure gradient, suction, separation and stability theory. In: AGARD-CP-224

van Ingen JL (1991) Research on laminar separation bubbles at Delft University of Technology. In: Kozlov and Dovgal (1991), pp 537–556

Jones LE, Sandberg RD, Sandham ND (2008) Direct numerical simulations of forced and unforced separation bubbles on an airfoil at incidence. J Fluid Mech 602:175–207

Kaiktsis L, Karniadakis GE, Orszag SA (1996) Unsteadiness and convective instabilities in two-dimensional flow over a backward-facing step. J Fluid Mech 321:157–187

Kiya M (1989) Separation bubbles. In: Germain P, Piau M, Caillerie D (eds) Theoretical and Applied Mechanics, Elsevier, Amsterdam, pp 173–191

Klebanoff PS, Tidstrom KD (1972) Mechanism by which a two-dimensional roughness element induces boundary-layer transition. Phys Fluids 15(7):1173–1188

Kozlov VV, Dovgal AV (eds) (1991) Separated flows and Jets, IUTAM Symposium, Springer–Verlag, Berlin

Leblanc P, Blackwelder R, Liebeck R (1991) Experimental results on laminar separation on two airfoils at low Reynolds numbers. In: Twenty-Ninth Aerospace Sciences Meeting, Reno, U.S.A.

Marquet O, Lombardi M, Chomaz JM, Sipp D, Jacquin L (2009) Direct and adjoint global modes of a recirculation bubble: Lift-up and convective non-normalities. J Fluid Mech 622:1–21

Marquillie M, Ehrenstein U (2003) On the onset of nonlinear oscillations in a separating boundary-layer flow. J Fluid Mech 490:169–188

Marxen O, Lang M, Rist U, Wagner S (2003) A combined experimental/numerical study of unsteady phenomena in a laminar separation bubble. In: Flow, Turbulence, and Combustion, vol 71, Kluwer, Dordrecht, pp 133–146

Masad JA, Iyer V (1994) Transition prediction and control in subsonic flow over a hump. Phys Fluids 6(1):313–327

Masad JA, Malik MR (1994) On the link between flow separation and transition onset. AIAA Paper 94–2370

Masad JA, Nayfeh AH (1992) Stability of separating boundary layers. In: Fourth Internat. Conf. Fluid Mechanics, Alexandria, vol 1, pp 261–278

Masad JA, Nayfeh AH (1993) The influence of imperfections on the stability of subsonic boundary layers. In: Ashpis DE, Gatski TB, Hirsh R (eds) Instabilities and Turbulence in Engineering Flows, Kluwer, Dordrecht, pp 65–82

Michalke A (1991) On the instability of wall-boundary layers close to separation. In: Kozlov and Dovgal (1991), pp 557–564

Mueller TJ, Batill SM (1982) Experimental studies of separation on a two-dimensional airfoil at low Reynolds numbers. AIAA J 20(4):457–463

O'Meara MM, Mueller TJ (1987) Laminar separation bubble characteristics on an airfoil at low Reynolds numbers. AIAA J 25(8):1033–1041

Pauley LP, Moin P, Reynolds WC (1990) The structure of two-dimensional separation. J Fluid Mech 220:397–411

Rannacher J (1969) Untersuchung von geraden ebenen Flugelgittern im kritischen Reynoldszahlbereich. Kurzfassung in Maschinenbautechnik 18:2–10

Ripley MD, Pauley LL (1993) The unsteady structure of two-dimensional steady laminar separation. Phys Fluids A 5(12):3099–3106

Rist U (1994) Nonlinear effects of 2D and 3D disturbances on laminar separation bubbles. In: Lin SP, Phillips WRC, Valentine DT (eds) Nonlinear Instability of Nonparallel Flows, Springer–Verlag, Berlin, IUTAM Symposium, pp 324–333

Rist U, Maucher U (1994) Direct numerical simulation of 2D and 3D instability waves in a laminar separation bubble. In: AGARD-CP-551 Application of Direct and Large Eddy Simulation to Transition and Turbulence, Chania, Crete, Greece, pp 34–1347

Smith FT, Bodonyi RJ (1985) On short-scale inviscid instabilities in flow past surface-mounted obstacles and other nonparallel motions. J Roy Aeronaut Soc 89:205–212

Spalart PR, Strelets MK (2000) Mechanisms of transition and heat transfer in a separation bubble. J Fluid Mech 403:329–349

Tafti DK, Vanka SP (1991) A numerical study of flow separation and reattachment on a blunt plate. Phys Fluids A 3(7):1749–1759

Theofilis V (2000) Global linear instabilities in laminar separated boundary layer flow. In: Fasel and Saric (2000), pp 663–668

Theofilis V (2003) Advances in global linear instability analysis of nonparallel and three-dimensional flows. Progr Aerosp Sci 39:249–315

Theofilis V, Hein S, Dallmann U (2000) On the origins of unsteadiness and three-dimensionality in a laminar separation bubble. Philos Trans R Soc Lond A 358:3229–3246

Wang BJ, Bogucki DJ, Redekopp LG (2000) Transition in separated flows via global instability. In: Fasel and Saric (2000), pp 669–674

Ward JW (1963) The behaviour and effects of laminar separation bubbles on airfoils in incompressible flow. J Roy Aeronaut Soc 67:783–790

Watmuff JH (1999) Evolution of a wave packet into vortex loops in a laminar separation bubble. J Fluid Mech 397:119–169

Wee D, Yi T, Annaswamy A, Ghoniem AF (2004) Self-sustained oscillations and vortex shedding in backward-facing step flows: Simulation and linear instability analysis. Phys Fluids 16(9):3361–3373

Weibust E, Bertelrud A, Ridder SO (1987) Experimental investigation of laminar separation bubbles and comparison with theory. J Aircraft 24(5):291–297

Yang Z, Voke PR (2001) Large-eddy simulation of boundary-layer separation and transition at a change of surface curvature. J Fluid Mech 439:305–333

Part III
Special topics on linear stability

Chapter 8
Linear wave packets of instability waves

The ability to observe monochromatic instability waves in laboratory experiments is stipulated by controlled application of special disturbance sources, which excite small-amplitude oscillations at a single frequency with a predominant transverse wavenumber. In contrast, natural disturbances are mostly localized in space and/or time. In such cases, the wave packet concept discussed in this chapter can be used, which appears highly valuable for the general analysis of flow instability properties.

8.1 Classification of wave packets

A small-amplitude disturbance localized in space and time in the region of shear-flow instability represents usually a group of amplifying waves with different frequencies and wavenumbers. Near the source (or at small time intervals), decaying modes also contribute to the disturbance. Far from the source (or asymptotically with time), however, the growing waves are effectively selected by the flow. When the amplifying disturbance is localized only in space, it becomes a degenerate, purely spatial wave packet, which is also called a *wave train* (Fig. 8.1). Such a packet can be created, for example, with the help of a point-like periodic disturbance source located inside a shear flow. In the general case, the 'shape' of the wave packet changes, since the discrete waves constituting it are *dispersive*, i.e., they have different phase velocities and growth rates for different wave components. In turn, a degenerate case of the wave train is a disturbance consisting of a single wave excited by, for example, a long vibrating ribbon.

The sets of isolated periodic waves and wave packets are represented by discrete (power) and continuous (power-density) spectra, respectively, see Fig. 8.2. The difference between them relates to normalization, as the sets of isolated periodic waves are infinitely extended in space (time) by definition, while the wave packets are localized. As the Fourier components of the linear wave packet do not interact, each element of the spectrum develops independently in time. This means that the spectrum at any time instant is determined by its initial conditions. The initial-value

A.V. Boiko et al., *Physics of Transitional Shear Flows*,
Fluid Mechanics and Its Applications 98, DOI 10.1007/978-94-007-2498-3_8,

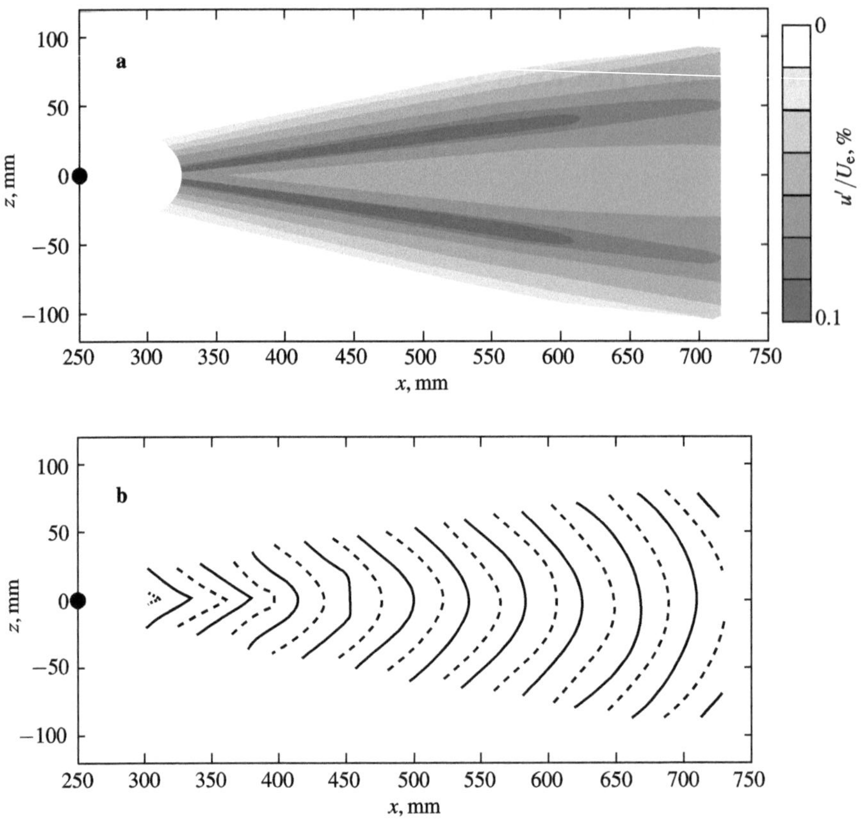

Fig. 8.1 Wave pattern of a typical wave train of the Tollmien–Schlichting waves in a flat-plate boundary layer: isolines of amplitude $u'_{max} = \max_y[u'(y)]$ (**a**); isolines of corresponding phase ψ_{max} with 180° shift (**b**): location of a point-like source of periodic disturbances (•) (Kozlov and Ryzhov, 1990)

problem from the viewpoint of disturbance excitation (the receptivity problem) is examined in Chap. 10. In what follows, we concentrate our attention on the general description of the flow far-field (asymptotic) response to perturbations localized in time and space, while the near-field response is considered in Chap. 9.

8.2 Group velocity

Since the wave packets are localized in space and time, it is possible to determine the transportation of the disturbance kinetic energy in terms of velocity, direction, and growth rate. Continuing the discussion of Sects. 1.1 and 2.2, we distinguish the wave packet development in time and/or space. Thus, if a growing localized disturbance

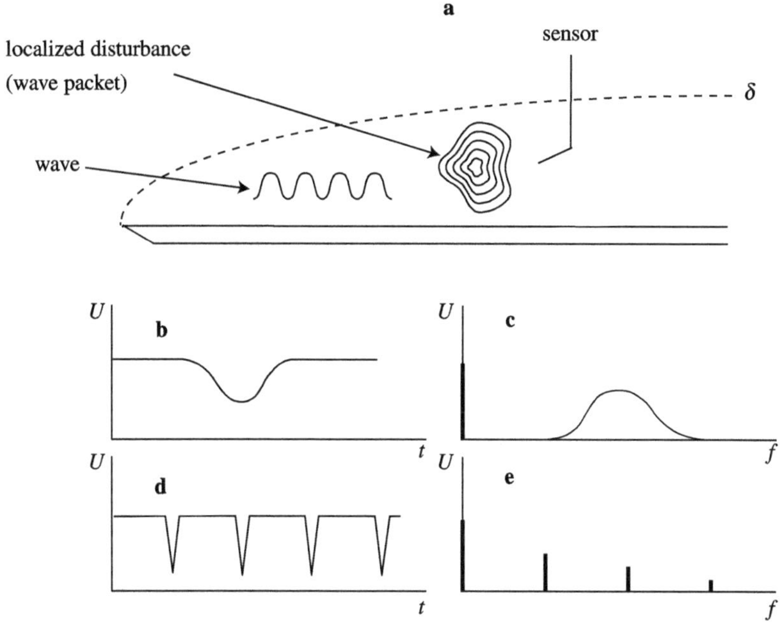

Fig. 8.2 To the definition of wave packet: sketch of a boundary layer with a localized disturbance and a wave sensed by a hot wire (**a**); oscilloscope traces of localized (**b**) and periodic (**d**) disturbances; power spectrum of the localized disturbance (**c**); linear spectrum of the periodic signal (**e**)

is concentrated permanently at the source and covers it, absolute instability takes place. If the disturbance propagates at least along one direction from the source and leaves the source position asymptotically with time, convective instability occurs.

The theory of dispersive waves was introduced for instability in plasma jets by Briggs (1964) and later was successfully applied for hydrodynamic wave packets. The theory states that the wave packet energy in the shear flows under consideration is transported with a group velocity vector

$$\mathbf{c}_{\mathrm{g}} = \nabla_{\kappa}\omega \,,$$

where, e.g., in the Cartesian coordinates

$$\nabla_{\kappa} = \left(\frac{\partial}{\partial\alpha}, \frac{\partial}{\partial\beta}\right)$$

is a gradient operator, provided that the Reynolds number is 'frozen' during differentiation.

Let us start with a simple guiding example of a wave packet consisting of two harmonic waves with close frequencies $\omega \pm \Delta\omega$ and wavenumbers $\alpha \pm \Delta\alpha$, whose amplitude A behaves as

$$A(x,t) = \cos\left[(\alpha - \Delta\alpha)x - (\omega - \Delta\omega)t\right] + \cos\left[(\alpha + \Delta\alpha)x - (\omega + \Delta\omega)t\right] \,.$$

Using a trigonometric identity we can rewrite it as

$$A(x,t) = 2\cos(\alpha x - \omega t)\cos(\Delta\alpha x - \Delta\omega t) \ .$$

This can be interpreted as a harmonic wave with the frequency ω and the wave number α modulated by the amplitude $2\cos(\Delta\alpha x - \Delta\omega t)$. As seen, the velocity of propagation of this modulation wave or envelope is $c_g = \Delta\omega/\Delta\alpha$. The propagation of energy in a wave always requires changes (modulations) in the wave frequency or amplitude. Hence, the actual speed of energy propagation is also c_g. This is the phase velocity of the envelope, but since it contains a group of waves, this speed is called the group velocity.

Let us now consider a continuous set of harmonic waves

$$A(x,t) = \int_{-\infty}^{+\infty} A_\alpha(\alpha) e^{\mathrm{i}[\alpha x - \omega(\alpha)t]} \mathrm{d}\alpha, \tag{8.1}$$

where the dependence $\omega(\alpha)$ is a particular form of the dispersion relation. We assume that the waveform $A(x,t)$ is a wave packet, i.e., it is narrow enough in x and t and, consequently, in α, so that the dispersion relation can be approximated by the leading terms in a Taylor expansion

$$\omega = \omega(\alpha_0) + \frac{\mathrm{d}\omega}{\mathrm{d}\alpha}(\alpha - \alpha_0) \ ,$$

where α_0 is some central wave number. Then, denoting $\omega_0 = \omega(\alpha_0)$ and $c_{g,0} = \mathrm{d}\omega(\alpha_0)/\mathrm{d}\alpha$,

$$\begin{aligned} A(x,t) &= \exp\left[\mathrm{i}\left(\alpha_0 c_{g,0} - \omega_0\right) t\right] \int_{-\infty}^{+\infty} A_\alpha(\alpha) \exp\left[\mathrm{i}\alpha\left(x - c_{g,0}t\right)\right] \mathrm{d}\alpha \\ &= \exp\left[\mathrm{i}\left(\alpha_0 c_{g,0} - \omega_0\right) t\right] A(x - c_{g,0}t, 0) \ . \end{aligned}$$

Thus, to within the phase factor $\exp\left[\mathrm{i}\left(\alpha_0 c_{g,0} x - \omega_0\right)\right]$ representing the carrier wave, the waveform $A(x,t)$ representing the wave packet is a function of a single combination of variables $t(x/t - c_{g,0})$, i.e., the packet propagates with the group velocity $c_{g,0}$.

To determine α_0 let us assume that $A(\alpha)$ varies slowly with α compared to the phase $\psi = \alpha x - \omega t$. Then, the phase ψ provides rapid oscillations of the integrand of (8.1) over the most of the integration range except the part (called the stationary point) for which $\mathrm{d}\psi/\mathrm{d}\alpha = 0$. Therefore, one can concentrate on such α_0, which satisfy the conditions

$$\frac{\partial\psi(\alpha_0)}{\partial\alpha} = \frac{x}{t} - c_{g,0} = 0$$

and make the main contribution to the integral. Hence, when the dependence $c_g(\alpha)$ is known, the condition determines α_0 for each ratio x/t.

In a similar way, consider now a linearized Navier–Stokes operator with a inhomogeneous right-hand side in a general form, cf. (2.3),

$$\mathscr{L}_{\mathrm{NS}}\mathbf{q} = \mathbf{G} \ , \tag{8.2}$$

where **G** represents a disturbance source and **q** represents the disturbance velocity and pressure. Using a Laplace transformation in t and a Fourier transformation in x and z, we reduce (8.2) to a system of ordinary differential equations in y:

$$\int_0^{\infty}\int_{-\infty}^{\infty}\int_{-\infty}^{\infty}\mathscr{L}_{\mathrm{NS}}\mathbf{q}\mathrm{e}^{-\mathrm{i}t\psi}\mathrm{d}x\mathrm{d}z\mathrm{d}t=\hat{\mathbf{G}}\,, \tag{8.3}$$

where $\hat{\mathbf{G}}(\alpha,\beta,\omega,y)$ is the Laplace–Fourier transformed disturbance source and $\psi = \alpha x/t+\beta z/t-\omega$ is the phase of oscillations.

Using the theory of the Green functions, it is possible to show that the solution of (8.3) in spectral space can be represented in the form $\hat{\mathbf{q}}=\mathbf{Z}/\mathcal{F}$, where the numerator $\mathbf{Z}$ depends on the particular forcing **G**, while the denominator $\mathcal{F}$, determined by the dispersion relation (2.8), depends on flow properties, but does not depend on **G**. The solution in physical space is obtained by inverse Fourier–Laplace transformation as

$$\mathbf{q}(x,y,z,t)=\frac{1}{8\pi^3}\int_{-\infty}^{\infty}\int_{-\infty}^{\infty}\int_{-\infty+\mathrm{i}\delta}^{\infty+\mathrm{i}\delta}\frac{\mathbf{Z}}{\mathcal{F}}\mathrm{e}^{\mathrm{i}t\psi}\mathrm{d}\omega\mathrm{d}\alpha\mathrm{d}\beta\,. \tag{8.4}$$

As we expect that at large t or far from the disturbance source only the growing leading modes make the most significant contribution to the integral, we may exclude the rest of the spectrum from consideration. Then, if initial conditions are given, Eq. (8.4) can be integrated directly. The integration requires typically, however, the evaluation of a several hundred normal modes for good resolution of the wave packet.

A different approach is to evaluate the integral using the method described above. The asymptotic behavior of the solution for the wave packet of the leading modes is determined by the wavevectors $\kappa_0=(\alpha_0,\beta_0)$, which satisfy the conditions

$$\frac{\partial\psi}{\partial\alpha}=\frac{x}{t}-\mathrm{Real}\left[c_{g\mathbf{x}}(\kappa_0)\right]=0\,,\quad \frac{\partial\psi}{\partial\beta}=\frac{z}{t}-\mathrm{Real}\left[c_{g\mathbf{z}}(\kappa_0)\right]=0\,.$$

Thus, an observer moving with the group velocity vector $\mathbf{c}_g=(c_{g\mathbf{x}},c_{g\mathbf{z}})^T$ follows the waves with the wavevector equal to κ_0, and with the frequency $\tilde{\omega}_0=\omega_0(\kappa_0)-\kappa_0\cdot\mathbf{c}_g(\kappa_0)$. Physically it means that the absolute instability (if any) is observed in this frame of reference with a maximum possible growth rate. In this moving coordinate system, if the 'neutral stability curve' $\omega_i(\mathbf{c}_g)=0$ in the $c_{g\mathbf{x}}$–$c_{g\mathbf{z}}$-space goes around the coordinate origin, the flow is absolutely unstable; otherwise, it is convectively unstable.

Such typical neutral stability curves for the Tollmien–Schlichting and crossflow instabilities are shown in Fig. 8.3. As seen, both instabilities have a convective character, since the coordinate origin is outside of region $\omega_i>0$. In the case of crossflow instability, the angular range formed by the tangents to the curve $\omega_i=0$ is relatively narrow and shifted downstream with respect to that of the Tollmien–Schlichting waves. The fundamental difference between the group and phase velocities is seen on comparing Fig. 8.3 and Fig. 6.7. The crossflow waves are either stationary or travel almost normal to the downstream direction. In contrast, the group velocity is directed almost parallel to it.

If the dispersion relation varies with spatial coordinates in a shear flow, appearance of additional critical Reynolds numbers corresponding to the transformation of

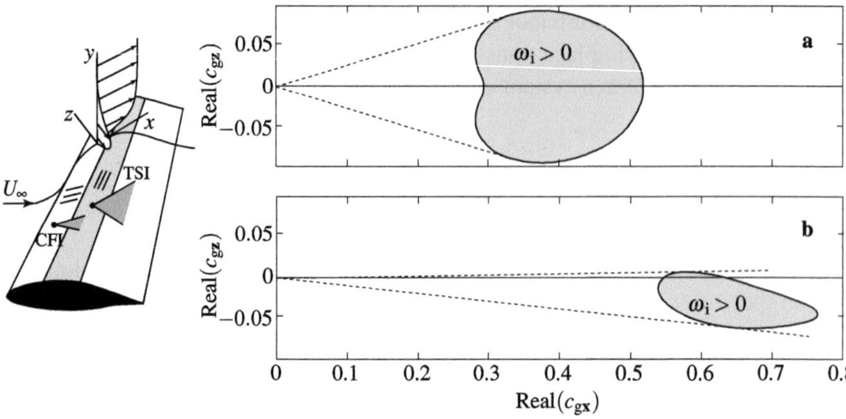

Fig. 8.3 Regions of the Tollmien–Schlichting (**a**) and crossflow (**b**) instability wave amplification in the group velocity plane (Oertel Jr. and Stank, 1999)

convective instability to absolute instability, or vice versa, can occur. In this case, the disturbance development depends on the disturbance source position. It is necessary to note, however, that the transition from one instability type to another in such a case is not stipulated by the appearance of a *qualitatively* new mechanism, but simply represents a *quantitative* shift in the balance between stabilizing and destabilizing forces.

Let us demonstrate a technique of determining the group velocity for a mode $\hat{v}$ satisfying the Orr–Sommerfeld equation (2.9). Multiplying it by the corresponding adjoint eigenfunction $\hat{v}^\dagger$ introduced in Sect. 2.5, integrating the resulting equation over y to remove functional dependencies, and then differentiating by α or β yields the following expressions for the group velocities:

$$c_{g\mathbf{x}} = \frac{\int \hat{v}^\dagger \left[\left(\mathcal{D}^2 + \beta^2\right) U - U_{yy} + \alpha \left(2\omega - 3\alpha U\right) \right] \hat{v} \mathrm{d}y}{\int \hat{v}^\dagger \left(\mathcal{D}^2 - \varkappa^2\right) \hat{v} \mathrm{d}y} - \frac{4\mathrm{i}\alpha}{\mathrm{Re}}, \tag{8.5a}$$

$$c_{g\mathbf{z}} = 2\beta \frac{\int \hat{v}^\dagger \left(\omega - \alpha U\right) \hat{v} \mathrm{d}y}{\int \hat{v}^\dagger \left(\mathcal{D}^2 - \varkappa^2\right) \hat{v} \mathrm{d}y} - \frac{4\mathrm{i}\beta}{\mathrm{Re}}. \tag{8.5b}$$

Note that the left-hand side of the orthogonality relation (2.15) coincides with the denominators in (8.5). Thus, proper normalization of the regular and adjoint eigenfunctions by requiring $C = 1$ in (2.15) reduces the task to calculation of the integrals in the numerators of (8.5).

8.3 Experimental investigations of wave packets

There are at least two main reasons for considering the development of the wave packets experimentally. One of them is to simulate a naturally occurring disturbance and to study the stages of its transformation to turbulence as a transition scenario.

The other is to consider only the initial (linear) stage of its development in order to extract wave characteristics of the waves constituting it, verify some theoretical results, and find their restrictions.

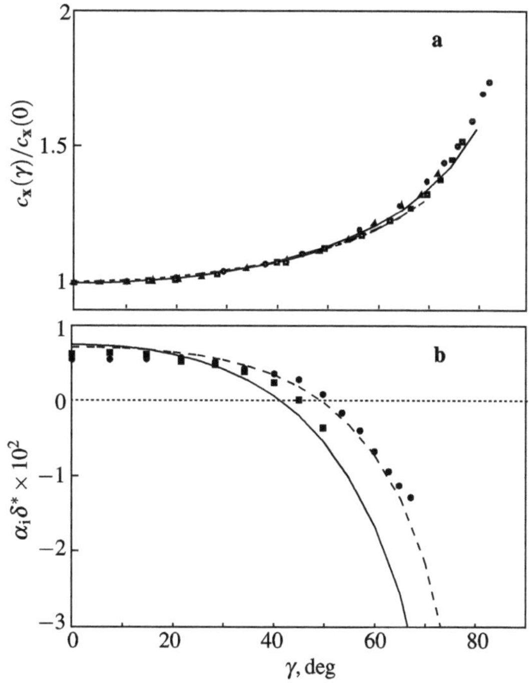

Fig. 8.4 Downstream phase velocity ratio as a function of the wave angle (**a**): $\mathrm{Re}_{\delta^*} = 827$–$1226$, $F = (63.6$–$138.7) \times 10^{-6}$; experimental results (□, △, ○); corresponding theoretical curves (*lines*). Normalized spatial wave growth rate as a function of the wave angle (**b**): experimental data for $F = 94.1 \times 10^{-6}$ at $\mathrm{Re}_{\delta^*} = 1048$; $y/\delta^* = 0.74$ (•); $y = 4.5$ mm (■); theoretical curves at $\mathrm{Re}_{\delta^*} = 1089$ (*solid line*) and $\mathrm{Re}_{\delta^*} = 963$ (*dashed line*) (Kachanov, 1985, Kachanov and Michalke, 1994)

Vasudeva (1967) was the first investigator, who studied the development of the Tollmien–Schlichting-wave packet in a boundary layer. He was not able to overcome certain technical difficulties, in particular, the small signal-to-noise ratio (dynamic range) of hot-wire measurements. The experiments of Gaster and Grant (1975) were an essential step forward in the investigation of wave packets. Periodic pulses of disturbances of quite a small amplitude were excited by periodic suction and blowing through a small hole in a flat plate. To reduce errors associated with the nonparallel flow effects, the measurements were performed in the region of the outer maximum of the Tollmien–Schlichting waves, whose location weakly depends on the streamwise coordinate and the frequency of disturbances. To increase the signal-to-noise ratio, computer-aided synchronous detection was applied. As a result, the pattern of downstream development of the excited wave train was obtained, which corresponded to the result of calculations of superposition of the Tollmien–Schlichting waves of the given frequency and different wavenumbers.

The experiments of Gaster and Grant (1975) were aimed at the general analysis of processes leading to the laminar–turbulent transition via instability phenomena in the wave packet, while the studies of Kachanov (1985) and Kachanov and Michalke (1994) were performed to study the characteristics of waves of a linear wave train.

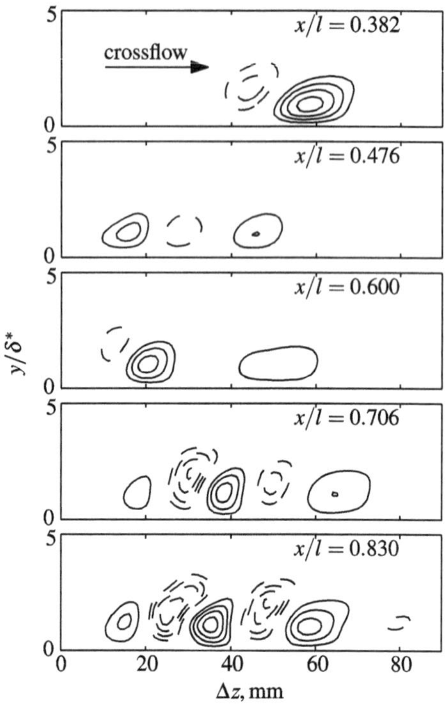

Fig. 8.5 Isolines of streamwise velocity of swept-wing stationary disturbances induced by a tip vortex in the free stream; wing chord length $l = 0.5$ m; equidistant isolines from -0.130 to 0.116 of external-flow velocity U_e (Boiko, 2002)

The main motivation of these experiments was the absence of detailed measurements of the dispersion and stability characteristics for oblique Tollmien–Schlichting waves. In the latter study, examination of the wave packet closely followed its theoretical description. The obtained periodic velocity distributions were subjected to a spatial Fourier transformation to obtain the dispersion characteristics of waves with different streamwise α and transverse β wavenumbers at a fixed frequency. As a result, the dispersion characteristics of oblique waves for a number of frequencies and angles of wave inclination $\gamma = \arctan(\beta/\alpha_r)$ up to 60–80° were obtained. Most of them appeared to be very close to the results predicted by the parallel linear stability theory. The general shape of the wave packet also agrees with the theoretical conclusions. However, the wave growth rates at large angles of inclination were larger than those obtained theoretically (Fig. 8.4). This fact was attributed to an increased role of the nonparallel flow effects for oblique waves. Another possible reason for the discrepancy are the transient phenomena described in Chap. 9.

A phenomenon accompanying the development of disturbances behind an isolated roughness element in a swept-wing boundary layer consists in apparent multiplication of the number of stationary vortices along the streamwise coordinate. This feature is characteristic for packets of unstable stationary crossflow vortices, see. Fig. 8.5. The main reason for this essentially linear phenomenon is small variations of the direction of the wavevector of vortices in the packet, and, as a consequence, variations of the direction of growth of modes with different wavenumbers.

Disturbances with smaller spanwise wavenumbers appear at the leeward side of the packet, while those with larger wavenumbers, at the windward side.

Exercises

8.1. Design an algorithm to obtain the group velocities of Squire modes (in temporal formulation). Write a program for finding the group velocities of Squire modes in the plane Poiseuille flow.

8.2. Write a program for finding the group velocities of Tollmien–Schlichting waves (in temporal formulation) in the Blasius boundary layer using finite differences and Eq. (8.5). Compare the obtained group velocities with corresponding phase velocities of the modes.

8.3. Modify the program from Exercise 8.2 to obtain the group velocities for the unstable leading crossflow modes in the Falkner–Skan–Cooke boundary layer. Compare the obtained group velocities with corresponding phase velocities of the modes. How does the result of the comparison differ from the result of similar comparison for the Blasius boundary layer in Exercise 8.2?

References

Boiko AV (2002) Swept-wing boundary layer receptivity to a steady free-stream vortex disturbance. Fluid Dyn 37(1):37–45

Briggs RJ (1964) Electron-stream interactions in plasmas. MIT, Cambridge, Mass.

Gaster M, Grant T (1975) An experimental investigation of the formation and development of a wave-packet in a laminar boundary layer. Proc R Soc Lond A 347:253–269

Kachanov YS (1985) Development of spatial wave packets in boundary layer. In: Kozlov VV (ed) Laminar–Turbulent Transition, Springer–Verlag, Berlin, IUTAM Symposium, pp 115–123

Kachanov YS, Michalke A (1994) Three-dimensional instability of flat-plate boundary layers: Theory and experiment. Eur J Mech B/Fluids 13(4):401–422

Kozlov VV, Ryzhov OS (1990) Receptivity of boundary layers: Asymptotic theory and experiment. Proc R Soc Lond A 429:341–373

Oertel Jr H, Stank R (1999) Dynamics of localized disturbances in transonic wing boundary layers. AIAA Paper 99–0551

Vasudeva BR (1967) Boundary-layer instability experiment with localized disturbance. J Fluid Mech 29:745–763

Further Reading

Breuer KS, Cohen J, Haritonidis JH (1997) The late stages of transition induced by a low-amplitude wave packet in a laminar boundary layer. J Fluid Mech 340:395–411

Cohen J, Breuer KS, Haritonidis JH (1991) On the evolution of a wave packet in a laminar boundary layer. Part 1. Weak disturbances. J Fluid Mech 225:575–606

Gaster M (1975) A theoretical model of a wave-packet in the boundary layer on a flat plate. Proc R Soc Lond A 347:271–289

Huerre P, Monkewitz PA (1990) Local and global instabilities in spatially developing flows. Ann Rev Fluid Mech 22:473–537

Joslin RD, Streett CL (1994) The role of stationary crossflow vortices in boundary-layer transition on swept wings. Phys Fluids 6(10):3442–3453

Monkewitz PA (1990) The role of absolute and convective instability in predicting the behavior of fluid systems. Eur J Mech B/Fluids 9(5):395–413

Schmid PJ, Henningson DS (2000) Stability and transition in shear flows. Springer–Verlag, Berlin

Chapter 9
Transient disturbances in shear flows

In the previous chapters, a far-field (asymptotic) response of shear flows to modal disturbances was considered. In this chapter, we concentrate on the response in the near-field of the disturbance source. As we know, the classical analysis of linear stability treats the disturbances as sets of separate modes of the linearized Navier–Stokes equations. However, the approach does not take into account the fact that these equations are not self-adjoint, i.e., the modes are not orthogonal. This may give rise to strong transient effects pronounced in the source near-field. It has been proven experimentally that transient phenomena close to the disturbance source can be drastic for the downstream transition to turbulence, even if the disturbances are asymptotically stable.

9.1 Lift-up effect

Ellingsen and Palm (1975) noticed that the inviscid stability theory allows a linear temporal growth of kinetic energy of disturbances independent of x (i.e., with a zero streamwise wavenumber α) in internal parallel shear flows. In this case, the linearized momentum equation (2.3a) in the absence of pressure gradient reduces to

$$\frac{\partial u}{\partial t} + vU_y = 0 ,$$

where the wall-normal disturbance velocity v is independent of time. Integration of the above equation gives

$$u(t) = u(0) - vU_y t ,$$

i.e., the amplitude of streamwise disturbance velocity u increases linearly in time. Such a mechanism of disturbance amplitude and kinetic energy growth is called the algebraic instability as the growth is linear rather than exponential or, more physically, it is called the lift-up effect as the growth can be interpreted as a consequence of the effect of the wall-normal velocity v in the presence of shear (i.e., when $U_y \neq 0$) on a fluid particle.

A.V. Boiko et al., *Physics of Transitional Shear Flows*,
Fluid Mechanics and Its Applications 98, DOI 10.1007/978-94-007-2498-3_9,

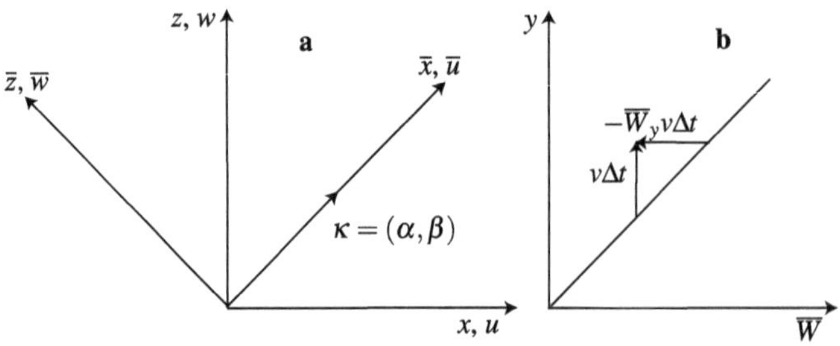

Fig. 9.1 Illustration of the lift-up effect: definition of the coordinate system fitted to wavevector (**a**); generation of the horizontal velocity defect in the direction $\bar{z}$ by lift-up of the fluid element preserving horizontal momentum in this direction (**b**) (Henningson et al, 1994)

To analyze the behavior of an initial perturbation with arbitrary fixed wavenumbers α and β in the inviscid limit, it is necessary to analyze the linearized Navier–Stokes equations at $\mathrm{Re} \to \infty$. Similarly to the normal vorticity equation (2.12) for normal modes, the normal vorticity equation for such initial disturbances can be written in the form

$$\left(\frac{\partial}{\partial t} + \mathrm{i}\alpha U\right) \tilde{\eta}(y,t) = -\mathrm{i}\beta U_y v(y,t) .$$

Let us transform the coordinate system so that the axis $\bar{x}$ is directed along the wavevector κ, as in Fig. 9.1, and the axis $\bar{z}$ is perpendicular to it. Such a transformation is called the Stuart transformation (cf. the Squire transformation in Sect. 6.4.2). The basic flow in this new wavevector-aligned coordinate system has the components

$$\overline{U} = \frac{\alpha}{\varkappa} U , \quad \overline{W} = -\frac{\beta}{\varkappa} U .$$

The components of the disturbance along the new coordinate axes are

$$\bar{u} = \frac{\mathrm{i}}{\varkappa} v , \quad \bar{w} = -\frac{\mathrm{i}}{\varkappa} \tilde{\eta} .$$

Combining the equations for $\tilde{\eta}(t)$, $\overline{W}$, and $\bar{w}$ given above yields

$$\Delta \bar{w} = -\overline{W}_y v \Delta t ,$$

where the observer is assumed to move together with the disturbance, and terms of the order of $\mathcal{O}(\Delta t^2)$ are dropped. Similarly to the $\alpha = 0$ case, the obtained expression for $\Delta \bar{w}$ corresponds to a disturbance of horizontal velocity induced by lift-up of a fluid particle due to wall-normal velocity.

From the viewpoint of the linear modal approach, the growth can be observed when the amplitude functions of the nonorthogonal modes constituting the perturbation cancel each other initially, but then—due to dispersion of phase velocities and amplification rates—induce a large transient growth of the disturbance kinetic energy (cf. the discussion of the loss of monotonic stability in Chap. 1). The phenomenon is related to the selection of certain initial conditions or, more correctly, to such a superposition of the nonorthogonal modes at the initial moment, which would trigger the processes of establishment in the hydrodynamic system accompanied by a growth of the total disturbance kinetic energy.

9.2 Growth of optimal disturbances

Viscosity prevents the unlimited growth of the disturbance kinetic energy, i.e., it becomes transient. Additionally, according to the discussion in Sect. 1.2, the growth can occur only at $\mathrm{Re} > \mathrm{Re}_\mathrm{E}$ when the flow loses monotonic stability. Evidently, such a rather slow transient growth is significant first of all in studying the transition to turbulence at subcritical Reynolds numbers $\mathrm{Re}_\mathrm{E} < \mathrm{Re} < \mathrm{Re}_\mathrm{L}$ and for flows in which the modal instability mechanism does not work (i.e., when $\mathrm{Re}_\mathrm{L} = \infty$).

A simple geometrical analogy of the phenomenon is the behavior of two initially opposite vectors with an equal length l on a plane. Accordingly, the length of the resulting vector is zero. If the angle between the vectors changes in time, the length of resulting vector begins to grow, even if the length of each vector becomes smaller with time, i.e., $l(t) < l(0)$. However, if $l(t) \to 0$ as $t \to \infty$, there exist a certain instant of time after which the resulting length begins to decrease.

The lift-up effect accounting for viscosity in plane channel flow was considered theoretically for the first time by Gustavsson (1991) and Henningson and Schmid (1994). It was found that the transient behavior of disturbance kinetic energy depends substantially on the initial conditions. The amplification rates and 'the time of survival' appeared to be maximum for waves with small streamwise wavenumbers α. For example, at $\mathrm{Re}_h = 3000$, as seen in Fig. 9.2, perturbations with $\alpha = 0$ and $\beta \approx 2.5$ corresponding to stationary streamwise vortices, are the most intensively growing ones in a plane channel flow.

The initial disturbances displaying the maximum energy growth at a fixed Reynolds number Re and wavenumbers β and α (or frequency ω instead of α for instability in space) are called optimal. The time interval or the downstream region in which the kinetic energy of transient disturbance generated by a particular disturbance source is increased characterizes the near-field of the source for these disturbances. It appears that there is a big pool of disturbances with $\alpha \approx 0$, which can amplify for a long time or downstream distance, so that the source 'near-field' with respect to them becomes very extended.

The possibility of algebraic instability and transient growth is found in calculations for most typical shear flows, for example, in a round pipe (Schmid and Henningson, 1994), free-shear (Farrell and Ioannou, 1993) and two-phase flows (Schmid and Kytomaa, 1994). Breuer and Kuraishi (1994) extended the analysis to three-dimensional and Hanifi et al (1994) to compressible boundary layers.

Gustavsson (1979), Hultgren and Gustavsson (1981) theoretically showed the necessity of accounting for the continuous spectrum in the boundary layer by considering three-dimensional initial-value problem. The studies indicate that the corresponding eigenfunctions at $\alpha \approx 0$ can have large amplitude inside the boundary

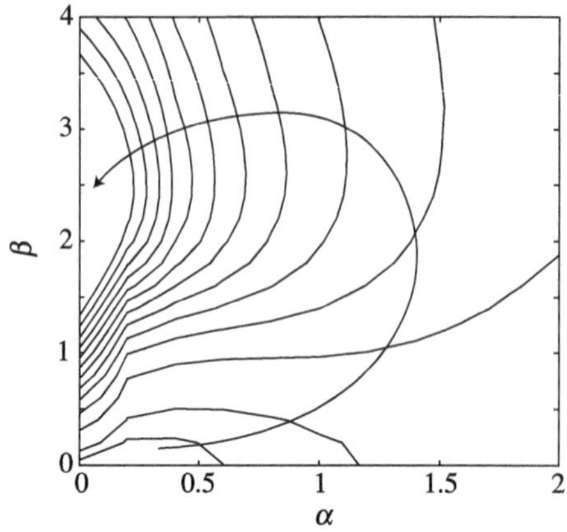

Fig. 9.2 Isolines of the maximum growth of kinetic energy $G(t)$ in the wavenumber α–β-plane for the Poiseuille flow at $\mathrm{Re}_h = 3000$. The transition from one curve to another in the direction indicated by the arrow corresponds to a change in $G(t) = 10, 25, 100, 200, \ldots, 1100$ (calculated by A. V. Boiko)

layer and stipulate a considerable temporal growth of the disturbance kinetic energy (Fig. 9.3). Moreover, at $\alpha = 0$ only three-dimensional disturbances consisting exclusively of components of the continuous spectrum are allowed in the near-wall region.

Most theoretical and numerical studies of transient perturbations use the optimal disturbances as their basis. They found their place in building some modern laminar–turbulent transition scenarios and in developing techniques for robust flow control. To determine the characteristics of disturbances that experience the maximum growth of kinetic energy $G_{\max}(t)$, it is necessary to solve an optimization problem for $G(t)$. Let us outline the related procedure, which is valid for linear stability approaches not only in the parallel-flow approximation. We define formally a propagator $\mathscr{P}(t)$—an evolution operator that takes an initial condition $\mathbf{u}_0 = \mathbf{u}(t = 0)$ to the state $\mathbf{u}(t > 0) = \mathscr{P}(t)\mathbf{u}_0$. Then the expression for $G(t)$ in terms of the propagator is

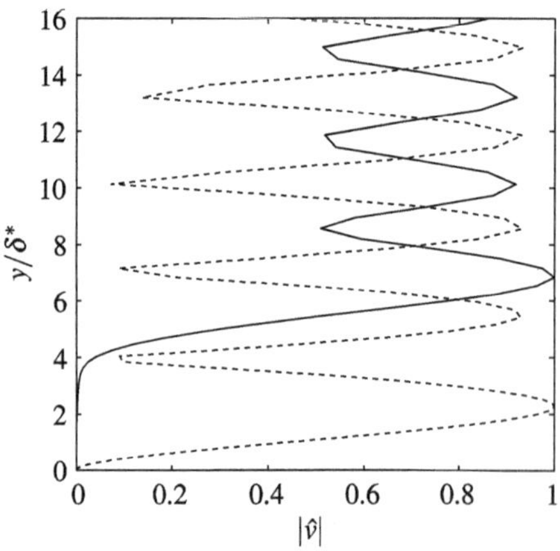

Fig. 9.3 Amplitude distribution of the wall-normal velocity for two modes of continuous spectrum, $\mathrm{Re}_{\delta^*} = 580$: $\alpha = 0$, $\beta = 1$, $k = 1$ (*solid line*) and $\alpha = 1$, $\beta = 0$, $k = 1$ (*dashed line*); k is defined in Sect. 2.6 (Hultgren and Gustavsson, 1981)

$$G(t) = \frac{E(t)}{E(0)} \equiv \frac{\mathbf{u}(t)\cdot\mathbf{u}(t)}{\mathbf{u}_0\cdot\mathbf{u}_0} = \frac{(\mathscr{P}(t)\mathbf{u}_0)\cdot(\mathscr{P}(t)\mathbf{u}_0)}{\mathbf{u}_0\cdot\mathbf{u}_0} = \frac{\mathbf{u}_0\cdot(\mathscr{P}(t)^\dagger\mathscr{P}(t)\mathbf{u}_0)}{\mathbf{u}_0\cdot\mathbf{u}_0}$$

which is the Rayleigh quotient, cf. (1.5). It means that $G_{\max}(t) = \max_{\mathbf{u}_0\neq 0} G(t)$ is positive and finite.

In calculations, finding $G_{\max}$ is equivalent to finding the largest singular value of the propagator $\mathscr{P}(t)$ approximated by a matrix $\mathbf{P}$ of a finite size $n\times n$. According to the singular value decomposition, we have

$$\mathbf{PU} = \mathbf{VS}\,,$$

where $\mathbf{V}$ and $\mathbf{U}$ are unitary matrices and $\mathbf{S} = \mathrm{diag}\{\sigma_0,\sigma_1,\ldots,\sigma_{n-1}\}$ is the matrix of singular values $\sigma_0 \geq \sigma_1 \geq \ldots \geq \sigma_{n-1}$; then

$$\mathbf{Pu}_0 = \sigma_0\mathbf{v}_0\,,$$

where $\mathbf{u}_0$ and $\mathbf{v}_0$ are the right and left singular vectors for σ_0. This equation represents a mapping of the initial conditions given by $\mathbf{u}_0$ to the final state at the time instant $t > 0$ given by $\mathbf{v}_0$ accompanied by the growth (mathematically, the stretching) factor σ_0, that is $G_{\max}(t) = \sigma_0$.

Assuming that the singular values are well separated, the general way to find the optimal initial perturbations is through the power iterations

$$\mathbf{u}_0^{k+1} = \mathbf{P}^H\mathbf{Pu}_0^k\,.$$

For quasi-parallel flows, this procedure can be specified as follows. In this case, the propagator is represented by the linearized equations (2.3). After eliminating the pressure, the solution of the linearized initial value problem to be found

$$\mathbf{u}(t) = \begin{pmatrix} u(t)\\ v(t)\\ w(t)\end{pmatrix}$$

can be expanded in the basis of normal modes $\hat{\mathbf{u}}_i$, $i = 1,2,\ldots$:

$$\mathbf{u}(t) = \sum_{i=1}^{\infty} k_i(t)\hat{\mathbf{u}}_i\,.$$

In this basis, the initial-value problem takes a particularly simple form:

$$\frac{\mathrm{d}\mathbf{k}}{\mathrm{d}t} = -\mathrm{i}\Upsilon\mathbf{k}\,,\quad \mathbf{k} = (k_1,k_2,\ldots)^T\,,\quad \Upsilon = \mathrm{diag}\{\omega_1,\omega_2,\ldots\}\,.$$

The formal solution

$$\mathbf{k}(t) = \exp(-\mathrm{i}\Upsilon t)\mathbf{k}_0\,,$$

gives the explicit form of the evolution operator $\mathscr{P}(t) = \exp(-\mathrm{i}\Upsilon t)$ in this basis.

The disturbance kinetic energy

$$E = \int_{y_1}^{y_2}(u^*u + v^*v + w^*w)\mathrm{d}y\,,$$

defines its measure (norm), which results in the following form of the scalar product of two vectors:

$$\mathbf{u}_1 \cdot \mathbf{u}_2 = \int_{-1}^{+1} \mathbf{u}_2^* \mathbf{u}_1 \, \mathrm{d}y \, .$$

When only a finite number n of modes (e.g., in numerical solution) is taken into account, this expression is approximated by

$$\mathbf{u}_1 \cdot \mathbf{u}_2 = \mathbf{k}_2^H \mathbf{W} \mathbf{k}_1 \, ,$$

where $\mathbf{W}$ is the weight matrix whose elements $W_{ij} = \hat{\mathbf{u}}_i \cdot \hat{\mathbf{u}}_j$ and $i, j = 1, \ldots, n$. The matrix $\mathbf{W}$ is Hermitian and positive definite, i.e., it can be decomposed (factored) as $\mathbf{W} = \mathbf{F}^H \mathbf{F}$. Then, the scalar product can be calculated as

$$\mathbf{u}_1 \cdot \mathbf{u}_2 = \mathbf{k}_2^H \mathbf{F}^H \mathbf{F} \mathbf{k}_1 = \|\mathbf{F}\mathbf{k}_1, \mathbf{F}\mathbf{k}_2\|_2^2 \, ,$$

where the subscript ‘2’ indicates the 2-norm. Thus, we have

$$G(t) = \max_{\mathbf{u} \neq 0} \frac{\|\mathbf{u}(t)\|^2}{\|\mathbf{u}_0\|^2} = \|\exp(-\mathrm{i}\Upsilon_n t)\|^2 = \|\mathbf{F}\exp(-\mathrm{i}\Upsilon_n t)\mathbf{F}^{-1}\|_2^2 \, ,$$

where $\Upsilon_n = \mathrm{diag}\{\omega_1, \ldots \omega_n\}$.

Function `Optimal` is an implementation of this method. To run it, one has to solve numerically in advance an eigenvalue problem for homogeneous linearized Navier–Stokes equations with homogeneous boundary conditions. The isolines of $G(t)$ obtained using this procedure for a plane channel flow in a certain range of streamwise and transverse wavenumbers at $\mathrm{Re}_h = 3000$ are shown in Fig. 9.2.

Profiles of disturbance velocities corresponding to the optimal disturbances at different time instants can be determined, if we take into account that

$$G(t) = \|\mathbf{F}\exp(-\mathrm{i}\Upsilon_n t)\mathbf{F}^{-1}\|_2^2 = \sigma_0^2 \, ,$$

where σ_0 is the leading (maximum) singular value of the matrix

$$\mathbf{B}(t) = \mathbf{F}\exp(-\mathrm{i}\Upsilon_n t)\mathbf{F}^{-1} \, ,$$

and

$$\mathbf{B}(t)\mathbf{k}_0 = \sigma_0 \mathbf{k}(t) \, ,$$

where $\mathbf{k}_0$ and $\mathbf{k}(t)$ are equal to the leading right and left singular vectors of $\mathbf{B}(t)$ corresponding to the disturbance states at time instants $t = 0$ and $t > 0$, respectively.

Figure 9.4 illustrates different characteristics of the optimal disturbances found with the procedure described above. A comparison of the growth of two-dimensional (homogeneous in the transverse direction) and three-dimensional (periodic in the transverse direction) optimal disturbances in the Blasius and the plane Poiseuille flow at subcritical Reynolds numbers is given in Fig. 9.4**a**. This figure clearly demonstrates that the transient growth of three-dimensional disturbances is potentially much more dangerous for destroying the initial flow than the two-dimensional growth. The corresponding initial and resulting (at large t) velocity patterns (so-called *streaks* or *streaky structures*) in y–z-planes for the time instants of the maximum value of G are shown in the remaining part of the figure. Additionally, the profile of the streamwise velocity amplitude for the optimal perturbation in the Blasius boundary layer is shown in Fig. 9.4**c**. Such a profile appeared to be one of

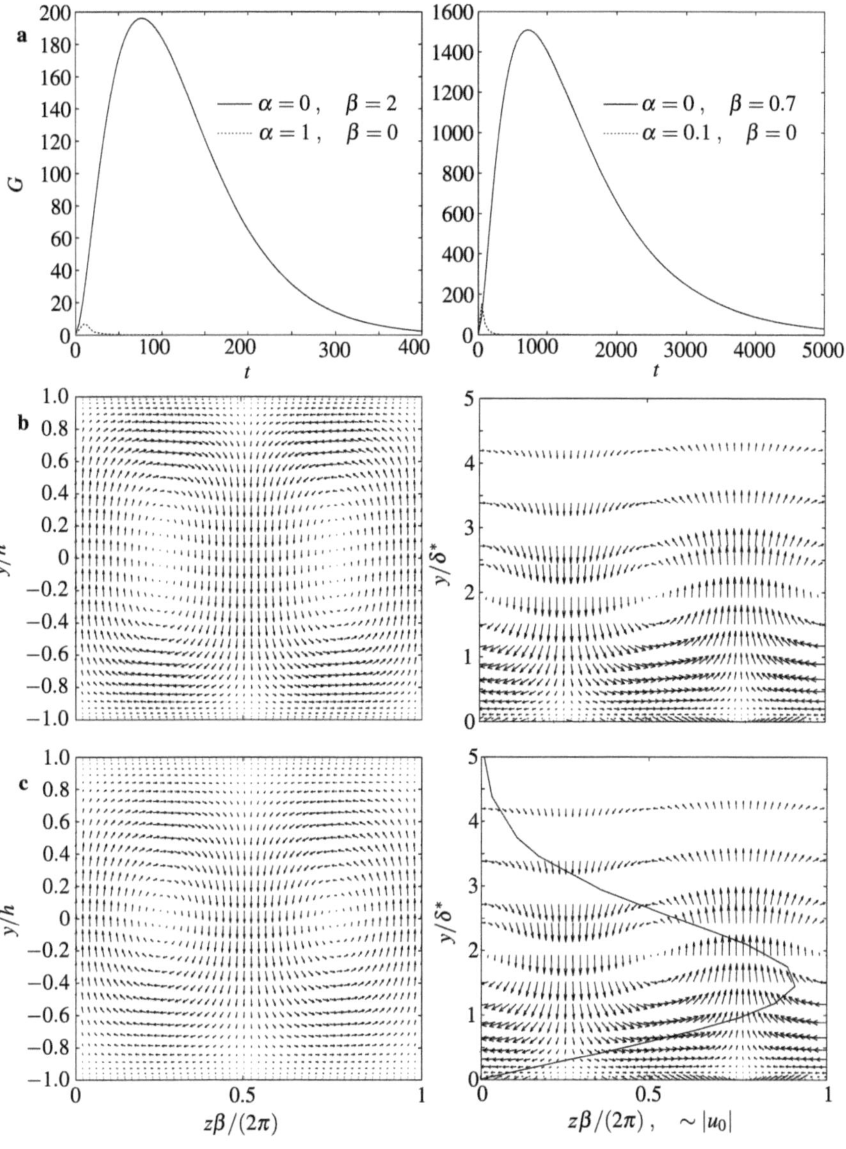

Fig. 9.4 Development in time of optimal disturbances for a plane Poiseuille flow at $\mathrm{Re}_h = 1000$ (*left*) and Blasius boundary layer at $\mathrm{Re}_{\delta^*} = 10000$ (*right*): amplification $G(t)$ of three-dimensional (*solid line*) and two-dimensional disturbance (*dotted line*) (**a**); initial optimal cross-sectional waveforms (**b**); the cross-sectional waveforms developed at the instant of the maximum $G(t)$ (**c**); streamwise disturbance amplitude profile $|u_0|$ (*solid line*) (calculated by A. V. Boiko)

```
function [t,G]=Optimal(Re,o,u,v,w,tmax,nodes,D)
% OPTIMAL calculates growth rates G(t)
% for optimal disturbances.
% Re     - Reynolds number;
% o      - eigenvalues of linearized NS-equation;
% u,v,w  - corresponding eigenfinctions
%          (disturbance velocities);
%
% tmax   - upper limit of time integration;
% nodes  - number of time steps;
% D      - a container of derivative matrices.

% form the energy matrix M
weight=sqrt(D{6});
vv=diag(weight)*v;
uu=diag(weight)*u;
ww=diag(weight)*w;
M=uu'*uu+ww'*ww+vv'*vv;

% factorize
[u0,s0]=svd(M);
s0=sqrt(diag(s0));
F=diag(s0)*u0'; iF=u0*diag(1 ./s0);

% form the nodes in time and calculate G(t)
t=linspace(0,tmax,nodes);
for k=1:nodes,
   G(k)=norm(F*diag(exp(-1i*o*t(k)))*iF,2).^2;
end
```

the most typical characteristics of experimentally observed boundary-layer streaks caused by a variety of sources.

Predominance of very oblique stationary modes in the optimal disturbances indicates that the nonparallel flow effects bias substantially their formation and development similarly to that observed for the Görtler vortices (Sect. 5.2.3). Thus, for a quantitative description of the quasi-stationary (low-frequency) optimal disturbances in growing boundary layers, it was proposed to use parabolic equations similar to (5.11) at Gö $=0$. Then, the equations take the form

$$\hat{u}_x+\hat{v}_y+\mathrm{i}\beta\hat{w}=0\,, \tag{9.1a}$$

$$(U_x-\mathrm{i}\omega)\hat{u}+U\hat{u}_x+V\hat{u}_y+U_y\hat{v}=\hat{u}_{yy}-\beta^2\hat{u}\,, \tag{9.1b}$$

$$(V_y-\mathrm{i}\omega)\hat{v}+U\hat{v}_x+V\hat{v}_y+V_x\hat{u}+\hat{p}_y=\hat{v}_{yy}-\beta^2\hat{v}\,, \tag{9.1c}$$

$$-\mathrm{i}\omega\hat{w}+U\hat{w}_x+V\hat{w}_y+\mathrm{i}\beta V\hat{p}=\hat{w}_{yy}-\beta^2\hat{w}\,. \tag{9.1d}$$

They are solved at $x_\mathrm{f}>x_0$ with the boundary conditions

$$y=0:\hat{u}=\hat{v}=\hat{w}=0\,, \tag{9.2a}$$

$$y \to \infty : \hat{u} = \hat{w} = \hat{p} = 0 \quad \text{or} \quad \hat{u} = \hat{v} = \hat{w} = 0\,, \tag{9.2b}$$

and the initial conditions $(\hat{u}_0(y), \hat{v}_0(y), \hat{w}_0(y))^T$ at $x = x_0$.

The growth or decay of disturbance kinetic energy is determined by the ratio

$$G(x_0, x_\mathrm{f}, \beta, \omega, \mathrm{Re}) = \frac{E(x_\mathrm{f})}{E(x_0)}\,.$$

To take into account different scaling of different disturbance velocity components, the energy in this case is defined as:

$$E(x) = \int_0^\infty (\mathrm{Re}|\hat{u}|^2 + |\hat{v}|^2 + |\hat{w}|^2)\mathrm{d}y\,.$$

Hence, at $\mathrm{Re} \gg 1$, the maximum energy growth occurs if the initial disturbance at x_0 does not contain the streamwise velocity. Using the same reasoning, at $x_\mathrm{f} \gg x_0$, where $\hat{u} \neq 0$, the terms $|\hat{v}|^2$ and $|\hat{w}|^2$ can be neglected. This leads to the following expression for G:

$$G = \mathrm{Re}\frac{\int_0^\infty |\hat{u}|^2 \mathrm{d}y}{\int_0^\infty (|\hat{v}|^2 + |\hat{w}|^2)\mathrm{d}y}\,.$$

As seen, at $\mathrm{Re} \gg 1$, the growth rate of the optimal disturbances scaled as $\overline{G} = G/\mathrm{Re}$ is independent of Re.

Let us define the norms of the disturbances as

$$\|\hat{u}\|^2 = \int_0^\infty |\hat{u}|^2 \mathrm{d}y\,,$$
$$\|\hat{\mathbf{q}}\|^2 = \int_0^\infty (|\hat{v}|^2 + |\hat{w}|^2)\mathrm{d}y\,,$$

where $\hat{\mathbf{q}} = (\hat{v}, \hat{w})^T$. Then,

$$\overline{G} = \frac{\|\hat{u}\|^2}{\|\hat{\mathbf{q}}\|^2}\,.$$

As the initial-boundary-value problem (9.1) with the homogeneous boundary conditions (9.2) is linear, its solution can be written through the propagator operator $\mathscr{P}$ as

$$\hat{u} = \mathscr{P}\hat{\mathbf{q}}\,.$$

Thus, we have

$$\overline{G}_{\max} = \max_{\hat{\mathbf{q}} \neq 0} \frac{(\mathscr{P}^\dagger \mathscr{P}\hat{\mathbf{q}}, \hat{\mathbf{q}})}{(\hat{\mathbf{q}}, \hat{\mathbf{q}})}\,.$$

Here $\mathscr{P}^\dagger$ means the adjoint of $\mathscr{P}$.

The maximum of $\overline{G}$ is reached at certain initial conditions $\hat{\mathbf{q}}$. From the theory of linear operators, it follows that such initial conditions are determined by the eigenvector corresponding to the leading eigenvalue of the problem

$$\mathscr{P}^\dagger \mathscr{P}\hat{\mathbf{q}} = \overline{g}\hat{\mathbf{q}}\,,$$

whereas $\overline{G}_{\max}$ is equal to this eigenvalue, which is non-negative. To calculate $\hat{\mathbf{q}}$, it is possible to utilize the simple iteration technique

$$\hat{\mathbf{q}}^{n+1} = \mathscr{P}^{\dagger}\mathscr{P}\hat{\mathbf{q}}^{n}\,.$$

9.3 Experimental studies

The mechanism of the disturbance energy growth under consideration is closely related to a specific choice of the initial conditions, which are, however, not atypical for shear flows. In a flow subject to a stochastic effect, e.g., at a high level of disturbances in the free stream or at the channel entrance, such initial perturbations can arise naturally.

In particular, at a high free-stream turbulence level, two different phenomena are usually distinguished inside the boundary layer, both of which can be responsible for the transition to turbulence: excitation of traveling waves with characteristics of Tollmien–Schlichting waves and generation of quasi-stationary longitudinal structures or ‘streaks’ with characteristics determined by external vortical and pressure disturbances. The streaks, which were named due to the visualization patterns produced by the flow structures, are experimentally observed as quasi-stationary deformations of the laminar boundary layer in transverse direction which grow downstream in amplitude and length. It is instructive to compare a smoke visualization of a flat-plate boundary layer at low ($\mathrm{Tu} \leq 0.02\%$) and high ($\mathrm{Tu} = 1.5\%$) free-stream turbulence (Fig. 9.5). In the first case, the boundary layer can be characterized as undisturbed (homogeneous smoke sheet), whereas the flow at $\mathrm{Tu} = 1.5\%$ is modulated by the streamwise structures.

It should be emphasized that the boundary-layer perturbations with a dominating streaky pattern, although random in time and space, are not what is usually called turbulence. The typical energy density spectra of streamwise velocity disturbances obtained in the free stream at $\mathrm{Tu} = 1.5\%$ and in the laminar boundary layer are shown in Fig. 9.6. They demonstrate the essential difference of free-stream turbulence from large-scale boundary-layer oscillations, where the low-frequency disturbances are much higher. Moreover, the transverse and the wall-normal scales of the low-frequency structures are of the order of the boundary-layer thickness, and the amplitude of wall-normal velocity fluctuations v' inside the boundary layer is several times smaller than u' (Westin et al, 1994).

At the same time, the profiles $u'(y)$ in the boundary layer are very different from the corresponding profiles of oblique Tollmien–Schlichting waves, see Fig. 9.7. In contrast, almost ideal coincidence of the measured amplitude distributions and the theoretical profile related to the streaks generated by optimal disturbances occurs. This indicates that disturbances close to optimal ones contribute substantially to the experimentally observed streaks. Thus, it can be assumed that the theory of optimal disturbances can describe the behavior of the most energy-intensive harmonics in wavenumber spectra of localized quasi-stationary deformations of boundary layers.

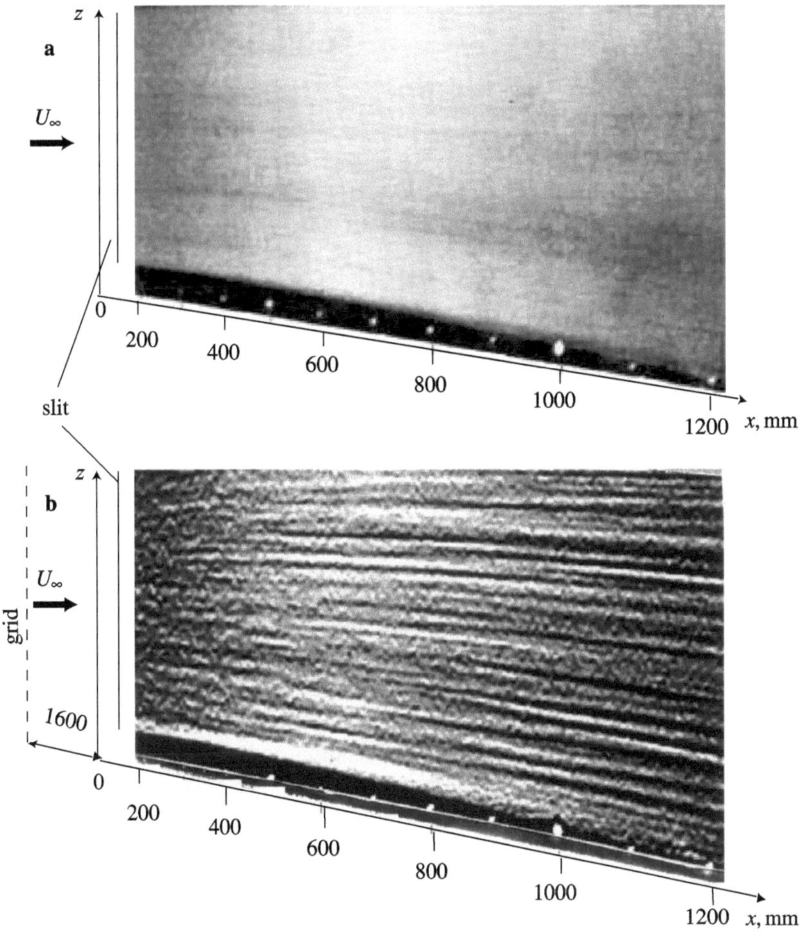

Fig. 9.5 Smoke visualization of a flat-plate boundary layer (Matsubara et al, 1996): Tu $\leq$ 0.02% (**a**) and Tu = 1.5% (**b**); $U_e = 6$ m/s

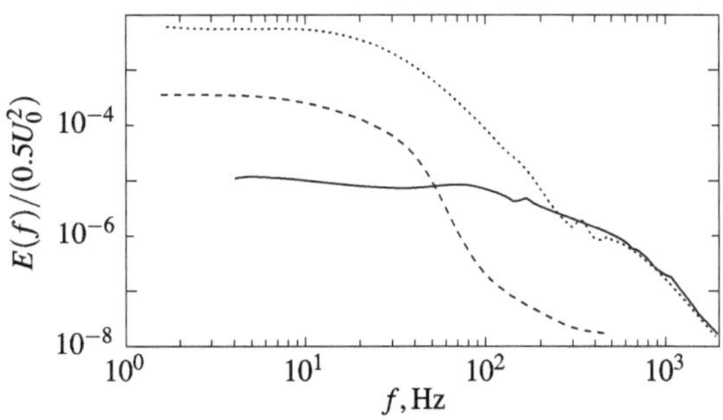

Fig. 9.6 Energy density spectra for u' at different wall-normal positions at $\mathrm{Re}_{\delta^*} = 890$ in the free stream (*solid line*), in the middle of the boundary layer (*dashed line*), and in the near-wall region (*dotted line*) (Westin et al, 1994)

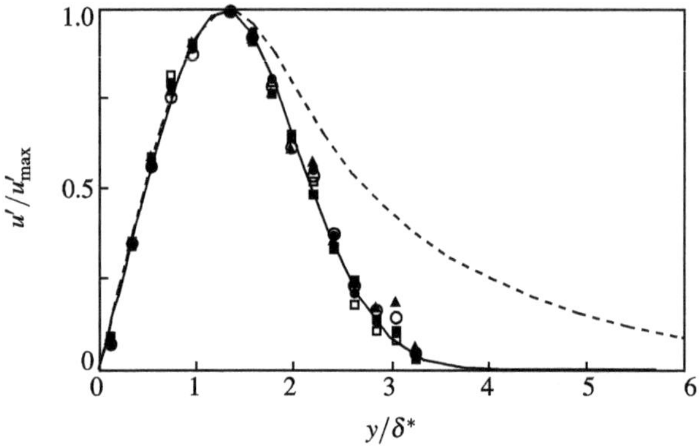

Fig. 9.7 Amplitude profile of boundary-layer streaks at $F = 27 \times 10^{-6}$, $\alpha = 0.022$ and $\beta = 0.419$ (■), 0.582 (□), 0.791 (•), 0.977 (○), 1.163 (△) (Boiko et al, 2003); Tollmien–Schlichting wave profile at $\alpha = 0.022$ and $\beta = 0.419$ (*dashed line*); profile of the stationary boundary-layer mode (*solid line*) (Luchini, 1996)

Experimentally, several types of controlled boundary-layer perturbations excited through a point-like source are found, including linear wave packets, turbulent spots (described in Chap. 13), and transverse-localized structures associated with the streaks, Fig. 9.8. These structures are deterministic, very localized (with respect to the Tollmien–Schlichting wave packets or turbulent spots) three-dimensional formations with velocities of propagation of the leading and trailing fronts approximately equal to 0.9 and 0.5 of external-flow velocity U_e, respectively (as opposed to the group velocity of the Tollmien–Schlichting wave packet of about 0.3–0.4 of U_e). They quickly elongate downstream, experience practically no expansion in the transverse direction, and the magnitude of the maximum velocity defect continuously decay downstream (Fig. 9.9). The spectral analysis by Westin et al (1998) shows that, indeed, the majority of harmonics, except for those at low frequencies, decay as the disturbance propagates downstream, while oblique waves dominate in the wavenumber spectra, Fig. 9.10. The maximum amplitude of disturbances is centered at zero frequency, which is interpreted as a quasi-stationary flow distortion. It is remarkable that the structures are observed at low Reynolds numbers in the region

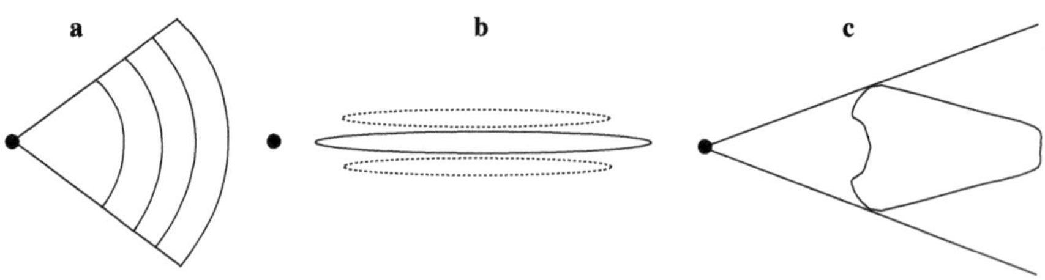

Fig. 9.8 Three types of localized boundary-layer disturbances (Grek et al, 1985): wave packet (**a**), streaky structure (**b**), and turbulent spot (**c**)

where the Tollmien–Schlichting waves are highly damped. At present, it is firmly established that the cause of their appearance is the lift-up effect discussed above.

Klingmann (1991, 1992) was the first who studied in detail the development of three-dimensional transient disturbances in a channel flow experimentally and conducted a comparison with the results of the viscous linear analysis of the corresponding initial-value problem. The disturbances were excited by injection of fluid into the channel through a point-like hole in one of the channel walls. At the initial stage, despite the disturbance amplitude reaching 10–20% of U_a, good correspondence with the theory was obtained: appearance of streamwise streaks of retarded and accelerated fluid with strong velocity shears in the transverse direction, which propagated with velocities between 0.5 and 0.9 of U_a; formation of new streaks at downstream disturbance propagation; linear dependence of the disturbance kinetic energy growth in time; and proportionality of amplification time to the Reynolds number. However, due to nonlinear effects at late stages of the amplification, the disturbances grew longer than the theory predicts and downstream, depending on the initial intensity of excitation, they either decayed or led to turbulence through a stage of nonlinear development.

Breuer and Haritonidis (1990) applied the inviscid linear stability theory to the analysis of experimental data concerning the development of small localized disturbances in a flat-plate boundary layer excited by transverse oscillations of a membrane installed on the plate surface. The disturbances, which they generated, were similar to other experimentally observed localized perturbations: the streak moved faster than the Tollmien–Schlichting waves, without strong attenuation and transverse expansion. It was shown that the initial stage of the streak development is satisfactorily described by the inviscid lift-up mechanism. In accordance with the linear stability theory and experimental data, the disturbance is divided into two parts. One of them is a broadening wave packet described by the solutions of the Rayleigh equation with waves propagating at their own phase velocities. The other is the advective part propagating at a local basic flow velocity, which, due to the downstream elongation, forms a streak of strong transverse velocity shear. However, the results of the experiment show natural dissipation of the disturbance by viscosity downstream, which, obviously, cannot be obtained in the framework of the inviscid analysis.

The initial stage of development of transient disturbances in a flat-plate boundary layer was considered in detail by Boiko et al (2003). The free-stream disturbances were induced locally at the leading edge of the experimental model, Fig. 9.11. As a result, ranges of streamwise coordinate and transverse wavenumbers were found, in which the amplitude of low-frequency disturbances experiences the growth, while high-frequency perturbations decay continuously, Fig. 9.12**a**. The values of the downstream propagation velocity $c_{\mathbf{x}}(\beta)$ for Tollmien–Schlichting waves corresponding to the experimental and the optimal (using the model of Luchini, 2000) conditions are shown in Fig. 9.12**b**. As seen, the model quite accurately predicts the velocities of low-frequency disturbances; the deviation is less than 4.5%. Also, good quantitative conformity exists between the theoretical results on the phase velocities of the Tollmien–Schlichting waves and the experimental data at $F = 80.6 \times 10^{-6}$.

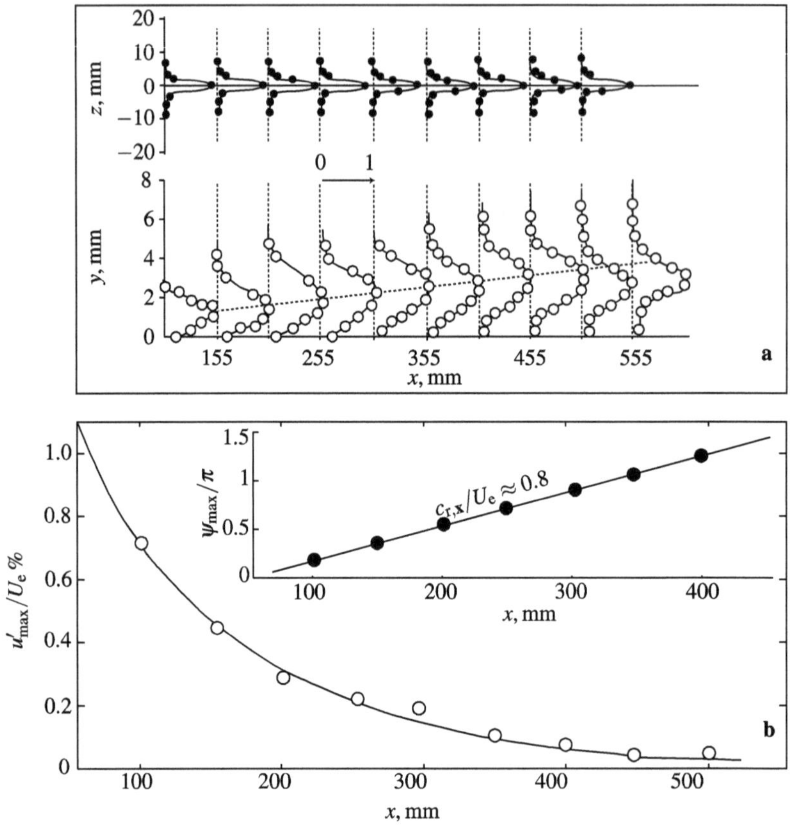

Fig. 9.9 Integral characteristics of the streaky structure in the boundary layer far from the source: transverse and wall-normal distributions of streamwise disturbance velocity $u'(y)/u'_{max}$ along with the streamwise coordinate (**a**), a distribution of the amplitude $u'_{max} = \max_y[u'(y)]$ and the corresponding phase ψ_{max} of the disturbance velocity as functions of the streamwise coordinate (**b**) (Grek et al, 1985)

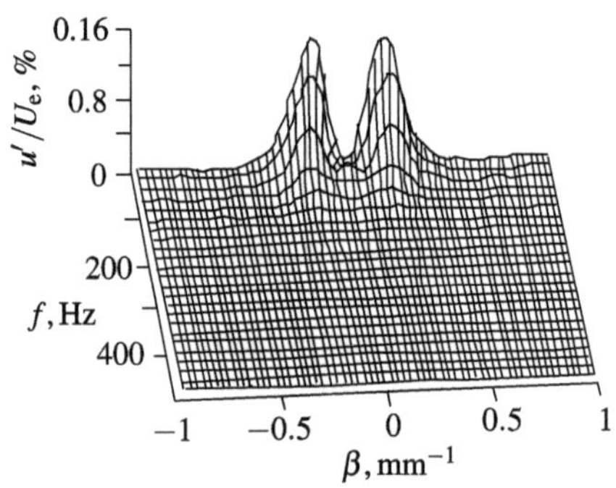

Fig. 9.10 Typical spectral content of the localized disturbance in a flat-plate boundary layer. Measurements at a subcritical Reynolds number close to the disturbance amplitude maximum (Westin et al, 1998)

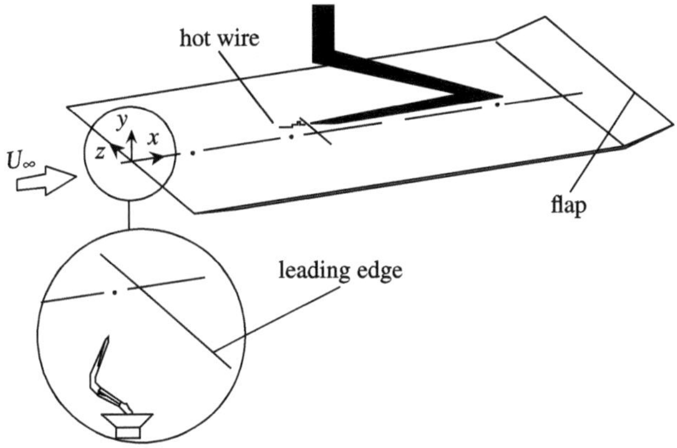

Fig. 9.11 Principal scheme of the experimental setup (Boiko et al, 2003)

Note that the disturbance growth shown in Fig. 9.12**a** cannot be compared with the numerical results, as the nonoptimal disturbances excited in the experiment cannot experience the optimal growth. Meanwhile, the dominant transverse wavenumbers as well as the amplitude and phase profiles of disturbances across the boundary layer are close to those of the optimal ones.

One problem concerning the concept of optimal disturbances is the dependence of the transverse scale of excited natural streaks on flow parameters, which exhibits a large scatter in different facilities. From dimensional considerations, it may be assumed that the transverse extent of the structures, as a rule, is close to the boundary-layer thickness. However, there is an experimental evidence of excitation of the

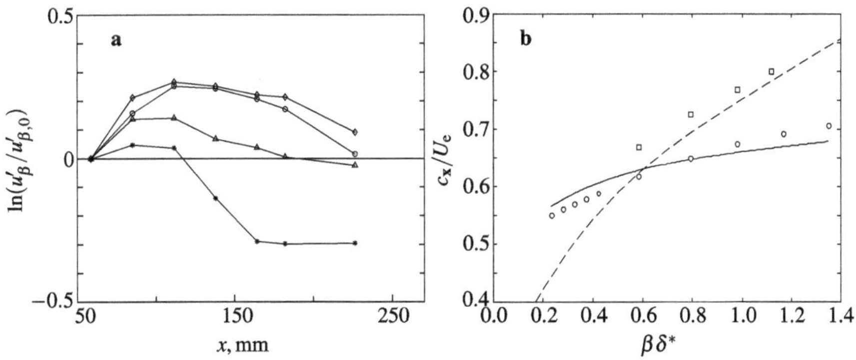

Fig. 9.12 Downstream behavior of normalized amplitudes u'_β of spectral components of the localized disturbance (**a**): $F = 26.9 \times 10^{-6}$; $\beta = 0.442$ (∗); 0.614 (△); 0.834 (○); 1.031 (◇) rad/mm; downstream propagation velocity of the disturbance as a function of the transverse wavenumber at $\mathrm{Re}_{\delta^*} = 420$ (**b**): experimental disturbances at $F = 26.9 \times 10^{-6}$ (○), 80.6×10^{-6} (□) vs. corresponding numerical results of Luchini (private communication) for $F = 26.9 \times 10^{-6}$ (*solid line*) and the data of linear stability calculations for $F = 80.6 \times 10^{-6}$ (*dashed line*)

boundary-layer perturbation at the same transverse scale as the free-stream vortices. The difference between the experimental data can be explained by some theoretical and numerical results, which show that the mechanism of scale selection is rather weak.

Exercises

9.1. Using the programs for solving the linear stability equations for the plane channel flow developed in Chap. 3 and function `Optimal` given in this chapter, reproduce Fig. 9.2.

9.2. Write programs to calculate velocities of the optimal disturbances for the plane channel flow and self-similar boundary layer. Reproduce the data given in Fig. 9.4.

9.3. Write programs to solve the linear stability equations for the plane channel flow and the self-similar boundary layer with respect to disturbances developing in space. Modify function `Optimal` to find $G(x)$ at $\mathrm{Re} < \mathrm{Re_L}$. Hint: remove all modes with $\alpha_\mathrm{i} > 0$ from consideration as they correspond to the waves propagating upstream (see, e.g., Tumin and Reshotko, 2001).

References

Boiko AV, Grek GR, Sboev DS (2003) Spectral analysis of localized disturbances in boundary layer at subcritical Reynolds numbers. Phys Fluids 15(12):3613–3624

Breuer KS, Haritonidis JH (1990) The evolution of a localized disturbance in a laminar boundary layer. Part 1. Weak disturbances. J Fluid Mech 220:569–594

Breuer KS, Kuraishi T (1994) Transient growth in two- and three-dimensional boundary layers. Phys Fluids A 6(6):1983–1993

Ellingsen T, Palm E (1975) Stability of linear flow. Phys Fluids 18(4):487–478

Farrell BF, Ioannou PJ (1993) Optimal excitation of three dimensional perturbations in viscous constant shear flow. Phys Fluids A 5(6):1390–1400

Grek GR, Kozlov VV, Ramazanov MP (1985) Three types of disturbances from the point source in the boundary layer. In: Kozlov VV (ed) Laminar–Turbulent Transition, Springer–Verlag, Berlin, IUTAM Symposium, pp 267–272

Gustavsson LH (1979) Initial-value problem for boundary layer flows. Phys Fluids 22(9):1602–1605

Gustavsson LH (1991) Energy growth of three-dimensional disturbances in plane Poiseuille flow. J Fluid Mech 224:241–260

Hanifi A, Schmid PJ, Henningson DS (1994) Transient growth in compressible boundary layer flow. In: Henningson DS (ed) Bypass Transition – Proceedings from a Mini-Workshop, Royal Institute of Technology, Stockholm, pp 51–65

Henningson DS, Schmid PJ (1994) A note on measures on disturbance size in spatially evolving flows. Phys Fluids A 6(8):2862–2864

Henningson DS, Gustavsson LH, Breuer KS (1994) Localized disturbances in parallel shear flows. Appl Sci Res 53:51–97

Hultgren LS, Gustavsson LH (1981) Algebraic growth of disturbances in a laminar boundary layer. Phys Fluids 24(6):1000–1004

Klingmann BGB (1991) Laminar–turbulent transition in plane Poiseuille flow. PhD thesis, Department of Gasdynamics, Stockholm

Klingmann BGB (1992) On transition due to three-dimensional disturbances in plane Poiseuille flow. J Fluid Mech 240:167–195

Luchini P (1996) Reynolds-number-independent instability of the boundary layer over a flat surface. J Fluid Mech 327:101–115

Luchini P (2000) Reynolds-number-independent instability of the boundary layer over a flat surface: Optimal perturbations. J Fluid Mech 404:289–309

Matsubara M, Kozlov VV, Alfredsson PH, Bakchinov AA, Westin KJA (1996) On flat plate boundary layer perturbations at high free stream turbulence level. In: Proc. Eighth Internat. Conf. on Methods of Aerophysical Research, RAS. Sib. Branch Inst. Theoret. Appl. Mech., Novosibirsk, vol 1, pp 174–179

Schmid PJ, Henningson DS (1994) Optimal energy density growth in Hagen-Poiseuille flow. J Fluid Mech 277:197–225

Schmid PJ, Kytomaa HK (1994) Transient and asymptotic stability of granular shear flow. J Fluid Mech 264:255–275

Tumin A, Reshotko E (2001) Spatial theory of optimal disturbances in boundary layers. Phys Fluids 13(7):2097–2104

Westin KJA, Boiko AV, Klingmann BGB, Kozlov VV, Alfredsson PH (1994) Experiments in a boundary layer subjected to free stream turbulence. Part 1. Boundary layer structure and receptivity. J Fluid Mech 281:193–218

Westin KJA, Bakchinov AA, Kozlov VV, Alfredsson PH (1998) Experiments on localized disturbances in a flat plate boundary layer. Part 1: The receptivity and evolution of a localized free stream disturbance. Eur J Mech B/Fluids 17(6):823–846

Further Reading

Andersson P, Berggren M, Henningson DS (1999) Optimal disturbances and bypass transition in boundary layers. Phys Fluids 11(1):134–150

Berlin S, Henningson DS (1994) A study of boundary layer receptivity to disturbances in the free stream. In: Henningson DS (ed) Bypass Transition – Proceedings from a Mini-Workshop, Royal Institute of Technology, Stockholm, pp 43–49

Boiko AV, Westin KJA, Klingmann BGB, Kozlov VV, Alfredsson PH (1994) Experiments in a boundary layer subjected to free stream turbulence. Part 2. The role of TS-waves in the transition process. J Fluid Mech 281:219–245

Butler KM, Farrell BF (1992) Three-dimensional perturbations in viscous shear flow. Phys Fluids A 4(8):1627–1650

Corbett P, Bottaro A (2000) Optimal perturbations for boundary layers subject to stream-wise pressure gradient. Phys Fluids 12(1):120–130

Henningson DS (1991) An eigenfunction expansion of localized disturbances. In: Johansson AV, Alfredsson PH (eds) Advances in Turbulence 3, Springer–Verlag, Berlin, pp 162–169

Herbert T, Lin N (1993) Studies of boundary-layer receptivity with parabolized stability equations. AIAA Paper 93–3053

Reddy SC, Schmid PJ, Henningson DS (1993) Pseudospectra of Orr-Sommerfeld operator. SIAM J Appl Math 53(1):15–47

Schmid PJ, Henningson DS (2000) Stability and transition in shear flows. Springer–Verlag, Berlin

Chapter 10
Excitation of shear flow disturbances

Turbulence in convectively unstable shear flows subjected to extrinsic dynamics results from amplification of their perturbations, which are generated by external disturbances and usually start to grow far upstream of the turbulent flow region. In the previous chapters, we considered consecutively the transitional events in far-field and near-field regions of disturbance sources and emphasized the importance of the regions for different laminar–turbulent transition scenarios. Now we concentrate on the disturbance excitation in shear layers. This process is referred to as 'receptivity' and is the main concern in this chapter.

10.1 Receptivity problem

In the linear stability theory, the laminar flow perturbations are described by the solution of an eigenvalue problem, which excludes their excitation from consideration. As a consequence, the primary question in the course of the receptivity investigation formulated for the first time by Morkovin (1968) is the search, selection, and experimental as well as theoretical modeling of those mechanisms that are most responsible for the formation of the near-wall disturbance field, followed by the transition to turbulence in a particular flow.

From the practical viewpoint, several kinds of instability modes generated in boundary layers are of primary interest. They are the Tollmien–Schlichting waves in two-dimensional flows, crossflow instabilities appearing as stationary vortices and traveling waves in three-dimensional configurations, as well as the Görtler instability vortices at concave walls. External acoustics, vortical disturbances in the form of localized flow modulations or free-stream turbulence, surface vibrations and roughness elements are those factors that most frequently contribute to the boundary-layer perturbations (Nishioka and Morkovin, 1986, Saric et al, 2002). A general sketch of the modern view for receptivity in the presence of both different types of external disturbance fields and different physical mechanisms of instability is given in Fig. 10.1.

A.V. Boiko et al., *Physics of Transitional Shear Flows*,
Fluid Mechanics and Its Applications 98, DOI 10.1007/978-94-007-2498-3_10,

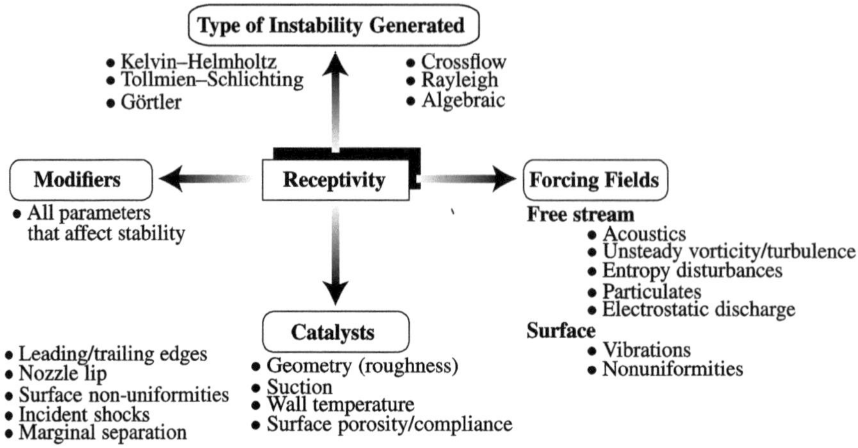

Fig. 10.1 Various aspects of the receptivity problem; based on the figure by Choudhari and Streett (1990)

The excitation process can be illustrated by a simple mechanical model of a forced harmonic oscillator with a slowly varying frequency, as given by Schmid and Henningson (2000):

$$\frac{d^2u}{dx^2} + \omega^2(x)u = \cos \Omega x \, .$$

At a constant frequency ω, the solution to this equation reads

$$u(x) = A\cos(\omega x + \phi) + B\cos\Omega x$$

where $B = 1/(\omega^2 - \Omega^2)$. The first term represents the homogeneous solution with the amplitude A and phase ϕ, and the second term is the particular solution due to forcing. In the case of slowly varying $\omega(x)$, A and ϕ also vary slowly with x. Let us assume that the initial amplitude of the homogeneous solution is $A(0) = 0$ and that there is certain initial mismatch between the frequencies $\omega(0) \neq \Omega$. Then the functions $A(x)$ and $\phi(x)$, where $x > 0$, can be found by marching as, e.g., in function `RecModel`, where a MATLAB initial value problem solver performs the job.

As can be seen in Fig. 10.2, the behavior of such a linear system is dominated initially by the particular solution, and the system oscillates with the frequency Ω. If the frequency is $\omega(x_0) = \Omega$ at a certain x_0, then a resonance occurs. At $x > x_0$, the amplitude of the particular solution decays to zero as $1/(\omega(x) - \Omega)$, but the amplitude $A(x)$ saturates slowly, and the system starts to oscillate with the 'natural' frequency ω.

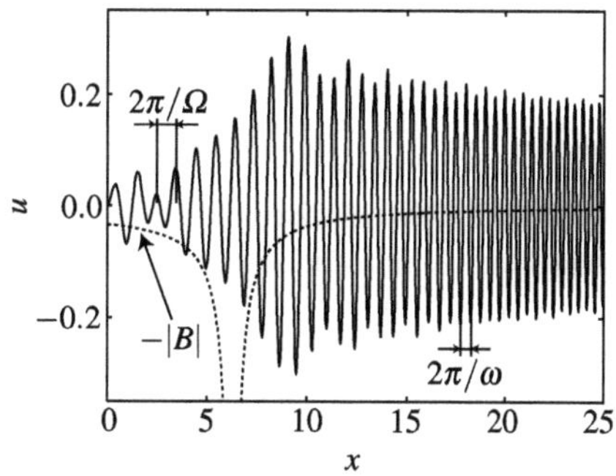

Fig. 10.2 Excitation of an 'eigenmode' of a forced quasi-harmonic oscillator: (*solid line*), full solution; (*dotted line*), negative amplitude of the particular solution. System parameters: $\omega = \pi + x/2$; $\Omega = 2\pi$

```
function [x,u,o]=RecModel(Omega)
% RECMODEL solves equation:
% u''+omega(x)^2*u=cos(Omega*x),
% where omega(x)=pi+x/2.

xspan=[0 25];
y0=[0;0];

[x,Y]=ode45(@(x,y) fcn(x,y,Omega),xspan,y0);
u=Y(:,1);
o=omega(x);
% -------------------------------------------------------
function dy=fcn(x,y,Omega)

dy=[y(2); cos(Omega*x)-omega(x).^2.*y(1)];
% -------------------------------------------------------
function o=omega(x)

o=pi+x/2;
```

10.2 Localized and distributed generation of laminar flow disturbances

For the receptivity problem, the linearized Navier–Stokes equations play the role of a spatial-temporal hydrodynamic resonator system with external forces as the driving right-hand-side term and inhomogeneous boundary conditions.

Two limit cases of excitation of the boundary-layer perturbations can be distinguished, that is, their distributed generation in an extended area and localized generation by a concentrated impact on the boundary layer. In practice, distributed excitation of, e.g., Tollmien–Schlichting waves means their continuous origination in a flow region of at least several wavelengths in the streamwise direction. Localized excitation occurs in a smaller domain (less than a wavelength), the disturbances developing downstream in the absence of the generator.

For effective excitation of the instability wave, forcing is needed, which should have not only the same frequency, but also comparable spatial scales. If the perturbations are introduced from inside the near-wall region (e.g., by mechanical vibration, oscillatory air blowing, or periodic heating of surface strips), these scales are easily matched. Otherwise, the free-stream disturbances (acoustical and vortical) are controlled by inviscid dynamics and their wavelengths in subsonic flows are quite different from that of the instability waves. Hence, for the transformation of some classes of external-flow perturbations to boundary-layer oscillations, there should be a mechanism of wavelength reduction. As the two-dimensional quasi-parallel stationary boundary layer has no short scales necessary for the effective process of wavelength reduction, the transformation can be observed when the basic flow experiences 'rapid' changes in the region of generation. This mechanism of wavelength reduction is called the 'scattering' of external disturbances at a boundary-layer nonuniformity.

Mean-flow variations contributing to the receptivity process are associated with pressure gradients, drastic changes in the boundary conditions caused by surface roughness, porosity, suction ports, etc. Local stationary distortions of the surface geometry or of the pressure field result in the basic flow nonuniformity inside the boundary layer possessing a broad wave spectrum. Time-dependent external perturbations modulate this stationary disturbance, resulting in a spectrum of traveling forced waves, which resonantly excite modes matched with the instability waves of interest (Goldstein and Hultgren, 1989). Generally, more localized boundary-layer variations cause an increase of the energy scattered to the short-wave part of the spectrum, which leads to higher amplitudes of the excited boundary-layer oscillations. An opposite example of effective distributed generation of the instability waves is a boundary layer with an extended region of small-scale distortions such as wall waviness with its spacing tuned to the wavelength of a particular instability mode.

10.3 Methodology of receptivity studies

The mechanisms of generation of shear flow disturbances can be characterized by the response of the boundary layer to the perturbations imposed on it, i.e., by a receptivity function, which should be determined in the region or at points of interest in a parameter space of the boundary-layer instability. For localized modal excitation (e.g., for the Tollmien–Schlichting or crossflow instability), the problem reduces to finding the receptivity function at a fixed Reynolds number as a complex ratio of the amplitude of the external harmonic disturbance and the 'initial' amplitude of the instability wave, with the phase lag between them characterizing the 'delay' in the transformation.

10.3.1 Theoretical approaches

In theoretical studies, the shear layer receptivity is investigated in different ways including numerical integration of the complete two-dimensional Navier–Stokes equations with vortical or sound forcing terms, application of asymptotic methods, and perturbation theory. Details of the available approaches and their limitations are discussed in a number of original studies and reviewing papers.

A straightforward and physically justified approach describing the generation of linear instability waves (such as the Tollmien–Schlichting and crossflow ones) in parallel and quasi-parallel flows is based on a consideration of adjoint equations of a corresponding linear problem.

Let us start with a simple guiding example. As known, a set of trigonometric functions $\phi_\alpha(x) = \exp(\mathrm{i}\alpha x)$, where α is integer, is an orthogonal system over the interval $x \in [0; 2\pi]$, that is,

$$\int_0^{2\pi} \phi_k(x)^* \phi_l(x) \mathrm{d}x = 2\pi \delta_{kl} ,$$

where δ_{kl} is the Kronecker delta and the asterisk denotes the complex conjugate. Moreover, the set of the functions is complete in the sense that any integrable complex-valued function $h(x)$ defined on $[0; 2\pi]$ can be decomposed into its constituent harmonics. Thus, we can write the formal expansion of $h(x)$ in terms of the Fourier orthogonal system

$$h(x) = \sum_{\alpha=-\infty}^{-\infty} \hat{h}_\alpha \phi_\alpha ,$$

where the coefficients of the decomposition or the Fourier coefficients are

$$\hat{h}_\alpha = \int_0^{2\pi} \phi_\alpha(x)^\dagger h(x) \mathrm{d}x = \int_0^{2\pi} \mathrm{e}^{-\mathrm{i}\alpha x} h(x) \mathrm{d}x , \quad \alpha = 0, \pm 1, \pm 2, \dots ,$$

where $\phi_\alpha(x)^\dagger = \phi_\alpha(x)^*$ is the function adjoint to $\phi_\alpha(x)$. If we now assume that the norm of $h(x)$ is unity, i.e., $h(x)$ has been somehow scaled, for example, using the scalar product to ensure that

$$\int_0^{2\pi} h(x)^* h(x) \mathrm{d}x = 1 ,$$

then the relative content of a particular harmonic is given by 'dimensionless' coefficients $\hat{h}_\alpha$. For functions, that differ from $h(x)$ only by a constant, the coefficients differ also by the same constant.

We know that the set of the normal modes constituting the solutions to the linearized Navier–Stokes equations is also complete (see Sect. 2.6 for details). Hence, if we find a method of the modal decomposition of the initial disturbance, then the coefficient of decomposition related to a particular mode (usually to the leading one) normalized by the norm of the disturbance provides the dimensionless receptivity coefficient for the mode. The set of the coefficients for the modes in a certain range of governing parameters (e.g., frequency) gives the required receptivity function. The technical difficulty in determining the coefficients is related to the absence of a simple orthogonality condition for the modes. In contrast to the trigonometric functions, the modes under consideration are bi-orthogonal. As a consequence, derivation of the receptivity coefficients requires a more involved procedure.

Let us turn to an inhomogeneous system of the linearized hydrodynamic equations, assuming harmonic sources of momenta and mass independent of z, as well as accepting inhomogeneous boundary conditions (Hill, 1995):[1]

$$
\begin{aligned}
-\mathrm{i}\omega\bar{\mathbf{u}} + \mathscr{L}\bar{\mathbf{u}} + \nabla\bar{p} &= \mathbf{g}(x,y;\omega)\,, \\
\nabla\cdot\bar{\mathbf{u}} &= \varphi(x,y;\omega)\,, \\
\bar{\mathbf{u}}(y_b) &= \mathbf{u}_b(x;\omega)\,,
\end{aligned}
$$

where y_b is the coordinate of one of the two boundaries (equal to either y_1 or y_2), any of which can be infinity (free stream). The sources are assumed to be localized in x, i.e., there exist such coordinates x_1 and x_2 that $x_2 > x_1$ and $\mathbf{g} = \varphi = \mathbf{u}_b = 0$ at $x < x_1$ and $x > x_2$. In the general form, the solution of the inhomogeneous equations can be written as $\mathbf{q} = \bar{\mathbf{q}}\exp(-\mathrm{i}\omega t)$, where $\bar{\mathbf{q}} = (\bar{\mathbf{u}}, \bar{p})^T$. Recalling that the Lagrange identity defined in Sect. 2.5 is valid for any suitably differentiable fields $\mathbf{q}$ and $\mathbf{q}^\dagger$, we choose $\mathbf{q}^\dagger = \hat{\mathbf{q}}^\dagger \mathrm{e}^{-\mathrm{i}(\alpha x-\omega t)}$, where $\hat{\mathbf{q}}^\dagger = (\hat{\mathbf{u}}^\dagger, \hat{p}^\dagger)^T$, to be adjoint to the normal mode, whose amplitude we wish to determine. This yields

$$
(\hat{\mathbf{u}}^\dagger \mathrm{e}^{-\mathrm{i}\alpha x})\cdot\mathbf{g} + (\hat{p}^\dagger \mathrm{e}^{-\mathrm{i}\alpha x})\varphi = \nabla\cdot\mathbf{j}(\hat{\mathbf{q}}^\dagger \mathrm{e}^{-\mathrm{i}\alpha x}, \mathbf{q})\,,
$$

where $\mathbf{j}$ is defined by (2.14). If the source is present only at one of the boundaries, say, y_1, the right-hand side term can be further itemized. In this case, from direct integration over y between the boundaries it follows that the wall-normal component of $\mathbf{j}$ at $y = y_1$ is

$$
(\mathscr{S}^\dagger \mathrm{e}^{-\mathrm{i}\alpha x})\cdot\mathbf{u}_b\,,
$$

where

$$
\mathscr{S}^\dagger = \left(\frac{1}{\mathrm{Re}}\frac{\partial\hat{u}^\dagger}{\partial y}\mathbf{e_x} + \hat{p}^\dagger\mathbf{e_y}\right)_{y=y_b}
$$

is called the adjoint stress ($\mathbf{e_x}$ and $\mathbf{e_y}$ are the unit vectors in the directions x and y, respectively). Finally, integrating between x_1 and x_2, the normalized amplitude of the mode excited by the sources is obtained

$$
A = \int_{x_1}^{x_2}\int_{y_1}^{y_2} \mathbf{g}\hat{\mathbf{u}}^\dagger \mathrm{e}^{-\mathrm{i}\alpha x}\mathrm{d}y\mathrm{d}x + \int_{x_1}^{x_2}\int_{y_1}^{y_2} \varphi\hat{p}^\dagger \mathrm{e}^{-\mathrm{i}\alpha x}\mathrm{d}y\mathrm{d}x + \int_{x_1}^{x_2} \mathbf{u}_b\mathscr{S}^\dagger \mathrm{e}^{-\mathrm{i}\alpha x}\mathrm{d}x\,.
$$

Such a receptivity analysis provides a clear and direct tool for solving a wide class of receptivity problems suitable for modal analysis based on the linearized Navier–Stokes equations. In this context, the adjoint eigensolutions to the equations define the efficiency with which a particular unit-power forcing excites the corresponding normal modes. In what follows, we outline some particular disturbance sources in more detail.

---

[1] Oblique disturbances can be considered in the same way.

10.3.1.1 Vibrating ribbon

A problem that can be treated with the last equation is the generation of boundary-layer waves by a vibrating ribbon placed near the wall and oscillating at a given frequency. It is approximated by the boundary-layer source of wall-normal velocity of unit amplitude $\mathbf{u}_b = (0, v_b)^T$, where

$$v_b(x,t) = \delta(x)\mathrm{e}^{-\mathrm{i}\omega t} ,$$

and $\delta(x)$ is the Dirac delta-function. It follows directly that the excitation amplitude is

$$A = \int_{x_1}^{x_2} \mathbf{u}_b \mathscr{S}^{\dagger} \mathrm{e}^{-\mathrm{i}\alpha x} \mathrm{d}x = \hat{p}^{\dagger}(0) .$$

10.3.1.2 Periodic blowing-suction at the wall through a transverse slot

The problem is similar to the previous one, but now we have

$$v_b(x,t) = v_{\mathrm{w}}(x)\mathrm{e}^{-\mathrm{i}\omega t}, \quad -d/2 \le x \le d/2 ,$$

where d is the dimensional slot width. After renormalization with the length scale d, the response is given by

$$A = \hat{p}^{\dagger}(0)\frac{\mathrm{Re}_d}{\mathrm{Re}_l}\int_{-1/2}^{1/2} v_{\mathrm{w}}(\chi)\exp\left(-\mathrm{i}\alpha\frac{d}{l}\chi\right)\mathrm{d}\chi ,$$

where l is the original scale used to normalize the linearized Navier–Stokes equations.

10.3.1.3 Wall oscillations

Let us suppose that a wall section oscillates about y_1 with a small velocity $\mathbf{u}_d(x,\omega)$. Then, the boundary displacement is given by $\mathrm{i}\mathbf{u}_d(x,\omega)/\omega$. Linearization of the boundary conditions in a way similar to that for the boundary conditions in considering the compliant coatings in Sect. 6.7 yields

$$\mathbf{u}_b = \mathbf{u}_d - \mathrm{i}\frac{\mathbf{u}_d}{\omega}(\nabla \cdot \mathbf{V})_{y=y_1}$$

or, componentwise,

$$u_b = u_d - \mathrm{i}\frac{u_d}{\omega}\left(\frac{\mathrm{d}U}{\mathrm{d}y}\right)_{y=y_1} ,$$
$$v_b = v_d .$$

Thus, we obtain

$$\int_{x_1}^{x_2} \mathbf{u}_b \mathscr{S}^{\dagger} \mathrm{e}^{-\mathrm{i}\alpha x} \mathrm{d}x = \int_{x_1}^{x_2} \left[\frac{\hat{u}_d}{\mathrm{Re}}\frac{\partial \hat{u}\dagger}{\partial y} + \hat{v}_d\left(\hat{p}\dagger - \frac{\mathrm{i}}{\omega \mathrm{Re}}\frac{\mathrm{d}U}{\mathrm{d}y}\frac{\partial \hat{u}\dagger}{\partial y}\right)\right]_{y=y_1} \mathrm{e}^{-\mathrm{i}\alpha x}\mathrm{d}x .$$

10.3.1.4 Scattering of a free-stream acoustic wave by surface roughness

The effect of a stationary roughness element with a shape $h(x) \ll 1$ ($h(x) = 0$ at $x > x_2$ and $x < x_1$) on scattering of a free-stream acoustic wave given by a planar fluctuation $\sim \exp(-\mathrm{i}\omega t)$ is considered. In this case, $y_1 = 0$ and $y_2 = \infty$. Thus, the Stokes boundary layer with the basic flow velocity profile

$$u(y) = 1 - \exp\left[-\sqrt{\omega \mathrm{Re}/2}(1-\mathrm{i})y\right]$$

is superimposed on the steady boundary layer with the streamwise $U(y)$ and wall-normal $V(y)$ basic flow velocities. In the presence of roughness, the modulated boundary layer and pressure field are distorted. After evaluation of corresponding integrals, the normalized amplitude of the instability mode can be expressed as

$$A = -\frac{1}{\mathrm{Re}}\left(\frac{\mathrm{d}U}{\mathrm{d}y}\frac{\mathrm{d}U^{\dagger}}{\mathrm{d}y} + \frac{\mathrm{d}u}{\mathrm{d}y}\frac{\mathrm{d}\hat{u}^{\dagger}}{\mathrm{d}y}\right)_{y=0}\int_{x_1}^{x_2} h(x)\mathrm{e}^{-\mathrm{i}\alpha x}\mathrm{d}x\,.$$

Here $U^{\dagger}$ is the solution of the auxiliary adjoint problem

$$\mathscr{L}[\mathbf{V}^{\dagger}(y)\mathrm{e}^{-\mathrm{i}\alpha x}] + \nabla[P(y)^{\dagger}\mathrm{e}^{-\mathrm{i}\alpha x}] = \mathbf{G}^{\dagger}(y)\mathrm{e}^{-\mathrm{i}\alpha x}\,,$$
$$\nabla\cdot[\mathbf{V}^{\dagger}(y)\mathrm{e}^{-\mathrm{i}\alpha x}] = 0\,,$$

where

$$\mathbf{G}^{\dagger}(y) = \left(\mathrm{i}\alpha\hat{u}^{\dagger}u\,, \mathrm{i}\alpha\hat{v}^{\dagger}u + \hat{u}^{\dagger}\frac{\mathrm{d}u}{\mathrm{d}y}\right)^{T}\,,$$

and the corresponding boundary conditions read

$$y = 0 : \mathbf{V} = 0 \quad \text{and} \quad y \to \infty : \mathbf{V} = 0\,.$$

10.3.2 *Experimental strategy*

To achieve the goals of a receptivity experiment, i.e., to find, select, and model a particular receptivity mechanism, it is reasonable to consider such situations that allow one to 'purify' an effect under consideration and reveal the underlying physical mechanisms governing the process. This can be obtained through a choice of simplified flow geometries and eliminating possible 'side effects,' e.g., by noise reduction along with subjecting the flow to controllable disturbances. That is why the idea of experimental modeling, i.e., performing an experiment under controlled conditions, appears to be a powerful tool for receptivity investigations. In contrast to the data obtained in a 'natural' environment, the results can be directly compared with calculations without any significant assumptions about the physical nature of the disturbances under investigation.

For modal excitation, the receptivity functions are considered as the main outcome of such experiments. To obtain the receptivity functions in the case of localized excitation, one needs to determine the initial forced field of disturbances and to measure or estimate the field of the excited normal modes at the position of the source. Corresponding experimental data and numerical results are discussed in more detail in the following sections.

10.4 Receptivity of two-dimensional boundary layers

In the case of viscous boundary-layer instability, e.g., of the flow over a flat plate, the receptivity to free-stream disturbances is dominated by the mechanism of wavelengths reduction. The transformation of external perturbations to the Tollmien–Schlichting waves occurs preferably alongside with sudden variations of the basic flow. In what follows, we begin with this path of receptivity.

10.4.1 Leading-edge receptivity

A strong coupling between external forcing, i.e., acoustic and vortical oscillations, and the boundary-layer instability is found in the leading-edge region of a flat plate. The transformation of harmonic two-dimensional vortical disturbances of the oncoming flow into a two-dimensional Tollmien–Schlichting wave is illustrated by wind-tunnel data in Fig. 10.3. For convenience, the amplitude and phase isolines of the velocity perturbations are presented mirrored with respect to the x-axis.

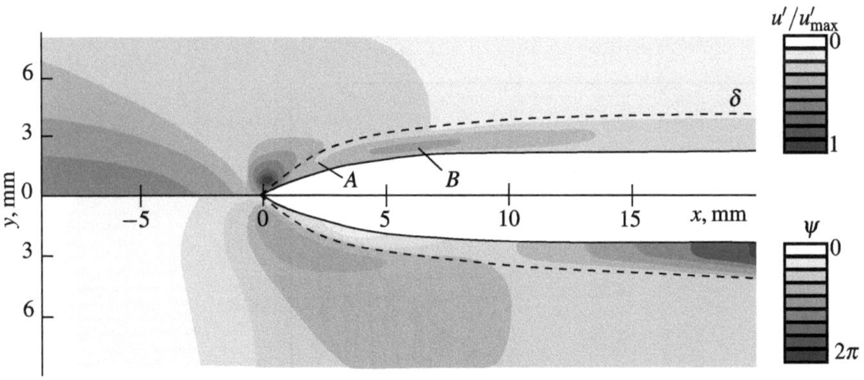

Fig. 10.3 Amplitude and phase isolines of streamwise disturbance velocity about the leading edge of a flat plate (Kachanov et al, 1979)

The size of the generation region is characterized by the distance between the leading edge and the disturbance minimum close to the wall (point A in Fig. 10.3). Downstream of this point, a wavy perturbation of the boundary layer with growing amplitude is observed. Starting from point B, the excited wave decays at subcritical Reynolds numbers, and its propagation velocity stabilizes. In this example, the measure of the excitation effectiveness is the ratio of the streamwise velocity of the perturbations in the leading-edge region and in the external flow. At rather weak forcing, this ratio does not change with the amplitude of oscillations, which means linearity of the disturbance transformation. However, the receptivity depends on the excitation frequency and the basic flow velocity.

As is found experimentally and numerically, the most important for the Tollmien–Schlichting wave generation in the present configuration is the wall-normal velocity v' of the external disturbances. The perturbed motion streamlines converge in front of the leading edge with an increase of the wall-normal velocity and a sharp change of the streamline direction. As a result, the wall-normal disturbance velocity v' 'transforms' to the streamwise one u', and a localized vortex is generated, as is seen in Fig. 10.3. At the same time, the streamwise velocity of the external oscillations does not make any significant contribution to the receptivity.

10.4.2 Excitation of instability waves at local boundary-layer nonuniformities

Local stationary distortions of a developed boundary layer are important agents of receptivity, not only because of their general occurrence, but also due to their possible proximity to the instability region. In contrast to the Tollmien–Schlichting waves originating near the leading edge, which are subject to strong decay before they start to grow, the disturbances generated close to the critical Reynolds number can exhibit immediate downstream amplification.

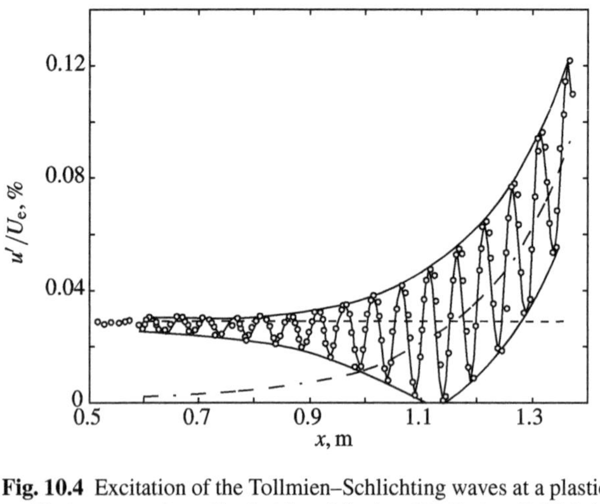

Fig. 10.4 Excitation of the Tollmien–Schlichting waves at a plastic strip on a flat plate by acoustic oscillations: sound amplitude (*dotted line*) and Tollmien–Schlichting wave amplitude (*dash-dotted line*) separated from the hot-wire signal (∘), the strip position is $x = 0.565$ m (Aizin and Polyakov, 1979)

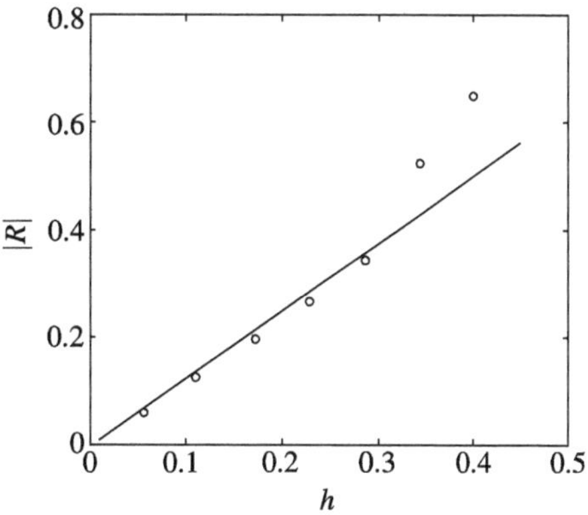

Fig. 10.5 Variation of acoustic receptivity amplitude $|R|$ with the rectangular bump height h at $F = 49.34 \times 10^{-6}$, $\mathrm{Re}_{\delta^*} = 582$, bump width $d = 31.6$: theoretical results (*solid line*) of Crouch (1992) and experimental data (○) of Saric et al (1991)

An example of the localized boundary-layer receptivity at a two-dimensional roughness element on a flat plate to streamwise acoustic waves is given in Fig. 10.4. When traversing a hot-wire probe along the boundary layer, characteristic beatings of the anemometer signal caused by interference of the excited vorticity perturbations with the acoustic oscillations are observed. The amplifying Tollmien–Schlichting wave is detected at a short distance (about one wavelength) downstream of the roughness element. In this case, with micron-size nonuniformities of the plate surface and a boundary-layer thickness of several millimeters, the coefficient of the transformation of acoustic disturbances into the Tollmien–Schlichting waves is about 10^{-2}.

In similar conditions, the receptivity coefficient grows with the roughness height, and the Tollmien–Schlichting wave generation can be described in a linear approximation until the basic flow disturbance produced by the surface variation exceeds a certain threshold value. A comparison of theoretical and experimental data for the acoustic receptivity amplitude $|R|$ on a two-dimensional bump on flat-plate surface as a function of the dimensionless height h is shown in Fig. 10.5. As seen, the agreement is very good for $h \lesssim 0.3$. As the the surface height irregularity becomes 'large,' the linear analysis fails to capture the experimentally observed behavior.

10.4.3 Receptivity to localized unsteady disturbances

When a developed boundary layer is forced locally by a non-stationary perturbation, its broad wave spectrum contains scales that are comparable with characteristic instability wavelengths. In this case, the mechanism of wavelength reduction on local nonuniformities of the basic flow is not necessarily required for excitation of normal modes. Through experimental and theoretical studies, the boundary-layer receptivity to different types of perturbations both external and originating at the wall was examined. Apparently, the more concentrated forcing, the wider its wave-number spectrum, facilitating excitation of instability waves.

Figure 10.6 shows the boundary-layer response to two-dimensional periodic in time disturbances generated locally above a flat plate by a vibrating ribbon. The amplitude distributions $u'(y)$ of the vorticity oscillations measured in the near-wall region during the excitation are compared with those of the Tollmien–Schlichting wave induced with the same technique from inside the boundary layer. In the upstream sections, the amplitude profiles of the boundary-layer perturbation generated by external forcing is superposed on the peripheral part of the vortex street behind the vibrating ribbon. The vortex street decays whilst propagating downstream, most significantly near the surface, so that the Tollmien–Schlichting wave that follows the boundary-layer stability characteristics is found in the right-hand side of Fig. 10.6. In the present case, the boundary layer is receptive to the strong nonuniformity of external disturbances along the streamwise coordinate in the vicinity of the ribbon.

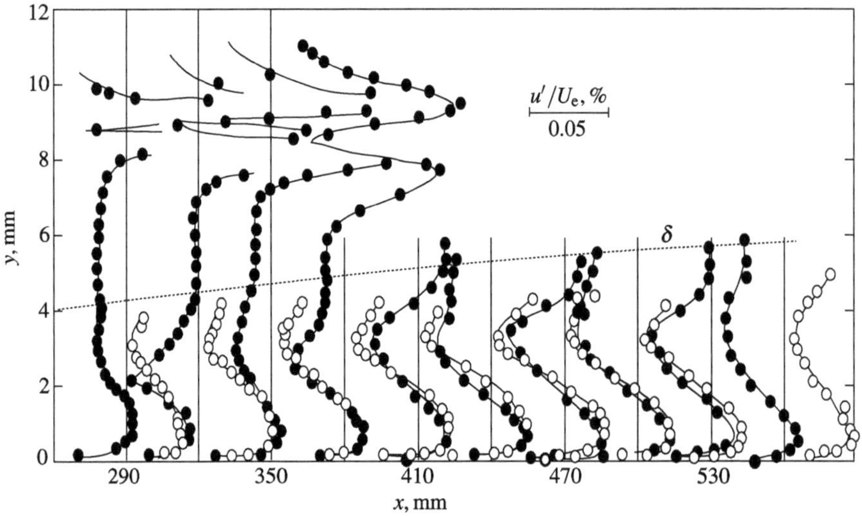

Fig. 10.6 Harmonic perturbations generated by a vibrating ribbon from outside (•) and inside (○) of the flat-plate boundary layer; the ribbon is positioned at $x = 251$ mm (Dovgal et al, 1980)

Otherwise, there is a number of possibilities of exciting the shear flow locally at the body surface, e.g., by vibration, periodic heating, and suction and blowing. The boundary-layer receptivity to oscillations generated by a three-dimensional vibrating membrane on a flat plate is illustrated in Fig. 10.7. The ratio of the initial wavenumber spectrum of the Tollmien–Schlichting waves to the corresponding resonant spectrum of the surface vibrations gives the receptivity function $R = |R|e^{i\varphi}$.

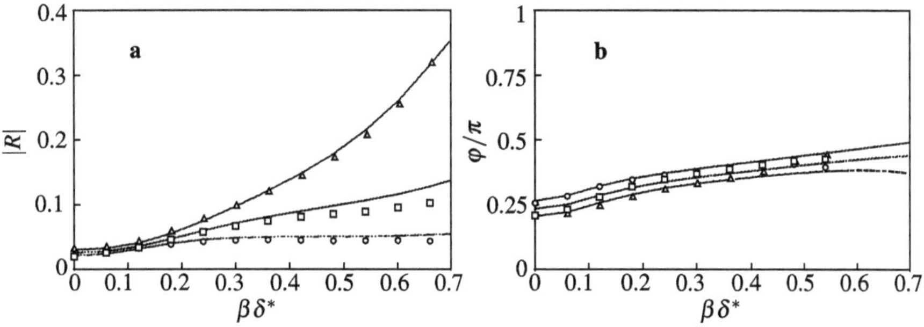

Fig. 10.7 Amplitude (**a**) and phase (**b**) of the vibration receptivity function for the Blasius boundary layer depending on the transverse wavenumber $\beta\delta^*$ and the frequency (○, □, △) of oscillations: experiments (*symbols*) and calculations (*lines*) (Ivanov et al, 1998)

10.4.4 Excitation of instability waves at laminar boundary-layer separation

Depending on the basic flow configuration, one can distinguish between two paths of receptivity of convectively unstable separated flows. These are the transformation of oscillations generated somehow in the upstream boundary layer into the separated flow perturbations and excitation of instability waves just at separation.

The first one is expected at a slow streamwise variation of the basic flow characteristics preventing a strong coupling between environmental perturbations, such as free-stream turbulence and acoustic waves, and the separated layer instability. An example is excitation of a boundary layer separating in the adverse pressure gradient region over an airfoil, Fig. 10.8. In this case, shear layer disturbances originating at the minimum static pressure along the wing chord grow to the largest amplitude before entering the separation bubble with the main contribution to the laminar flow perturbations. At the same time, generation of instability waves is not found or, at most, is negligible in the vicinity of the separation region.

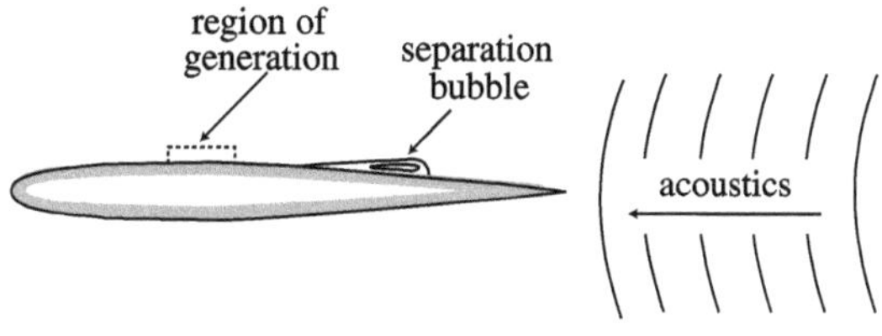

Fig. 10.8 Acoustic excitation of instability waves at the separation on an airfoil (Dovgal and Kozlov, 1983)

The second path applies to separated flows induced by local basic flow variations, which provide a mechanism of wavelength reduction, binding the separation with the receptivity process. This can happen when the surface shape changes rapidly

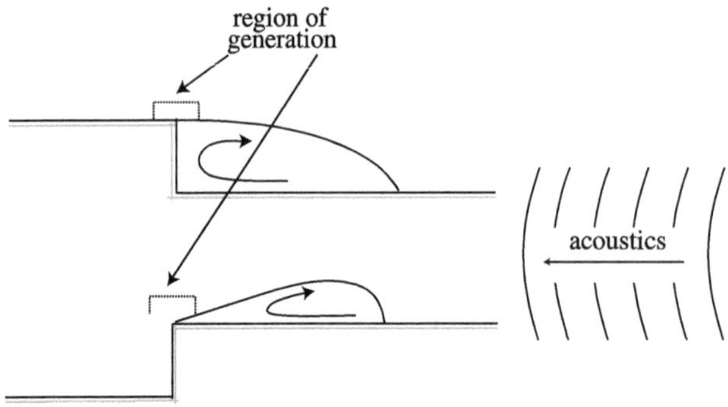

Fig. 10.9 Acoustic excitation of instability waves at the separation induced by two-dimensional steps on a flat plate (Boiko et al, 1990)

(e.g., on two-dimensional steps on a smooth wall), as is sketched in Fig. 10.9. The instability waves are generated close to the point of separation, while they evolve downstream due to instability of separation bubbles only. In such conditions, the initial amplitudes of the excited disturbances are strongly dependent on the size and shape of surface variations (Asai and Kaneko, 1998, Bodonyi et al, 1989). Furthermore, along with the perturbations excited in proximity to the separation point, the boundary-layer oscillations coming from upstream may contribute to the laminar flow breakdown as an interplay between two receptivity paths described above.

10.5 Receptivity of a swept-wing boundary layer

At low sweep angles, Tollmien–Schlichting instability is usually dominating in wing boundary layers. Starting from sweep angles of 30–40° and at rather low Reynolds numbers, the crossflow instability modes develop in the form of streamwise vortices and traveling waves. As is found in experimental studies, the stationary crossflow vortices are more pronounced at a low turbulence level due to a relatively large amplitude of the surface roughness seeding the vortices (Bippes, 1999). The contribution of traveling waves to boundary-layer perturbations gets higher with the growth of the oncoming-flow turbulence. Since the crossflow instability modes are governed by the inviscid mechanism, their receptivity characteristics are quite different from that for the Tollmien–Schlichting waves. Moreover, stationary boundary or flow nonuniformities have scales necessary for effective excitation of stationary crossflow vortices, and an additional mechanism of scale reduction is not needed, which is in contrast to excitation of traveling waves.

The generation of the crossflow instability modes in swept-wing boundary layers was in focus of extensive experimental and theoretical studies (Kachanov, 2000). In particular, much has been done on the receptivity to laminar flow perturbations originating at the wall. Research data reported in original and reviewing papers clarify

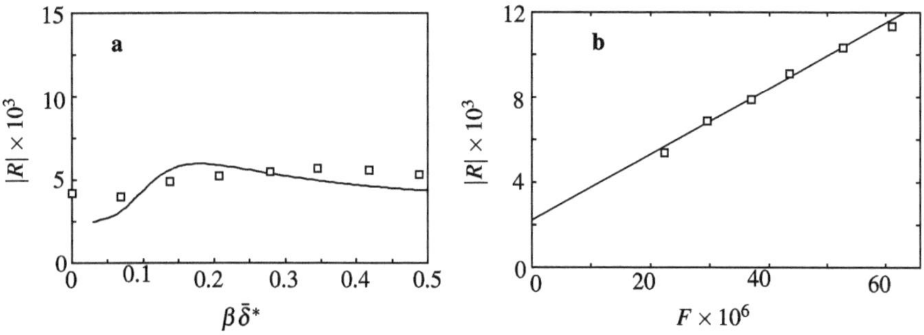

Fig. 10.10 Vibration receptivity coefficients for a swept-wing boundary layer at sweep angle $\chi = 25°$, $\mathrm{Re}_{\bar{\delta}^*} = 498$: as a function of spanwise wavenumber at $F = 22.2 \times 10^{-6}$; experimental (□) and theoretical (*solid line*) data (**a**) and as a function of frequency of vibrations at $\beta\bar{\delta}^* = 0.278$; experimental (□) and linearly approximated (*solid line*) data (**b**). The displacement thickness $\bar{\delta}^*$ was calculated along the local external-flow streamline (Gaponenko et al, 2002)

the contribution to the receptivity of the surface roughness and mean-flow disturbances induced by flow suction.

In experiments of Gaponenko et al (2002), the complex coefficients of the boundary-layer receptivity to local surface vibrations were obtained as functions of the disturbance frequency and the transverse wavenumber. A quantitative comparison of the wind-tunnel data with the results of the linear receptivity theory was also performed. The absolute values of the vibration receptivity function are shown in Fig. 10.10**a**. The theory is able to predict correctly the initial spectrum of the crossflow waves including the most unstable modes, which are close in this case to $\beta\bar{\delta}^* = 0.4$, where $\bar{\delta}^*$ is the boundary-layer displacement thickness at the position of the vibrator. The experimentally obtained dependence of the receptivity amplitude on the oscillation frequency is shown in Fig. 10.10**b** for a fixed spanwise wavenumber, where the receptivity coefficient is extrapolated to surface roughness at $F = 0$. It appeared that the linear receptivity theory predicts correctly the initial amplitudes of the stationary crossflow vortices in a wide range of spanwise wavenumbers including those of the most amplifying disturbances. As seen, both in the theory and experiment, the roughness receptivity coefficients turned out to be significantly lower than vibration receptivity coefficients.

10.6 Excitation of the Görtler vortices

The Görtler vortices at a concave wall—except for short wavelength ones—evolve over the same streamwise length scale as that of the undisturbed boundary layer, i.e., the nonparallel flow effects are large. Consequently, the local linear stability analysis based on the quasi-parallel-flow approximation is invalid in this case. To solve the problem for long-wavelength vortices, one has to take into account terms that are dependent on both wall-normal and streamwise coordinates as it is discussed in Sect. 5.2.3. Being essentially an initial-value problem, the instability analysis should be always accompanied by consideration of receptivity.

Below we outline a general approach to the problem using the Lagrange identity for parabolic systems. For a parabolic equation given by a positive-definite operator $\mathscr{L}$ as

$$\mathbf{q}_x = \mathscr{L}\mathbf{q}\,,$$

the adjoint operator $\mathscr{L}^\dagger$ is introduced so that

$$\mathbf{q}^\dagger_x = -\mathscr{L}^\dagger\mathbf{q}^\dagger\,,$$

with the property

$$(\mathbf{q}^\dagger\cdot\mathbf{q})_x = 0\,.$$

The Görtler equations can be written in a similar form as

$$(\mathscr{M}\bar{\mathbf{q}})_x = \mathscr{L}\bar{\mathbf{q}}\,,$$

where

$$\mathscr{M} = \begin{bmatrix} I & 0 & 0 & 0 \\ U & 0 & 0 & 0 \\ V & U & 0 & 0 \\ 0 & 0 & U & 0 \end{bmatrix}\,, \quad \mathscr{L} = \begin{bmatrix} 0 & -\mathcal{D} & I & 0 \\ \check{\Delta}-V_y & -U_y - U\mathcal{D} & 0 & 0 \\ -2\text{Gö}^2 U & \check{\Delta}-2V_y-2V\mathcal{D} & V & -\mathcal{D} \\ 0 & 0 & \check{\Delta} & -\beta^2 \end{bmatrix}\,,$$

$$\check{\Delta} = \mathcal{D}^2 - \beta^2 + \mathrm{i}\omega\,, \quad \bar{\mathbf{q}}(x,y) = (\bar{u}(x,y), \bar{v}(x,y), \tilde{w}(x,y), \bar{p}(x,y))^T\,, \quad \tilde{w} = \mathrm{i}\beta\bar{w}\,,$$

and the boundary conditions $\bar{u} = \bar{v} = \tilde{w} = 0$ at $y = 0$ and $y \to +\infty$. Using a suitably differentiable field

$$\bar{\mathbf{q}}^\dagger(x,y) = (\bar{u}^\dagger(x,y), \bar{v}^\dagger(x,y), \tilde{w}^\dagger(x,y), \bar{p}^\dagger(x,y))^T\,,$$

the following Lagrange identity can be derived:

$$\begin{aligned} \bar{\mathbf{q}}^\dagger\cdot\frac{\partial(\mathscr{L}\bar{\mathbf{q}})}{\partial x} - \bar{\mathbf{q}}^\dagger\cdot\mathscr{M}\bar{\mathbf{q}} + \bar{\mathbf{q}}\cdot\frac{\partial\left(\mathscr{L}^\dagger\bar{\mathbf{q}}^\dagger\right)}{\partial x} - \bar{\mathbf{q}}\cdot\mathscr{M}^\dagger\bar{\mathbf{q}}^\dagger \\ = \big(\bar{u}^\dagger\bar{v} + \bar{v}^\dagger V\bar{u} + \bar{v}^\dagger\bar{u}_y + \bar{v}^\dagger_y\bar{u} + 2\tilde{w}^\dagger V\bar{v} + \tilde{w}^\dagger\bar{v}_y \\ + \tilde{w}^\dagger\bar{p} + \tilde{w}^\dagger_y\bar{v} + \bar{p}^\dagger V\tilde{w} - \bar{p}^\dagger\tilde{w}_y + \bar{p}^\dagger_y\tilde{w}\big)_{y=0}^{y=\infty}\,. \end{aligned}$$

Here, the last two lines are the bilinear concomitant due to the boundary conditions, and the adjoint equations have the form

$$\begin{aligned} \bar{u}^\dagger_x + U\bar{v}^\dagger_x + V\tilde{w}^\dagger_x + (V\bar{v}^\dagger)_y + \check{\Delta}\bar{v}^\dagger - 2\text{Gö}^2 U\tilde{w}^\dagger &= 0\,, \\ U\tilde{w}^\dagger_x + \bar{u}^\dagger_y - U_y\bar{v}^\dagger + 2V\tilde{w}^\dagger_y + \check{\Delta}\tilde{w}^\dagger &= 0\,, \\ U\bar{p}^\dagger_x + \bar{u}^\dagger + V\tilde{w}^\dagger + V\bar{p}^\dagger_y + \check{\Delta}\bar{p}^\dagger &= 0\,, \\ \tilde{w}^\dagger_y - \beta^2\bar{p}^\dagger &= 0, \end{aligned}$$

with the boundary conditions $\bar{v}^\dagger = \tilde{w}^\dagger = \bar{p}^\dagger = 0$ at $y = 0$ and $y \to +\infty$ chosen to evaluate the concomitant to zero in the absence of forcing, and the scalar product

$$s = \bar{\mathbf{q}}^{\dagger} \cdot \mathscr{M} \bar{\mathbf{q}}$$

is independent of x.

In contrast to the local approach described in Sect. 10.3.1, to find the adjoint eigenfunctions at an initial position x_0, one has to integrate upstream from a suitable downstream position x_{f}, where the boundary-layer response to external forcing is observed. This, however, is not too complicated, if the equations are discretized and the operators $\mathscr{L}$ and $\mathscr{M}$ are approximated by matrices. Recall, for example, a simple scheme for x-integration used in (5.14):

$$\left(\mathbf{M}^{(i+1)} + \Delta x \mathbf{L}^{(i+1)}\right) \bar{\mathbf{q}}^{(i+1)} = \mathbf{M}^{(i)} \bar{\mathbf{q}}^{(i)} ,$$

where $i = 1, 2, \ldots, n-1$, $n\Delta x = x_{\mathrm{f}} - x_0$. Multiplying it scalarly by the vector of the adjoint functions $\bar{\mathbf{q}}^{\dagger(i)}$ to be found and subtracting terms

$$s^{(i+1)} = \bar{\mathbf{q}}^{\dagger(i+1)T} \mathbf{M}^{(i+1)} \bar{\mathbf{q}}^{(i+1)}$$

from both sides, which is equivalent to integration by parts, yields

$$\bar{\mathbf{q}}^{\dagger(i)T} \left(\mathbf{M}^{(i+1)} + \Delta x \mathbf{L}^{(i+1)}\right) \bar{\mathbf{q}}^{(i+1)} - \bar{\mathbf{q}}^{\dagger(i+1)T} \mathbf{M}^{(i+1)} \bar{\mathbf{q}}^{(i+1)} =$$
$$\bar{\mathbf{q}}^{\dagger(i)T} \mathbf{M}^{(i)} \bar{\mathbf{q}}^{(i)} - \bar{\mathbf{q}}^{\dagger(i+1)T} \mathbf{M}^{(i+1)} \bar{\mathbf{q}}^{(i+1)} .$$

Thus, for $s^{(i)}$ to be independent of i the following equality, which appears to be a finite-difference analog of the continuous adjoint Görtler equations, should be satisfied:

$$\left(\mathbf{M}^{(i+1)} + \Delta x \mathbf{L}^{(i+1)}\right)^T \bar{\mathbf{q}}^{\dagger(i+1)T} = \mathbf{M}^{(i)T} \bar{\mathbf{q}}^{\dagger(i)T} .$$

As soon as the adjoint functions are known, excitation of the Görtler vortices by a particular disturbance source can be considered.

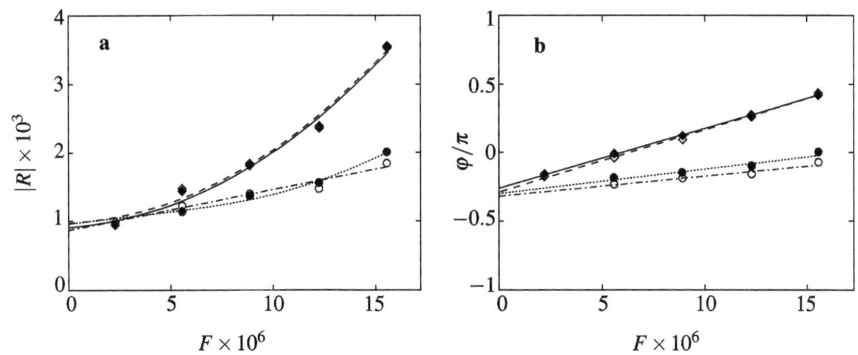

Fig. 10.11 Amplitude (**a**) and phase (**b**) of the vibration receptivity function for the non-stationary Görtler vortices in a flat-plate boundary layer at $\mathrm{G\ddot{o}}_{\delta^*} = 8.16$ as functions of frequency: experimental data for $\Lambda = 149$ ($\circ, \bullet$) and $\Lambda = 274$ ($\Diamond, \blacklozenge$) and their approximation by second-degree polynomials (*lines*) (Mischenko, 2010)

As is found theoretically and numerically, the vortices are excited effectively by both surface roughness and free-stream disturbances. A streamwise localized roughness promotes the vortex growth at higher Görtler numbers with respect to the free-stream disturbances, while a distributed roughness (with streamwise extent comparable with the evolution scale of the boundary layer) can excite the vortices at arbitrary small values of Gö. In the case of the short scale vortices, their development can be described by local linear stability approach, while their origination can be analyzed by inhomogeneous linearized Navier–Stokes equations. Since these modes are localized close to the wall, they can be more effectively excited by roughness than by free-stream vortical disturbances.

At present, reliable experimental data on excitation of the Görtler vortices in shear flows by free-stream disturbances are still lacking. First results on the vibration receptivity of a flat-plate boundary layer to the Görtler vortices were obtained only recently. Experimentally obtained dependencies of the receptivity amplitude and phase on the oscillation frequency are shown in Fig. 10.11 for two fixed spanwise wavelengths. Additionally, the values are extrapolated to surface roughness at $F = 0$. It appears that, similarly to the crossflow vortices, the roughness receptivity coefficients are significantly lower than those for the vibration receptivity.

10.7 Excitation of streaky structures

As it has been shown, the development of streaks is usually attributed to the lift-up effect, which is strongly dependent on the forcing disturbances, resulting in the receptivity rather than the stability problem. Different aspects of this problem are discussed below.

10.7.1 Localized generation of streaks

The first careful experiments on leading edge receptivity of the laminar boundary layer to free-stream vortical disturbances localized in time and space were carried out by Grek et al (1991); further details were given by Westin et al (1998). The perturbations were introduced by isolated pulses of air suction or blowing through a narrow tube located in front of the flat-plate leading edge. The isolines of the streamwise velocity shown in Fig. 10.12 give an impression of the initial stage of the disturbance development as rapid decay of a fairly symmetric formation.

The structure of the velocity fluctuations induced in the boundary layer is illustrated in Fig. 10.13**a**. In the plane parallel to the wall, it consists of two symmetric regions of a velocity defect separated by a region of a velocity excess (Fig. 10.13**b**). Approaching the wall, these regions grow in size and the intensity of perturbations inside them increase, the perturbed motion reaching 15% of U_∞.

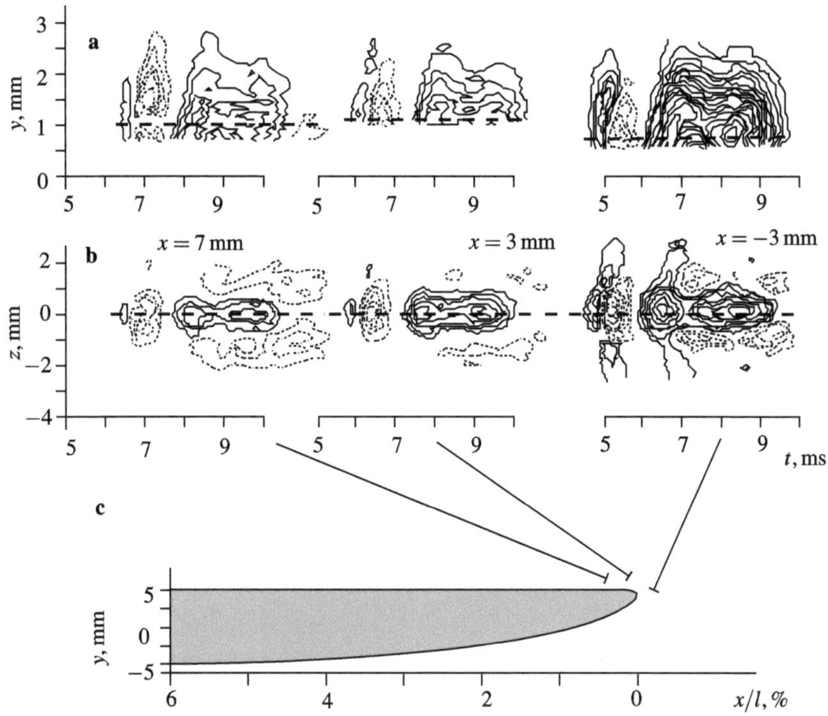

Fig. 10.12 Streamwise velocity u' distribution in y–t-plane at disturbance centreline (**a**) and z–t-plane (**b**) at the leading-edge (**c**): velocity excess (*solid lines*); velocity defects (*dotted lines*); corresponding sections through normal planes (*dashed lines*). Isoline spacing: 0.01 of U_e; plate length $l = 2$ m (Westin et al, 1998)

At the next streamwise position (Fig. 10.13**c, d**), most of the free-stream part of the perturbation has already damped, and the disturbance develops mainly inside the boundary layer. Thus, a decaying localized free-stream vortical disturbance generates a streamwise structure in the boundary layer with the phenomenological characteristics similar to those of the streaks observed at a high free-stream turbulence level.

The main features of receptivity of the flat-plate boundary layer to localized disturbances generated from the wall are the same as discussed above. The response of the boundary layer to such an excitation consists of generating streaks irrespective of the disturbance sources (by single pulses of suction or blowing through a hole, or transverse slot, or by surface vibration) over a wide range of excitation amplitudes. Smoke visualization of the disturbance generated by blowing through a transverse slot shows that narrow streamwise streaks appear in the boundary layer, Fig. 10.14. Detailed measurements indicated that the inner structure of the boundary-layer perturbations is qualitatively unaffected by the excitation magnitude, free-stream velocity, and parameters of the disturbance source.

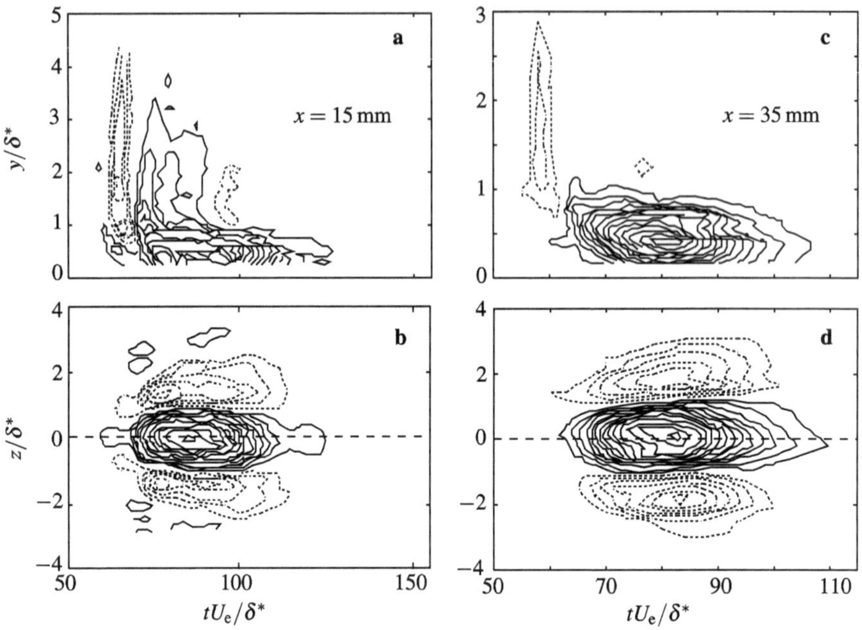

Fig. 10.13 Formation of the streak streamwise velocity in the boundary layer: (**a, c**) in the y–t-plane at the disturbance centerline; (**b, d**) in z–t-plane: velocity excess (*solid lines*); velocity defects (*dotted lines*); corresponding sections through normal planes (*dashed lines*). Isoline spacing: 0.01 of U_e (Westin et al, 1998)

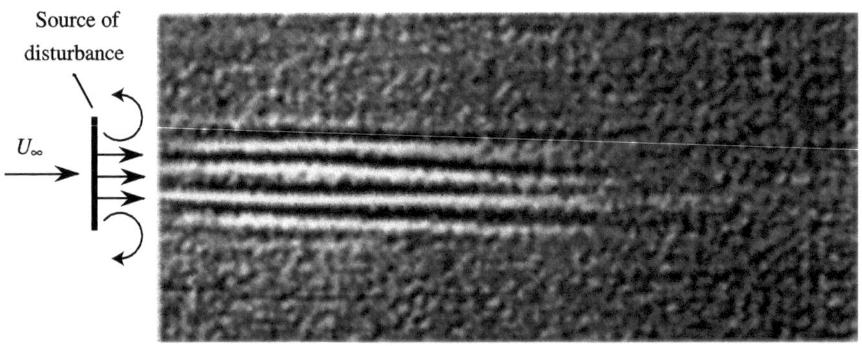

Fig. 10.14 Smoke visualization of the streaky structure generated in the flat-plate boundary layer by blowing air through a transverse slot; Tu = 0.02% (Alfredsson et al, 1996)

10.7.2 Distributed generation of streaks

The identity of the observed response of the boundary layer to three-dimensional free-stream disturbances over a broad range of their amplitudes, experimental models, and facilities indicates the generality of the mechanism of the boundary-layer distortion. However, the continuous decay of the locally introduced disturbance in quiet environmental conditions is in contrast to the downstream growth of the streamwise disturbance velocity observed in the presence of free-stream turbulence. The latter supports the idea of distributed receptivity of the boundary layer with respect to free-stream vortices at a high level of turbulence.

It is noteworthy that the three-dimensional inviscid free-stream disturbances are assumed to have scales comparable with the streaks, i.e., the mechanism of wave-scale reduction is not necessary or is of lower significance than for Tollmien–Schlichting wave excitation. It means that quite effective distributed forcing of the streaks in an extended region of the boundary layer may occur.

An experimental way to introduce certain controlled 'representative' disturbances to investigate the distributed receptivity from the free stream was found by Bertolotti and Kendall (1997). The idea is to produce a free-stream vortex at the tip of a micro-wing. By varying the wing's angle of attack and the free-stream velocity, the vortex strength can be controlled. Such an approach provides the fundamental possibility of investigating in detail the development of stationary disturbances independently of the traveling-wave modes. The successful comparison of the theoretical and experimental data on the response of the boundary layer indicated that the receptivity of the flow to quasi-stationary disturbances could be modeled in such a way.

In the experiment of Boiko (2000), the micro-wing was located above the plate so that the leading edge region was skipped. It was found that the excited stationary boundary-layer disturbance has the same phenomenological characteristics as those that appeared in the boundary layer under the effect of high free-stream turbulence. The free-stream velocity distributions downstream of the micro-wing as obtained by Boiko (2000) are illustrated in Fig. 10.15. The distributions measured at different downstream locations indicate a slow downstream decay of the vortex accompanied by little expansion of its core, which is in accordance with previous observations and the theoretical model of Batchelor (1964).

The boundary-layer response after a spatial Fourier transformation along the transverse coordinate at the same downstream positions is shown in Fig. 10.16. It can be seen that the disturbance maximum is concentrated at fixed $\beta = 0.082$ and $y/\delta^* \approx 1.3$. This indicates that, in the framework of the experimental conditions, the physical scale of the excited structure does not depend on the boundary-layer thickness in the region of generation and propagation of streaks ($\delta^* = 1.09$–1.99), which is in accordance with the known observations of the boundary-layer streaks excited by free-stream turbulence.

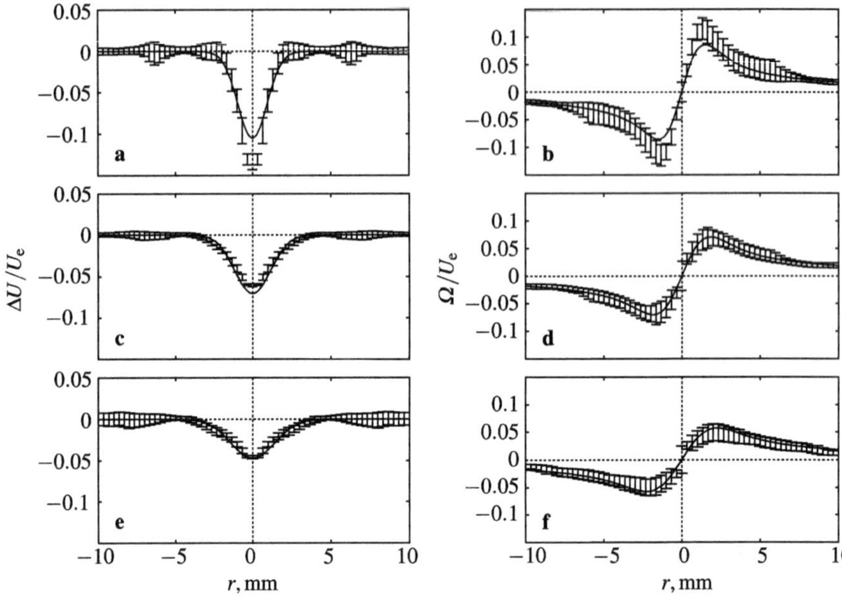

Fig. 10.15 Streamwise (**a, c, e**) and radial (**b, d, f**) velocity defects of the tip vortex at downstream distances from the micro-wing $\Delta x = 40$, 130 and 265 mm, respectively. Mean values (*points*) and error estimations (*bars*) obtained by Boiko (2000); data approximation by the model of Batchelor (1964) (*Solid lines*); $r = 0$ corresponds to the distance of 15 mm to the wall; the micro-wing is located at $x_0 = 215$ mm; $U_e = 5.9$ m/s

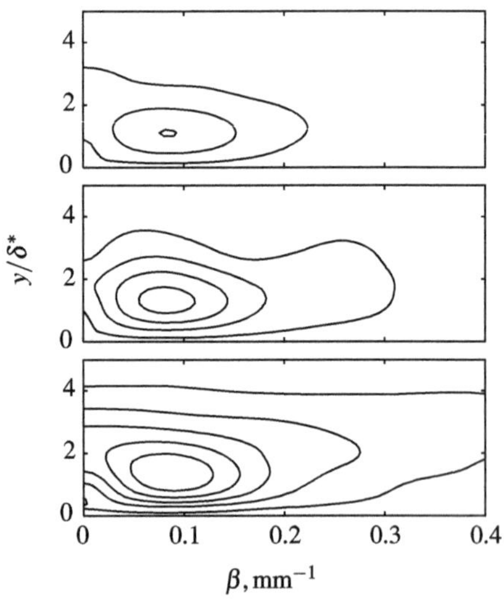

Fig. 10.16 Spectral distribution of ΔU at $U_e = 5.9$ m/s: $\Delta x = 40$, 130 and 265 mm (from *top* to *down*). Isolines with equidistant spacing from 0.1 to 1.1 (Boiko, 2000)

Additional material for this chapter is given in the multimedia file No. 7, which supplements the book, titled 'Response of shear flows to external disturbances.'

References

Aizin LB, Polyakov NF (1979) Generation of the Tollmien-Schlichting wave by sound at an isolated surface roughness. Preprint 17–79, RAS. Sib. Branch Inst. Theoret. Appl. Mech., Novosibirsk. In Russian.

Alfredsson PH, Bakchinov AA, Kozlov VV, Matsubara M (1996) Laminar-turbulent transition at a high level of a free stream turbulence. In: Duck and Hall (1996), pp 423–436

Asai M, Kaneko M (1998) Experimental investigation of the receptivity of separated flow. In: Proc. Third Internat. Conf. Fluid Mechanics. Beijing Institute of Technology, Beijing, pp 231–237

Batchelor GK (1964) Axial flow in trailing line vortices. J Fluid Mech 20:645–658

Bertolotti FP, Kendall JM (1997) Response of the boundary layer to controlled free-stream vortices of axial form. AIAA Paper 97–2018

Bippes H (1999) Basic experiments on transition in three-dimensional boundary layers dominated by crossflow instability. Progr Aerosp Sci 35(4):363–412

Bodonyi RJ, Welch WJC, Duck PW, Tadjfar M (1989) A numerical study of the interaction between unsteady free-stream disturbances and localized variations in surface geometry. J Fluid Mech 209:285–308

Boiko AV (2000) Receptivity of boundary layers to free stream axial vortices. IB 223–2000 A10, German Aerospace Center (DLR) – Institute for Fluid Mechanics, Göttingen

Boiko AV, Dovgal AV, Kozlov VV, Scherbakov VA (1990) Instability and receptivity of a boundary layer close to 2D surface imperfections. Izv Sib Otd Akad Nauk SSSR, Ser Tekhn Nauk 1:50–56

Choudhari M, Streett CL (1990) Boundary layer receptivity phenomena in three-dimensional and high-speed boundary layers. AIAA Paper 90–5258

Crouch JD (1992) Localized receptivity of boundary layers. Phys Fluids A 4(7):1408–1414

Dovgal AV, Kozlov VV (1983) Influence of acoustic perturbations on the flow structure in a boundary layer with adverse pressure gradient. Fluid Dyn 18(2):205–209

Dovgal AV, Kozlov VV, Levchenko VY (1980) Experimental investigation into the reaction of a boundary layer to external periodic disturbances. Fluid Dyn 15(4):602–606

Duck PW, Hall P (eds) (1996) Nonlinear instability and transition in three-dimensional boundary layers. Fluid Mechanics and its Application. Kluwer, Dordrecht

Gaponenko VR, Ivanov AV, Kachanov YS, Crouch JD (2002) Swept-wing boundary-layer receptivity to surface non-uniformities. J Fluid Mech 461:93–126

Goldstein ME, Hultgren LS (1989) Boundary layer receptivity to long wave free-stream disturbances. Ann Rev Fluid Mech 21:137–166

Grek GR, Kozlov VV, Ramazanov MP (1991) Laminar-turbulent transition at high free stream turbulence: Review. Izv. Sib. Otd. Akad. Nauk SSSR, Ser. Tekhn. Nauk 106–138

Hill DC (1995) Adjoint systems and their role in the receptivity problem for boundary layer. J Fluid Mech 292:183–204

Ivanov AV, Kachanov YS, Obolentseva TG, Michalke A (1998) Receptivity of the Blasius boundary layer to surface vibrations. Comparison of theory and experiment. In: Proc. Ninth Internat. Conf. on Methods of Aerophysical Research, vol 1. RAS. Sib. Branch Inst. Theoret. Appl. Mech., Novosibirsk, pp 93–98

Kachanov YS (2000) Three-dimensional receptivity of boundary layers. Eur J Mech B/Fluids 19(5):723–744

Kachanov YS, Kozlov VV, Levchenko VY (1979) Occurrence of Tollmien-Schlichting waves in the boundary layer under the effect of external perturbations. Fluid Dyn 13(5):85–94

Mischenko DA (2010) Experimental investigation of generation and development of boundary-layer unsteady Görtler-instability modes. Doctoral thesis, Institute of Theroretical and Applied Mechanics SB RAS, Novosibirsk

Morkovin MV (1968) Critical evaluation of transition from laminar to turbulent shear layers with emphasis on hypersonically traveling bodies. AFFDL TR 68–149

Nishioka M, Morkovin MV (1986) Boundary-layer receptivity to unsteady pressure gradients: Experiments and overview. J Fluid Mech 171:219–261

Saric W, Reed H, Kerschen E (2002) Boundary-layer receptivity to freestream disturbances. Ann Rev Fluid Mech 34:291–319

Saric WS, Hoos JA, Radeztsky RH (1991) Boundary-layer receptivity of sound with roughness. In: Reda DC, Reed HL, Kobayashi R (eds) Boundary Layer Stability and Transition to Turbulence. ASME, pp 17–22

Schmid PJ, Henningson DS (2000) Stability and transition in shear flows. Springer–Verlag, Berlin

Westin KJA, Bakchinov AA, Kozlov VV, Alfredsson PH (1998) Experiments on localized disturbances in a flat plate boundary layer. Part 1: The receptivity and evolution of a localized free stream disturbance. Eur J Mech B/Fluids 17(6):823–846

Further Reading

Bakchinov AA, Grek GR, Katasonov MM, Kozlov VV (1997) Experimental investigation of the structure and development characteristics of the localized disturbances in the flat plate boundary layer. Preprint 1–97, RAS. Sib. Branch Inst. Theoret. Appl. Mech., Novosibirsk. In Russian.

Balakumar P, Hall P, Malik MR (1990) On the receptivity and non-parallel stability of traveling disturbances in rotating disk flow. ICASE Rep. 90–89

Bassom AP, Hall P (1994) The receptivity problem for $O(1)$ wavelength Görtler vortices. Proc R Soc Lond A 446:449–516

Bassom AP, Seddougui SO (1995) Receptivity mechanisms for Görtler vortex modes. Theoret Comput Fluid Dyn 7:317–339

Berlin S, Henningson DS (1994) A study of boundary layer receptivity to disturbances in the free stream. In: Henningson DS (ed) Bypass Transition – Proceedings from a Mini-Workshop. Royal Institute of Technology, Stockholm, pp 43–49

Bertolotti FP (2000) Receptivity of three-dimensional boundary layers to localised wall roughness and suction. Phys Fluids 12(7):1799–1809

Bippes H (1977) Experimente zur Entwicklung der freien Wirbel hinter einem Rechteckflügel. Acta Mech 26:223–245

Bodonyi RJ, Duck PW (1992) Boundary layer receptivity to a wall suction control of Tollmien-Schlichting waves. Phys Fluids A 4(6):1206–1214

Boiko AV, Chung YM, Sung HJ (1997) Spatial simulation of the instability of channel flow with local suction/blowing. Phys Fluids 9(11):3258–3266

Chernorai VG, Grek GR, Katasonov MM, Kozlov VV (2000) Generation of the localized disturbances by the vibrating surface. Thermophys Aeromech 7(3):329–339

Choudhari M (1993) Boundary-layer receptivity due to distributed surface imperfections of a deterministic or random nature. Theoret Comput Fluid Dyn 4:101–117

Choudhari M, Streett CL (1992) A finite Reynolds-number approach for the prediction of boundary-layer receptivity in localized regions. Phys Fluids A 4(11):2495–2514

Collis SS, Lele SK (1999) Receptivity to surface roughness near a swept wing leading edge. J Fluid Mech 380:141–168

Crouch JD (1992) Non-localized receptivity of boundary layers. J Fluid Mech 244:567–581

Crouch JD (1993a) Receptivity and the evolution of boundary-layer instabilities over short-scale waviness. Phys Fluids A 5(3):561–567

Crouch JD (1993b) Receptivity of three-dimensional boundary layers. AIAA Paper 93–0074

Crouch JD (1994a) Distributed excitation of Tollmien-Schlichting waves by vortical free-stream disturbances. Phys Fluids A 6(1):217–223

Crouch JD (1994b) Theoretical study on the receptivity of boundary layers. AIAA Paper 94–2223

Crouch JD, Bertolotti FP (1992) Nonlocalized receptivity of boundary layers to three-dimensional disturbances. AIAA Paper 92–0740

Crouch JD, Spalart PR (1995) A study of nonlinear and nonparallel effects on the localized receptivity of boundary layers. J Fluid Mech 290:29–37

Denier JP, Hall P, Seddougui SO (1991) On the receptivity problem for Görtler vortices: Vortex motions induced by roughness. Philos Trans R Soc Lond A 335:51–85

Deyhle H, Bippes H (1996) Disturbance growth in an unstable three-dimensional boundary layer and its dependence on environmental conditions. J Fluid Mech 316:73–113

Dovgal AV, Kozlov VV (1990) Hydrodynamic instability and receptivity of small scale separation regions. In: Arnal D, Michel R (eds) Laminar-Turbulent Transition IUTAM Symposium. Springer, Berlin, pp 523–531

Dovgal AV, Kozlov VV, Michalke A (1996) On disturbances excited by a point source in an axisymmetric laminar separation bubble. Eur J Mech B/Fluids 15(4):651–664

Duck PW, Hall P (eds) (1996) Nonlinear Instability and Transition in Three-Dimensional Boundary Layers. Fluid Mechanics and its Application. Kluwer, Dordrecht

Elofsson PA, Alfredsson PH (2000) An experimental investigation of oblique transition in a Blasius boundary layer flow. Eur J Mech B/Fluids 19(5):615–636

Fasel H (1977) Reaktion von zweidimensionalen, laminaren, inkompressiblen Grenzschichten auf periodische Störungen in der Aussenströmung. Z Angew Math Mech 57:T180–T183

Fedorov AV (1984) Excitation of Tollmien-Schlichting waves in a boundary layer by a periodic external source located on the body surface. Fluid Dyn 19(6):888–893

Fedorov AV (1988) Excitation of waves of instability of the secondary flow in a boundary layer on a swept wing. J Appl Mech Tehn Phys 29(5):643–648

Gaponenko VR, Ivanov AV, Kachanov YS (1995) Experimental study of cross-flow instability of a swept-wing boundary layer with respect to travelling waves. In: Kobayashi R (ed) Laminar-Turbulent Transition Springer–Verlag, Berlin, IUTAM Symposium. pp 373–380

Gaponenko VR, Ivanov AV, Kachanov YS (1996) Experimental study of 3D boundary-layer receptivity to surface vibration. In: Duck and Hall (1996), pp 389–398

Gaster M (1965) On the generation of spatially growing waves in a boundary layer. J Fluid Mech 22:433–441

Gaster M, Gupta TKS (1993) The generation of disturbances in a boundary layer by wall perturbations: The vibrating ribbon revisited once more. In: Ashpis DE, Gatski TB, Hirsh R (eds) Instabilities and Turbulence in Engineering Flows. Kluwer, Dordrecht, pp 31–50

Gatski TB, Grosch C (1987) Numerical experiments in boundary-layer receptivity. In: Dwoyer DL, Hussaini MY (eds) Stability of Time Dependent and Spatially Varying Flows. Springer–Verlag, Berlin, pp 82–96

Gilev VM, Kozlov VV (1984) Excitation of Tollmien-Schlichting waves in the boundary layer on a vibrating surface. J Appl Mech Tehn Phys 25(6):874–877

Goldstein ME (1983) The evolution of Tollmien-Schlichting waves near a leading edge. J Fluid Mech 127:59–81
Goldstein ME (1984) Generation of instability waves in flows separating from smooth surfaces. J Fluid Mech 145:71–94
Goldstein ME (1985) Scattering of acoustic waves into Tollmien-Schlichting waves by small streamwise variations in surface geometry. J Fluid Mech 154:509–529
Goldstein ME, Hultgren LS (1989) Boundary layer receptivity to long wave free-stream disturbances. Ann Rev Fluid Mech 21:137–166
Goldstein ME, Leib SJ, Cowley SJ (1987) Generation of Tollmien-Schlichting waves on interactive marginally separated flows. J Fluid Mech 181:485–517
Hall P (1990) Görtler vortices in growing boundary layers: The leading edge receptivity problem, linear growth and the nonlinear breakdown stage. Mathematika 37:151–189
Heinrich RA, Choudhari M, Kerschen EJ (1988) A comparison of boundary layer receptivity mechanisms. AIAA Paper 88–3758
Herbert T, Lin N (1993) Studies of boundary-layer receptivity with parabolized stability equations. AIAA Paper 93–3053
Hultgren LS, Gustavsson LH (1981) Algebraic growth of disturbances in a laminar boundary layer. Phys Fluids 24(6):1000–1004
Ivanov AV, Kachanov YS, Koptsev DB (1994) An experimental investigation of instability wave excitation in three-dimensional boundary layer at acoustic wave scattering on a vibrator. Thermophys Aeromech 4(4):359–372
Kachanov YS (1996) Generation, development and interaction of instability modes in swept-wing boundary layer. In: Duck and Hall (1996), pp 115–132
Kachanov YS, Kozlov VV, Levchenko VY (1975) Generation and development of small-amplitude disturbances in a laminar boundary layer in the presence of an acoustic field. Izv Sib Otd Akad Nauk SSSR, Ser Tekhn Nauk 3(13):18–26
Kachanov YS, Kozlov VV, Levchenko VY (1979) The development of small-amplitude oscillations in a laminar boundary layer. Fluid Mech – Soviet Res 8(2):152–156
Kachanov YS, Kozlov VV, Levchenko VY, Maksimov VP (1979b) Transformation of external disturbances into the boundary layer waves. In: Proc. Sixth Intern. Conf. on Numerical Methods in Fluid Dyn., Springer–Verlag, Berlin, Lecture Notes in Physics, 19, pp 299–307
Kachanov YS, Kozlov VV, Levchenko VY (1982) Origin of Turbulence in Boundary Layer. Nauka. Sib. Otd., Novosibirsk
Kerschen EJ (1991) Linear and non-linear receptivity to vortical freestream disturbances. In: Reda DC, Reed HL, Kobayashi R (eds) Boundary Layer Stability and Transition to Turbulence. ASME, pp 43–48
King RA, Breuer KS (2001) Acoustic receptivity and evolution of two-dimensional and oblique disturbances in a Blasius boundary layer. J Fluid Mech 432:69–90
Kozlov VV, Ryzhov OS (1990) Receptivity of boundary layers: Asymptotic theory and experiment. Proc R Soc Lond A 429:341–373

Leehey P, Shapiro P (1980) Leading edge effect in laminar boundary layer excitation by sound. In: Eppler R, Fasel H (eds) Laminar–Turbulent Transition, Springer–Verlag, Berlin, IUTAM Symposium, pp 321–331

Lindors M, Laine S (1976) Numerical solution of the Navier-Stokes equations for unsteady boundary-layer flows past a wave-like bulge on a flat plate. Lect Notes Phys 59:293–299

Luchini P, Bottaro A (1998) Görtler vortices: A backward in time approach to the receptivity problem. J Fluid Mech 363:1–23

Maksimov VP (1979) Origin of Tollmien-Schlichting waves in oscillating boundary layers. In: Levchenko VY (ed) Development of Disturbances in Boundary Layer. RAS. Sib. Branch Inst. Theoret. Appl. Mech., Novosibirsk, pp 68–75. In Russian.

Michalke A (1993) On the receptivity of a compressible two-dimensional vortex sheet close to a wall for various types of excitation. Eur J Mech B/Fluids 12(4):421–445

Michalke A (1995) Receptivity of axisymmetric boundary layers due to excitation by a Dirac point source at the wall. Eur J Mech B/Fluids 14(4):373–393

Michalke A (1997) Excitation of small disturbances by a Dirac line source at the wall and their growth in a decelerated laminar boundary layer. Eur J Mech B/Fluids 16(1):17–37

Michalke A, Al-Maaitah AA (1992) On the receptivity of the unstable wall boundary layer along a surface hump excited by a 2-D Dirac source at the wall. Eur J Mech B/Fluids 11(5):521–542

Müller B, Bippes H (1988) Experimental study of instability modes in a three-dimensional boundary layer. In: AGARD-CP-438 Fluid Dynamics of Three-Dimensional Turbulent Shear Flows and Transition, Chismet, Turkey, pp 13.1–13.15

Murdock JW (1980) The generation of Tollmien-Schlichting waves by a sound wave. Proc R Soc Lond A 372:517–534

Ng LL, Crouch JD (1999) Roughness-induced receptivity to crossflow vortices on a swept wing. Phys Fluids 11(2):432–438

Polyakov NF (1975) Induction of hydrodynamic waves in laminar boundary layer by streamwise acoustic field. In: Proc. Symp. on Physics of Acousto-Hydrodynamic Phenomena. Nauka, Moscow, pp 216–223. In Russian.

Rogler HL, Reshotko E (1975) Disturbances in a boundary layer introduced by a low intensity array of vortices. SIAM J Appl Math 28(2):431–462

Rothmayer AP, Smith FT (1998) High Reynolds number asymptotic theories. In: Johnson RW (ed) The Handbook of Fluid Dynamics, Springer–Verlag, Berlin, chap 23–25 pp 23.1–25.26.

Ruban AI (1985) On the generation of Tollmien-Schlichting waves by sound. Fluid Dyn 19(5):709–716

Saric W, Reed H, Kerschen E (1994) Leading edge receptivity to sound: Experiments, DNS, and theory. AIAA Paper 94–2222

Saric WS (1990) Low-speed experiments: Requirements for stability measurements. In: Hussaini MY, Voight RG (eds) Instability and Transition, Springer–Verlag, Berlin, ICASE/NASA LaRC Series, vol 1, pp 162–172

Saric WS, White EB, Reed HL (1999) Boundary-layer receptivity to freestream disturbances and its role in transition. AIAA Paper 99–3788

Shapiro PJ (1977) The influence of sound upon laminar boundary layer instability. Tech. Rep. 83458-83560-1, Mass. Inst. Technol., Cambridge, Acoustic and Vibration Lab., See also NTIS AD-A046057.

Sidorenko NV, Erofeev EA (1985) Interaction of vortex disturbances and boundary layer on blunt bodies. In: Kozlov VV (ed) Laminar–Turbulent Transition, Springer–Verlag, Berlin, IUTAM Symposium, pp 261–266

Spalart PR (1993) Numerical study of transition induced by suction devices. In: So RMC, Speziale CG, Launder BE (eds) Near-Wall Turbulent Flows. Elsevier, Amsterdam, pp 849–858

Spiridonov AN, Chernoray VG (2000) Generation and development of disturbances by a surface vibration in a boundary layer with a pressure gradient. In: Proc. Seventh Internat. Conf. Stability of Homogeneous and Heterogeneous Fluids, Novosibirsk, pp 165–167. In Russian.

Takagi S, Saric WS, Radeztsky RH (1991) Effect of sound and micro-sized roughness on crossflow dominated transition. Bull Amer Phys Soc 36(10):2630

Tam CKW (1978) Excitation of instability waves in a two-dimensional shear layer by sound. J Fluid Mech 89:357–371

Tam CKW (1981) The excitation of Tollmien-Schlichting waves in low subsonic boundary layers by free stream sound waves. J Fluid Mech 109:483–501

Terent'ev ED (1984) A linear problem on a vibrator oscillating harmonically at supercritical frequencies in a subsonic boundary layer. Prikl Mat Mekh 48(2):264–272

Thomas ASW, Lekoudis SG (1978) Sound and a Tollmien–Schlichting wave in a Blasius boundary layer. Phys Fluids 21(11):2112–2113

Tumin A (1996) Receptivity of pipe Poiseuille flow. J Fluid Mech 315:119–137

Wu X (2001) Receptivity of boundary layers with distributed roughness to vortical and acoustic disturbances: A second-order asymptotic theory and comparison with experiments. J Fluid Mech 431:91–133

Yang ZY, Voke PR (1991) Numerical simulation of transition under turbulence. Tech. Rep. ME–FD/91.01, Department of Mechanical Engineering, University of Surrey, U.K.

Zavolskiy NA, Reutov VP, Rybushkina GV (1983) Excitation of Tollmien–Schlichting waves by acoustic and vortex disturbance scattering in boundary layer on a wavy surface. J Appl Mech Tehn Phys 24(3):355–361

Zhigulev VN, Tumin AM (1987) Onset of turbulence. Dynamical theory of excitation and development of instabilities in boundary layers. Nauka. Sib. Otd., Novosibirsk. In Russian.

Chapter 11
Secondary instabilities of shear layers

When a linear instability mode reaches a large-enough amplitude, it enters the region of its essentially nonlinear, but still deterministic development. Usually the disturbance amplitude saturates in this region, which resembles the formation of a new quasi-steady state that opens in some cases the door for mechanisms of secondary instabilities discussed below.

11.1 Secondary instability in a flow modulated by the Tollmien–Schlichting waves

The idea of the theory discussed in this section is a conjecture that the primary disturbance, whose amplitude becomes nearly constant due to nonlinear mechanisms, can be a forcing agent for parametric excitation of some other modes (Herbert, 1988). Such a secondary instability should be observed at relatively small amplitudes of the primary wave when its nonlinear distortions are small.

The starting point is to consider a basic flow formed from the initial stationary velocity profile with an imposed saturated primary periodic disturbance. Let us proceed with a two-dimensional Tollmien–Schlichting wave in mind. From the local viewpoint, i.e., in the coordinate system ($\chi = x - ct, y, z$) moving with the wave phase velocity $c = c_r$, the flow is steady. Then its linear stability to three-dimensional secondary disturbances is considered.

Denoting the streamwise and wall-normal velocities of the initial two-dimensional wave as $Au_{2D}(x,y)$ and $Av_{2D}(x,y)$, respectively, where A is a wave amplitude, and its phase velocity as c, the basic flow is described by

$$\mathbf{U} = (U(y) - c + Au_{2D}(\chi,y)\ ,\ Av_{2D}(\chi,y)\ ,\ 0)^T\ .$$

For swept bodies, the two-dimensional analysis can be easily extended by incorporating crossflow with the transverse primary disturbance to the equations.

Substituting this expression for the basic flow into the linearized Navier–Stokes momentum equations and the continuity equation yields the following system for

A.V. Boiko et al., *Physics of Transitional Shear Flows*,
Fluid Mechanics and Its Applications 98, DOI 10.1007/978-94-007-2498-3_11,

the linear secondary instability:

$$\frac{\partial u}{\partial t}(U-c)\frac{\partial u}{\partial \chi}+vU_y+\frac{\partial p}{\partial \chi}-\frac{1}{\mathrm{Re}}\nabla^2 u$$
$$=-A\left[\frac{\partial}{\partial \chi}(uu_{2\mathrm{D}})+\frac{\partial}{\partial y}(uu_{2\mathrm{D}}+uv_{2\mathrm{D}})-u_{2\mathrm{D}}\frac{\partial w}{\partial z}\right], \quad (11.1\mathrm{a})$$

$$\frac{\partial v}{\partial t}(U-c)\frac{\partial v}{\partial \chi}+vU_y+\frac{\partial p}{\partial y}-\frac{1}{\mathrm{Re}}\nabla^2 v$$
$$=-A\left[\frac{\partial}{\partial \chi}(uv_{2\mathrm{D}}+vu_{2\mathrm{D}})+\frac{\partial}{\partial y}(vu_{2\mathrm{D}})-v_{2\mathrm{D}}\frac{\partial w}{\partial z}\right], \quad (11.1\mathrm{b})$$

$$\frac{\partial w}{\partial t}(U-c)\frac{\partial w}{\partial \chi}+\frac{\partial p}{\partial z}-\frac{1}{\mathrm{Re}}\nabla^2 w=0, \quad (11.1\mathrm{c})$$

$$\frac{\partial u}{\partial \chi}+\frac{\partial v}{\partial y}+\frac{\partial w}{\partial z}=0. \quad (11.1\mathrm{d})$$

These equations with the boundary conditions $u=v=w=0$ at $y=0$ and $y\to\infty$ constitute the problem of secondary instability for a flow modulated by saturated Tollmien–Schlichting waves. It can be additionally simplified as in the case of derivation of the Orr–Sommerfeld equation, so that the problem reduces to solving two equations for u and v, decoupled from the solution for the third w-component, which is found from the continuity equation.

Since $u_{2\mathrm{D}}$ and $v_{2\mathrm{D}}$ are the solutions for the primary wave that is periodic with χ, then, in accordance with the Floquet theory, it is possible to present the solution as

$$\begin{pmatrix} u \\ v \\ w \\ p \end{pmatrix}(\chi,y,z,t)=\begin{pmatrix} \hat{u} \\ \hat{v} \\ \hat{w} \\ \hat{p} \end{pmatrix}(\chi,y)\mathrm{e}^{\mathrm{i}\bar{\alpha}\chi}\mathrm{e}^{\mathrm{i}(\beta z-\omega t)}, \quad (11.2)$$

where ω and $\bar{\alpha}$ are generally complex, while $\hat{u},\hat{v},\hat{w}$ and $\hat{p}$ are periodic disturbance amplitude functions in χ with period $2\pi/\alpha_\mathrm{r}$ and $\mathrm{Real}\bar{\alpha}\in[-\alpha_\mathrm{r}/2;\alpha_\mathrm{r}/2]$.

Since ω and $\bar{\alpha}$ are generally complex, one has to prescribe two of their four real components to solve the eigenvalue problem. The imaginary parts of ω and $\bar{\alpha}$ stand for the disturbance growth in time and space, respectively, so the choice can be based on consideration of the absolute or convective instability cases and the flow symmetry. The final reduction of the problem to compute eigenvalues of the secondary modes is fulfilled by substitution of (11.2) into (11.1). The limit cases give solutions for secondary instability of two types: harmonic ($\mathrm{Real}\bar{\alpha}=0$), where the period of the secondary mode coincides with that of the Tollmien–Schlichting wave, and subharmonic ($\mathrm{Real}\bar{\alpha}=\pm\alpha_\mathrm{r}/2$), where the period of the secondary mode is doubled. Between them, mixed or detuned cases exist.

Application of the theory to the Blasius boundary layer gives results that are in good accordance with experimental data, as is shown in Fig. 11.1 for the primary Tollmien–Schlichting wave and the subharmonic disturbance. It is interesting enough that both the nonlinear theory and the theory of linear secondary instability

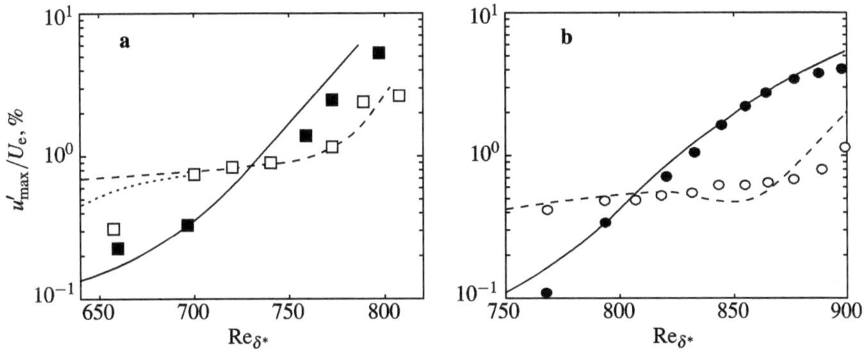

Fig. 11.1 Comparison of measured (*symbols*) and calculated (*curves*) amplification of primary (□, ∘, *dashed line*) and subharmonic (■, •, *solid line*) waves at the subharmonic breakdown [after Saric et al (1984), Zelman and Maslennikova (1993) (*left*) and Corke and Mangano (1989), Crouch and Herbert (1993) (*right*)]; growth of fundamental wave according to a nonparallel theory (*Dotted line*); $u'_{max} = \max_y[u'(y)]$

approach the experimental data quite well in this case, though the amplitudes of secondary disturbances exceed in some sections the amplitudes of primary waves by an order of magnitude. Usually one expects that the growing secondary disturbances should start to distort the primary periodic state, when the amplitudes of primary and secondary disturbances become of the same order (about 1% of free-stream velocity), as the latter need energy for their growth. This paradox is explained by a specific feature of the flat-plate boundary layer, where the energy for secondary disturbances is drawn directly from the two-dimensional basic flow. Nevertheless, the applicability of the theory of secondary instability is quite limited in this case as it cannot explain some effects of the laminar–turbulent transition in a plane boundary layer due to the presence of strong resonance phenomena discussed in Chap. 12.

11.2 Secondary instability in flows modulated by streamwise vortices and streaks

The streamwise vortices and streaky structures can arise in shear flows due to a variety of reasons. Centrifugal instability causes the appearance of the Taylor, Dean, or Görtler vortices, the crossflow over a swept wing initiates crossflow vortices, local roughness generates streaks in a boundary layer, etc. In contrast to the laminar–turbulent transition due to the Tollmien–Schlichting instability, nonlinear resonances in shear flows modulated by these streamwise vortices and streaks are weak if ever exist in typical situations. The primary disturbances can reach very large amplitudes (20–40% of free-stream velocity) before the final breakdown of the laminar flow. That provides a potential possibility of tracing the growth of secondary disturbances to quite large amplitudes using the secondary instability approach.

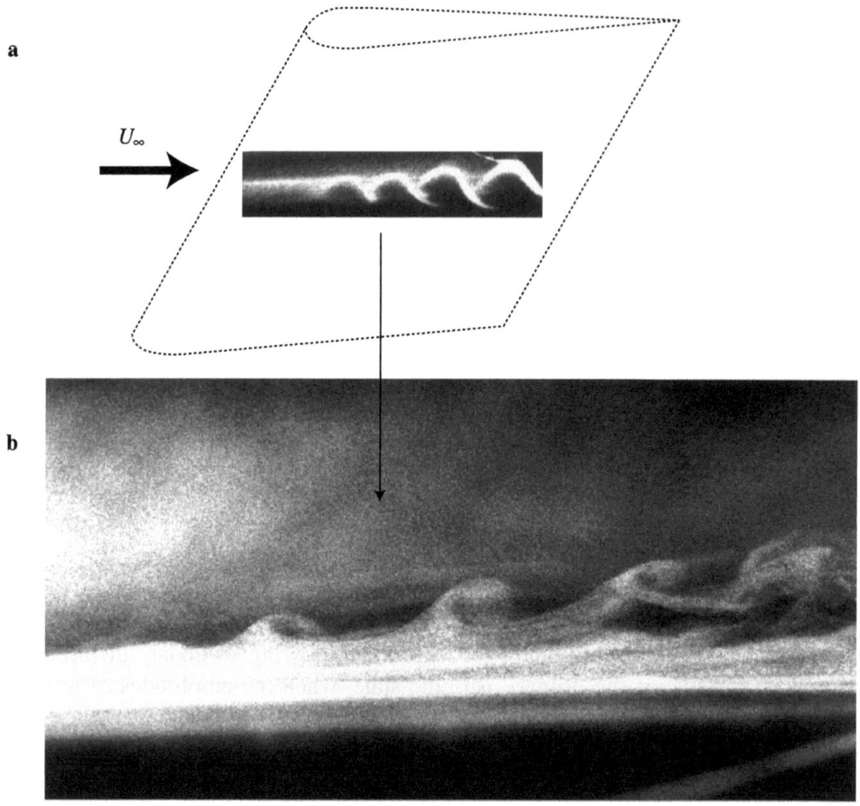

Fig. 11.2 Traveling disturbances arising on stationary vortices in a swept-wing boundary layer (Kohama et al, 1996): view parallel to the surface (**a**); view normal to the surface (**b**)

At a large-enough amplitude of the boundary-layer modulation, traveling disturbances propagating along the primary vortices, as, e.g., illustrated by flow visualization in Fig. 11.2, can appear. Alternatively, the large-amplitude vortices can experience certain irregularity or meandering (vortex splitting and coalescence) known as the Eckhaus instability. Both situations can be treated as caused by normal instability modes in the form of waves or streamwise vortices of the secondary flow composed of the primary basic state with a superimposed periodic in the transverse direction vortex pattern.

A typical example is the secondary instability of Görtler vortices, which are initiated with a characteristic periodicity in the transverse direction due to centrifugal instability (see Sect. 5.2.3 for details). The analysis of secondary instability is usually performed using the locally-parallel approach, i.e., the dependence of disturbances on wall-normal and transverse velocities of the basic flow modulated by the vortices is ignored. This leads to the system of equations formulated in Sect. 6.3. The main difference of the present problem is due to basic streamwise velocities and the shape

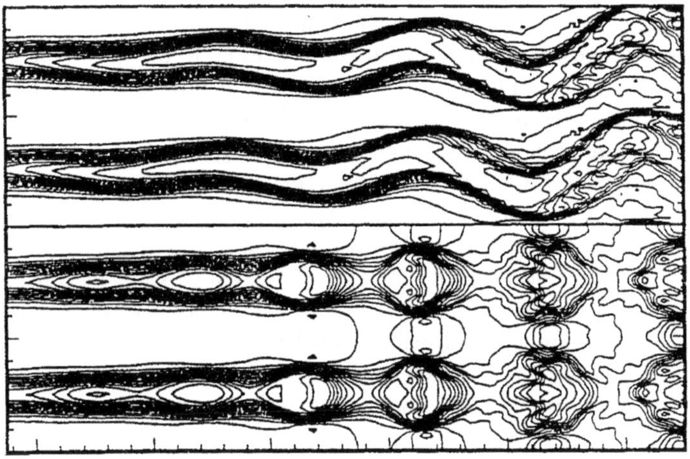

Fig. 11.3 Instantaneous velocities inside the Blasius boundary layer in the x–z-plane: *top*, secondary sinuous Görtler instability; *bottom*, secondary varicose Görtler instability (Li and Malik, 1995)

of the boundary. Additionally, assuming an inviscid nature of secondary instability, viscous terms can be dropped. This admits elimination of disturbance velocities, so that the equation for pressure fluctuations can be derived as

$$\left[(\alpha U-\omega)\left(\frac{\partial^2}{\partial y^2}+\frac{\partial^2}{\partial z^2}+2\mathrm{i}\beta\frac{\partial}{\partial z}-\varkappa^2\right)-2\alpha\left(U_y\frac{\partial}{\partial y}+U_z\frac{\partial}{\partial y}\right)\right]\hat{p}=0$$

with the boundary conditions

$$\hat{p}_y(y,z)=0 \quad \text{at} \quad y=0\,, \quad \hat{p}(y,z)\to 0 \quad \text{as} \quad y\to\infty$$

and the requirement of transverse periodicity of primary vortices.

To date, a detailed analysis of secondary instability of Görtler vortices both with and without account of viscosity has been performed, and the results have been compared with available experimental data. It was shown that the secondary instability approach is able to correctly describe the initial stage of the development of the traveling disturbances observed in experiments. It was also proved that the effect of viscosity is small. The flow is prone first to the growth of disturbances of sinuous type (see Fig. 11.3 and Table 6.3), the maximum of streamwise perturbations being located in the region of the maximum transverse velocity gradient. As is found, both the frequency and the streamwise wavenumber of the most amplifying disturbance are proportional to the basic flow velocity gradient in a wide range of parameters (Bottaro and Klingmann, 1996):

$$\left|\frac{\partial U}{\partial z}\right|_{\max}\approx 1.2\omega_{\mathrm{r}}\,, \qquad \left|\frac{\partial U}{\partial z}\right|_{\max}\approx 0.79\alpha\,.$$

The varicose mode usually becomes pronounced at larger amplitudes of the Görtler vortices.

Another wide-spread type of vortical structures in laminar flows is co-rotating stationary crossflow vortices. Growing downstream according to the linear stability theory, the vortices reach amplitudes of about 10–12% before saturation. Poll (1979, 1985) was probably the first to observe the high-frequency oscillations (in comparison to the low-frequency linearly unstable traveling crossflow modes) in the swept-wing boundary layer. Later these results were confirmed by other authors through detailed measurements and flow visualizations.

The analysis of secondary instability in this case is performed in the local frame of reference aligned with the vortices. An obvious distinction of secondary instability of the crossflow vortices from secondary instability of Görtler vortices is the presence of crossflow resulting in the absence of the basic flow symmetry. Hence, the classification of secondary modes based on their parity as sinuous and varicose modes makes no sense in this case. The analysis of secondary instability for this flow has been performed in a series of studies. A comparison of theoretical results with available experimental data showed their good correspondence to each other.

At free-stream turbulence $\mathrm{Tu} \lesssim 0.1\%$, stationary crossflow vortices usually dominate due to a large contribution of stationary background perturbations; as Tu grows, they are shaded by traveling crossflow disturbances. Such behavior demonstrates a competition of various transition mechanisms and introduces a considerable difficulty to the study of secondary instability of vortices in naturally developing flows.

This makes reasonable modeling of vortices in simplified cases to control their characteristics during experiments. With such an approach, a series of studies related to secondary instability of different flows was conducted. Boiko et al (1997) proved the linearity of the traveling disturbance at the initial stage of its growth inside an isolated vortex of large amplitude excited in a swept-wing boundary layer. This property indicates that the investigated waves represent a secondary linear instability of the flow, rather than a result of nonlinear interaction.

In Fig. 11.4, the growth curves of disturbances measured at two different initial amplitudes are shown, in comparison with exponential amplification. Another indication of the wave linearity is the amplitude independence of their development along the vortex. The superposition of the waves is demonstrated in Fig. 11.5, where

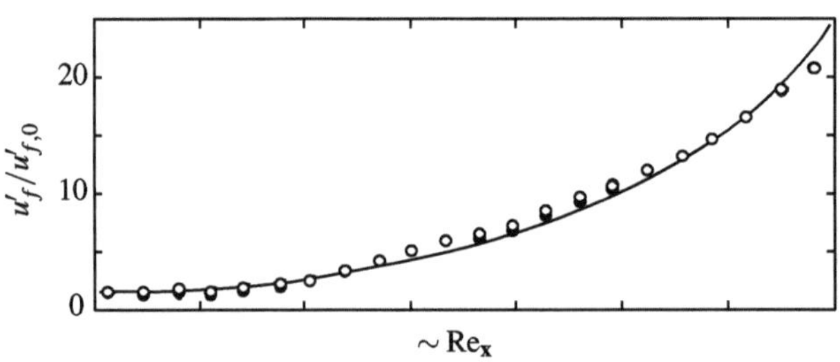

Fig. 11.4 Growth of the wave amplitude u'_f (measured in the disturbance maximum inside the boundary layer) approximately equal to 1% (∘) and 0.5% of U_e (•) at the end of the region shown; approximation by exponential growth (*solid line*). The wave was excited in a swept-wing boundary-layer stationary vortex by sound at fixed frequency f (Boiko et al, 1997)

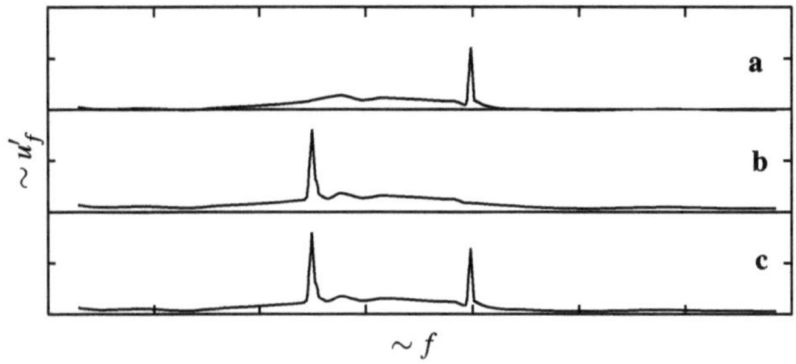

Fig. 11.5 Spectra of traveling disturbances in a swept-wing boundary-layer stationary vortex excited by periodic suction and blowing (**a**), sound (**b**) and simultaneously (**c**) (Boiko et al, 1997)

the spectra are shown at separate and combined excitation of disturbances with periodic suction and blowing through a hole in the surface of the experimental model and with acoustic oscillations from an external source. The spectra exhibit the absence of noticeable interaction of the waves (combinational modes) up to a total amplitude of the waves of about 1% of external-flow velocity U_e.

The inviscid nature of the secondary instability mechanism for this case is illustrated in Fig. 11.6. The distributions of the mean-velocity defect and the amplitude of growing disturbances, as well as the position of the layer where the local mean velocity is equal to the velocity of the wave propagation along the vortex (an analog of the critical layer) are shown. The core of the perturbed region is close to the critical layer, and the distribution of the traveling wave amplitude correlates well with the location of local maxima of the velocity gradient along the spanwise coordinate.

The visualization of a flat-plate boundary layer shown in Fig. 11.7 indicates that the streaks developing in a flat-plate boundary layer begin to oscillate in the trans-

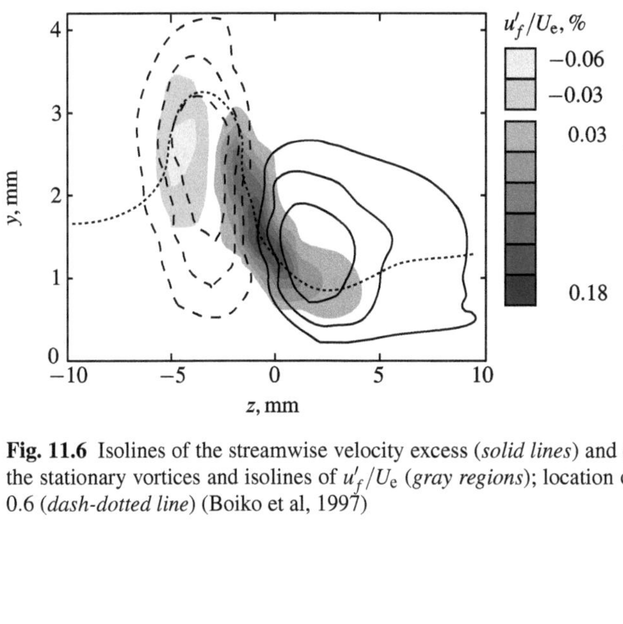

Fig. 11.6 Isolines of the streamwise velocity excess (*solid lines*) and defect (*dashed lines*) due to the stationary vortices and isolines of u'_f/U_e (*gray regions*); location of the critical layer $U/U_e =$ 0.6 (*dash-dotted line*) (Boiko et al, 1997)

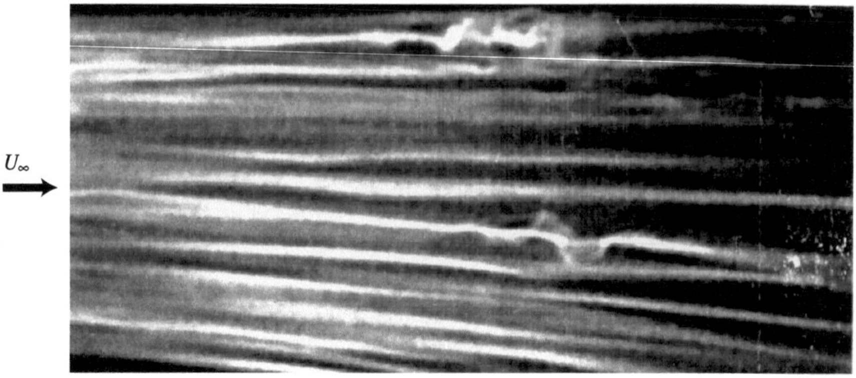

Fig. 11.7 Smoke visualization of high-frequency wave packet manifestation at laminar–turbulent transition in a flat-plate boundary layer at Tu = 1.5% (Alfredsson et al, 1996)

verse direction before the final breakdown to turbulence. A similar phenomenon is also observed at the late stages of the subcritical transition in the plane Couette and Poiseuille flows. It is conjectured that the breakdown of the streaks due to the activity of traveling disturbances at subcritical Reynolds numbers $\mathrm{Re} < \mathrm{Re}_\mathrm{L}$ is similar to the process of secondary instability of flows with the Görtler or crossflow vortices. At $\mathrm{Re} > \mathrm{Re}_\mathrm{L}$, a competitive mechanism is interaction of the streaks with the Tollmien–Schlichting waves, as discussed in Chap. 12.

Corresponding theoretical and numerical studies are mostly concentrated on secondary instability of flows modulated by streaks developed from initial optimal disturbances introduced locally. The downstream amplitude development of an optimal stationary disturbance is displayed in Fig. 11.8**a** for a set of upstream amplitudes. In Fig. 11.8**b**, a typical nonlinearly saturated streak is illustrated by contour plot of streamwise velocity distribution in boundary layer. As seen, regions of strong transverse shear are formed on the sides of the low-speed region, which is also displaced further away from the wall during the saturation process. Possible inviscid secondary instability is caused by inflectional profiles of the basic flow velocity. In numerical calculations, it appeared that the sinuous modes are the dominating instabilities in such a case. Varicose modes become amplifying only at very large amplitudes of the primary streaks (about 37% of external-flow velocity).

Note that such an approach prohibits a direct quantitative comparison with experimental results at a high free-stream disturbance level, where the streak generation is distributed over an extended region. Even at a local excitation of streaks in a controlled experiment, they cannot be excited optimally. As a consequence, in contrast to the theoretical predictions, both the varicose and the sinuous-like disturbances were observed in numerous experimental studies dealing with the secondary instability of streaks in different flows. This indicates a deficiency of the traditional numerical approach without proper treating the receptivity process, which is, however, not an easy task. Despite the mentioned specifics of such computations, many characteristics of the secondary sinuous modes related to the optimally generated

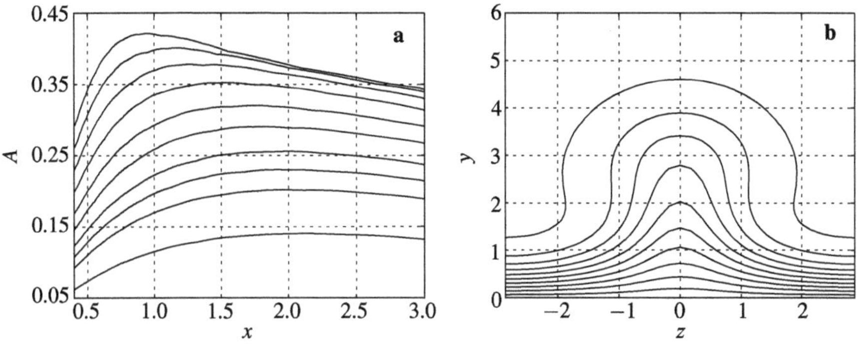

Fig. 11.8 Downstream development of the streak amplitude A versus the streamwise coordinate x for different upstream amplitudes (**a**) and isolines of boundary-layer streamwise velocity in a y–z-plane at $x = 2$, $A = 0.36$ (**b**). The coordinates are non-dimensionalized using the local Blasius length scale. Maximum contour level is 0.98, contour spacing is 0.1 (Brandt et al, 2003)

streaks (localization, phase velocities, etc.) were found to be quite close to the characteristics of disturbances determined experimentally.

11.3 Local high-frequency secondary instability

There is another possible mechanism of secondary instability of shear flows, which differs essentially from that described above. It can manifest itself in the region, where the instantaneous velocity distributions have strong shear layers along wall-normal or transverse directions, appearing in some situations at late stages of the laminar–turbulent transition. The corresponding (local in time and space) velocity profiles give rise to inviscid flow instability to high-frequency perturbations if quasi-stationarity and quasi-two-dimensionality of the former is assumed.

Landahl (1972) proposed a kinematic treatment of local high-frequency secondary instability, in which the coincidence of the phase velocity of the primary wave with the group velocity of the secondary disturbances is a necessary condition of instability. This requirement is needed for effective amplification of secondary perturbations in the unstable region, which propagates with the phase velocity of the primary wave. In other words, the wave packet of high-frequency secondary disturbances must be absolutely unstable in the frame of reference moving with the crests of the primary wave. The fact that flow destruction occurs on scales approximately equal to one wavelength of the primary instability wave testifies to feasibility of the local mechanism.

In the past, this was the most wide-spread theoretical idea on the laminar flow breakdown in the K-regime of the laminar–turbulent transition in a flat-plate boundary layer, occurring in the region of the high-shear layer formed by a Λ-structure (see Chap. 12 for a detailed description of the K-regime). It has been shown experimentally, however, that the criteria for realization of local high-frequency secondary instability are not satisfied in this case. In particular, the propagation velocities of the high-shear layer and high-frequency disturbances are different. Besides, the dominant disturbance frequencies are rather low, being centered around the subharmonic frequency, and the growth of irregular fluctuations is quite gradual, rather than explosive.

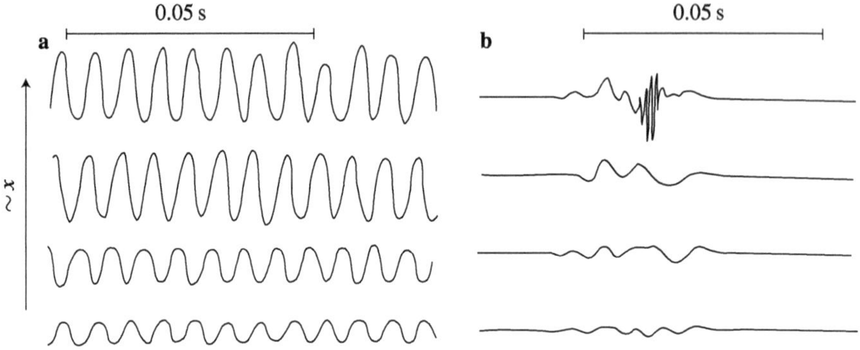

Fig. 11.9 Evidence of early appearance of an outburst of secondary instability in oscilloscope traces of the Tollmien–Schlichting wave packet (**b**) in comparison with the wave harmonic in time (**a**) (Gaster, 1979)

Nevertheless, the regions of high shear also appear in the absence of Λ-like vortices or even without previous amplification of the two-dimensional Tollmien–Schlichting wave, the regions being induced by an isolated wave packet or by a pair of oblique waves. It is not unlikely that the local high-frequency secondary instability approach is valid in these cases at a certain stage of flow breakdown. Figure 11.9 shows oscilloscope traces of a wave packet and a harmonic wave for several downstream positions. While the harmonic primary wave (at the left) of much higher amplitude does not lead to secondary instability, as there is no layer of strong shear in the transverse direction, it occurs at quite small amplitudes in the wave packet (at the right). An outburst of high-frequency oscillations probably associated with local secondary instability is clearly seen.

The results of flow visualizations also show that the transition to turbulence in boundary layers modulated by the Görtler vortices and streaks can be determined by instability mechanisms that produce growing waves independently in each vortical pair, so that the neighboring pairs can experience different types of secondary motions. In such cases, the explosive growth of traveling disturbances can be attributed to a local high-frequency instability mechanism.

Additional material for this chapter is given in the multimedia files Nos. 8–10, which supplement the book, titled 'Sinusoidal and varicose instabilities of streaks in a boundary layer', 'Streak instability in adverse pressure gradient boundary layer', and 'Secondary instability of streamwise vortices in a 3D flow.'

References

Alfredsson PH, Bakchinov AA, Kozlov VV, Matsubara M (1996) Laminar–turbulent transition at a high level of a free stream turbulence. In: Duck and Hall (1996), pp 423–436

Boiko AV, Kozlov VV, Syzrantsev VV, Scherbakov VA (1997) A study of the influence of internal structure of a streamwise vortex on the development of traveling disturbances inside it. Thermophys Aeromech 4(4):343–354

Bottaro A, Klingmann BGB (1996) On the linear breakdown of Görtler vortices. Eur J Mech B/Fluids 15(3):301–330

Brandt L, Cossu C, Chomaz JM, Huerre P, Henningson DS (2003) On the convectively unstable nature of optimal streaks in boundary layers. J Fluid Mech 485:221–242

Corke TC, Mangano RA (1989) Resonant growth of three-dimensional modes in transitioning boundary layers. J Fluid Mech 209:93–150

Crouch JD, Herbert T (1993) Nonlinear evolution of secondary instabilities in boundary-layer transition. Theoret Comput Fluid Dyn 4:151–175

Duck PW, Hall P (eds) (1996) Nonlinear Instability and Transition in Three-Dimensional Boundary Layers, Fluid Mechanics and its Application, Kluwer, Dordrecht

Gaster M (1979) The propagation of linear wave packet in laminar boundary layers — asymptotic theory for non-conservative wave systems. AIAA Paper 79–1492

Herbert T (1988) Secondary instability of boundary layers. Ann Rev Fluid Mech 20:487–526

Kohama Y, Onodera T, Egami Y (1996) Design and control of crossflow instability field. In: Duck and Hall (1996), pp 147–156

Landahl ML (1972) Wave mechanism of breakdown. J Fluid Mech 56:775–802

Li F, Malik MR (1995) Fundamental and subharmonic secondary instabilities of Görtler vortices. J Fluid Mech 297:77–100

Poll DIA (1979) Transition in the infinite swept attachment line boundary layer. Aeronaut Quart 30:607–629

Poll DIA (1985) Some observations on the transition process on the windward face of a long yawed cylinder. J Fluid Mech 150:329–356

Saric WS, Kozlov VV, Levchenko VY (1984) Forced and unforced subharmonic resonance in boundary-layer transition. AIAA Paper 84–0007

Zelman MB, Maslennikova II (1993) Tollmien–Schlichting-wave resonant mechanism for subharmonic-type transition. J Fluid Mech 252:449–478

Further Reading

Berlin S, Wiegel M, Henningson DS (1999) Numerical and experimental investigations of oblique boundary layer transition. J Fluid Mech 393:23–57

Betchov R (1960) On the mechanism of turbulent transition. Phys Fluids 3(6):1026–1027

Bippes H (1990) Instability features appearing on swept wing configurations. In: Arnal D, Michel R (eds) Laminar–Turbulent Transition, Springer–Verlag, Berlin, IUTAM Symposium, pp 419–430

Boiko AV, Kozlov VV, Syzrantsev VV, Scherbakov VA (1995) Experimental investigation of high-frequency secondary disturbances in a swept wing boundary layer. J Appl Mech Tehn Phys 36(3):385–393

Boiko AV, Kozlov VV, Syzrantsev VV, Scherbakov VA (1995) Experimental study of the transition to turbulence at a single stationary disturbance in the boundary layer on an oblique airfoil. J Appl Mech Tehn Phys 36(1):67–77

Borodulin VI, Kachanov YS (1994) Experimental study of nonlinear stages of a boundary layer breakdown. In: Lin SP, Phillips WRC, Valentine DT (eds) Nonlinear Instability of Nonparallel Flows, Springer–Verlag, Berlin, IUTAM Symposium, pp 69–80

Dryganets SV, Kachanov YS, Levchenko VY, Ramazanov MP (1990) Resonance flow randomization in the K-regime of boundary-layer transition. J Appl Mech Tehn Phys 31(2):239–249

Fischer TM, Dallmann U (1991) Primary and secondary stability analysis of a three-dimensional boundary-layer flow. Phys Fluids A 3(10):2378–2391

Floryan JM (1991) On the Görtler instability of boundary layers. J Aerosp Sci 28:235–271

Gaster M (1979) The propagation of linear wave packet in laminar boundary layers — asymptotic theory for non-conservative wave systems. AIAA Paper 79–1492

Guo Y, Finlay WH (1994) Wavenumber selection and irregularity of spatially developing nonlinear Dean and Görtler vortices. J Fluid Mech 264:1–40

Henningson DS (1995) Bypass transition and linear growth mechanisms. In: Benzi R (ed) Advances in turbulence, Kluwer, Dordrecht, pp 190–204

Klebanoff PS, Tidstrom KD, Sargent LM (1962) The three-dimensional nature of boundary-layer instability. J Fluid Mech 12:1–34

Koch W, Bertolotti FP, Stolte A, Hein S (2000) Nonlinear equilibrium solutions in a three-dimensional boundary layer and their secondary instability. J Fluid Mech 406:131–174

Lerche T (1997) Experimentelle Untersuchung nichtlinearer Strukturbildung im Transitionsprozess einer instabilen dreidimensionalen Grenzschicht. No. 310 in Reihe 7: Strömungstechnik, VDI Verlag GmbH, Düsseldorf

Nayfeh AH (1987) Nonlinear stability of boundary layers. AIAA Paper 87–0044

Nishioka M, Asai M, Iida S (1980) An experimental investigation of the secondary instability. In: Eppler R, Fasel H (eds) Laminar–Turbulent Transition, Springer–Verlag, Berlin, IUTAM Symposium, pp 37–46

Nitschke-Kowsky P, Bippes H (1988) Instability and transition of a three-dimensional boundary layer on a swept flat plate. Phys Fluids 31(4):786–795

Swearingen JD, Blackwelder RF (1987) The growth and breakdown of streamwise vortices in the presence of a wall. J Fluid Mech 182:255–290

Wu X, Stewart PA, Cowley SJ (2008) On the catalytic effect of resonant interactions in boundary layer transition. In: Bock HG, Hoog F, Friedman A, Gupta A, Neunzert H, Pulleyblank WR, Rusten T, Santosa F, Tornberg AK, Capasso V, Mattheij R, Neunzert H, Scherzer O, Bonilla LL, Moscoso M, Platero G, Vega JM (eds) Progress in Industrial Mathematics at ECMI 2006, Mathematics in Industry, vol 12, Springer, pp 146–156.

Part IV
Onset of turbulence

Chapter 12
Nonlinear effects during the laminar–turbulent transition

There is a variety of nonlinear processes taking place at flow breakdown to turbulence. Competing with each other, they occur more or less individually only with special adjustment of the initial conditions. Below we consider some prototypical mechanisms of the laminar flow breakdown originating from the preceding amplification of linear instability waves and streaks in boundary layers.

12.1 Onset of nonlinearity

Since in ordinary conditions two-dimensional Tollmien–Schlichting waves become unstable first in two-dimensional flows, it is assumed that they have better chances to trigger the nonlinear transition phenomena at high Reynolds numbers. A typical phenomenological sequence of stages of the laminar–turbulent transition in a boundary layer in such a case is shown in Fig. 12.1.

It is expected that nonlinear terms in the Navier–Stokes equations first become large in the critical layer close to the maximum of the disturbance wall-normal velocity. By transforming to a coordinate system moving with the wave, it is possible to introduce a stream function Ψ as

$$\Psi = \int (U-c)\mathrm{d}y + \varepsilon\psi$$

and to consider an equation for the disturbance stream function ψ of an arbitrary amplitude ε, derived from the Navier–Stokes equations (Zhigulev and Tumin, 1987):

$$\underbrace{\frac{1}{\varepsilon}\frac{\partial}{\partial t}(\varepsilon\nabla^2\psi)}_{D} + \underbrace{(U-c)\nabla^2\psi_x - \nabla^2 U\,\psi_x}_{A} + \underbrace{\varepsilon(\psi_y\nabla^2\psi_x - \psi_x\nabla^2\psi_y)}_{B} = \underbrace{\frac{1}{\mathrm{Re}_\delta}\nabla^4\psi}_{C},$$

where the values are normalized to the boundary-layer thickness δ and external-flow velocity U_e, and the subscripts denote differentiation with respect to corresponding coordinates. The linear terms A are minimal near the wave critical layer, where $U(y_\mathrm{c}) \approx c$ (Fig. 12.2). Therefore, it is possible to assume that the nonlinear terms

A.V. Boiko et al., *Physics of Transitional Shear Flows*,
Fluid Mechanics and Its Applications 98, DOI 10.1007/978-94-007-2498-3_12,

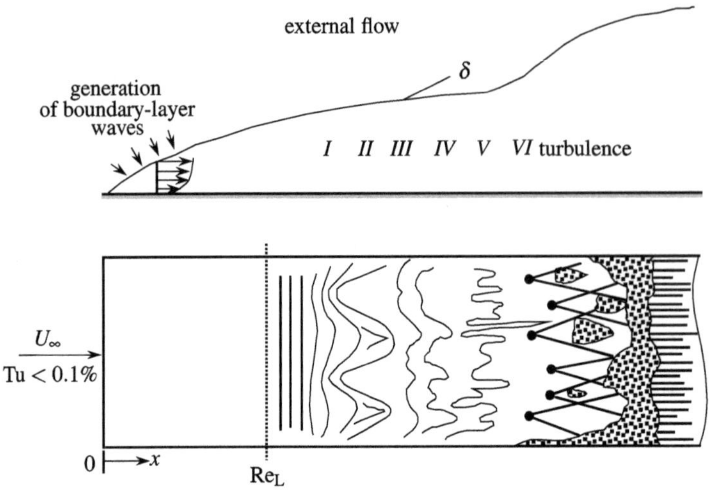

Fig. 12.1 Typical stages of the transition process at low free-stream turbulence (Schlichting and Gersten, (2000): linear instability stage of the Tollmien–Schlichting waves (*I*); development of three-dimensional Λ-structures (*II*); formation of streamwise vortical structures (*III*); appearance of strong shear layers (*IV*); region of origination of turbulent spots (*V*); interaction and merging of turbulent spots (*VI*)

B become comparable with the linear ones primarily at the critical layer, where the instability wave stream function has the maximum amplitude. Thus, the 'threshold' amplitude $\varepsilon_{\mathrm{ampl}}$, at which the nonlinear terms become pronounced, is determined at $A/B \sim 1$. It appears that, if the critical layer is well separated from the near-wall viscous region, then

$$\varepsilon_{\mathrm{ampl}} = [U_y(y_{\mathrm{c}})]^{1/3}(\alpha \mathrm{Re}_\delta)^{-2/3},$$

which usually gives $\varepsilon_{\mathrm{ampl}} \approx 1\%$ of U_{e}; if they are close to each other, then

$$\varepsilon_{\mathrm{ampl}} = [U_y(y_{\mathrm{c}})]^{4/3}(\alpha \mathrm{Re}_\delta)^{-2/3} c^{-1},$$

and it can reach 10% of U_{e}.

Fig. 12.2 Structure of the boundary layer at separation of critical and near wall viscous layers (Maslowe, 1985)

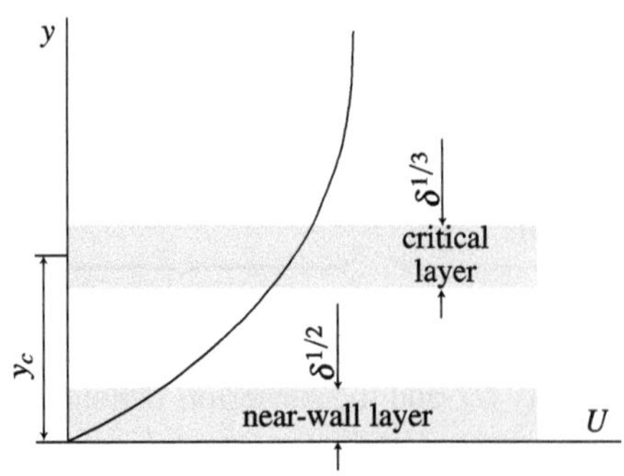

12.2 Basic nonlinear scenarios

Depending on the initial spectrum of laminar flow perturbations involved in wave interactions, different nonlinear scenarios are found. With predominance of two-dimensional waves at low free-stream turbulence levels, the K-regime (after *Klebanoff*) and the subharmonic one (also called the N-regime after its discoverers, the *Novosibirsk* group) are distinguished (Kachanov, 1994). The transition route originating from interaction of oblique waves is known as 'oblique breakdown'(Schmid and Henningson, 2000).

12.2.1 K- and N-regimes of the laminar–turbulent transition

These transition scenarios are featured by generation of the so-called Λ-structures at coalescence of strong shear layers directed at an angle to the flow, both in the streamwise and wall-normal directions. In the K-regime, the structures follow each other, while in the N-regime they are staggered, as is shown for a flat-plate boundary layer and the plane Poiseuille flow in Fig. 12.3. Mixed or combination types can also take place.

The transition in straight and swept-wing boundary layers is also connected with formation of Λ-structures (Fig. 12.4). In the straight-wing flow, a symmetric Λ-vortex pattern develops similar to that in the flat-plate boundary layer. Owing to the crossflow effect, it becomes asymmetric at the swept wing. The fact that the structures are observed in different external and internal flows indicates the universality of the mechanisms underlying their formation.

Normally, the K-type of laminar flow breakdown takes place when the initial wave amplitude is high enough. For the first time it was found in the Blasius boundary layer by Klebanoff et al (1962). By introducing controlled two-dimensional Tollmien–Schlichting waves with the vibrating-ribbon technique, they observed the appearance of stationary transverse modulations of their wavefronts. The transverse flow periodicity was accompanied by appearance of low-velocity '*spikes*' (intensive outbursts of the streamwise velocity at each period of the primary wave) on oscilloscope traces when measured by a hot-wire (Fig. 12.5). The subsequent transition seemed to be localized in the maxima (peaks) of the transverse basic flow velocity distribution and was characterized by formation of strong shear layers.

More recent experimental and numerical results revealed further details of the K-regime of breakdown. At the initial stage of the transition, the three-dimensional distortion of a plane Tollmien–Schlichting wave leads to formation of a system of Λ-vortices. The vortex 'head' traveling through a hot-wire probe produces phase-synchronized oscillations of higher harmonics of the primary wave in the oscilloscope signal sensed by an observer as the spikes. During downstream propagation, the vortex head is displaced in the direction of the boundary-layer edge and enters the region of higher velocities, forming a ring-like vortex. Figure 12.6 illustrates the formation and development of several such transitional structures. The right-most

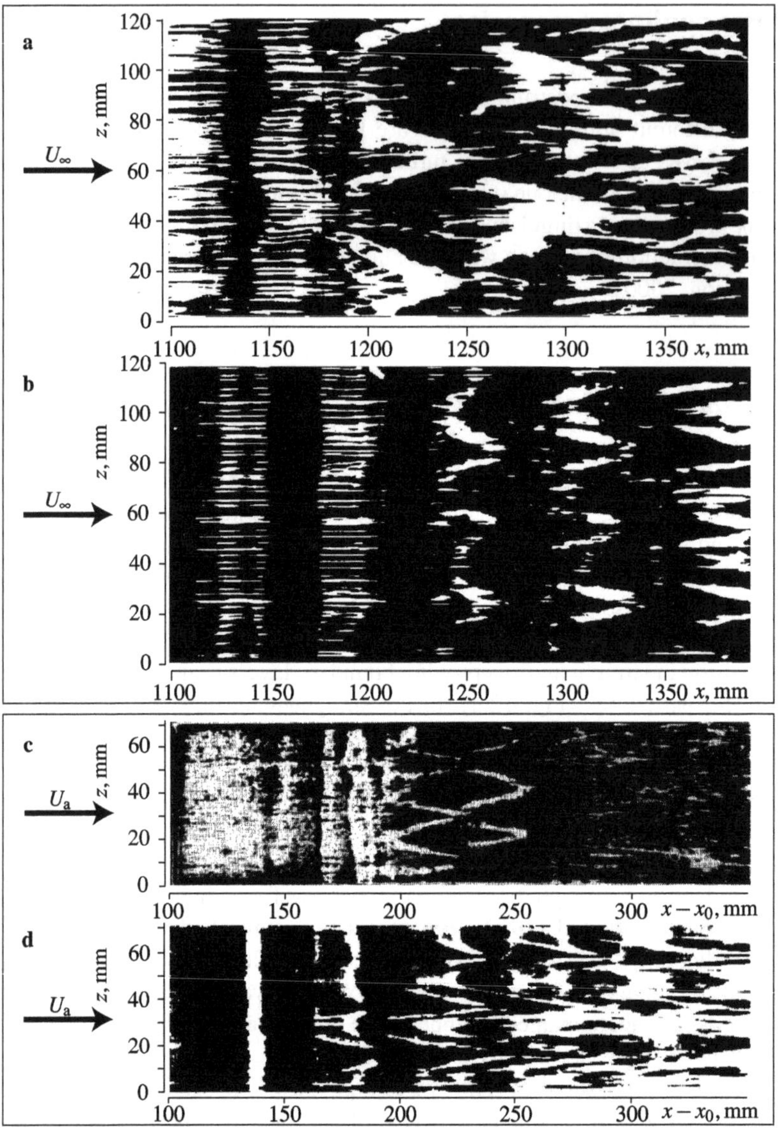

Fig. 12.3 Smoke visualization of Λ-structures in the plane parallel to the surface: N-regime in a flat-plate boundary layer (**a**) and K-regime in a flat-plate boundary layer (**b**) (Saric et al, 1984); N-regime in a plane Poiseuille flow (**c**) and K-regime in a plane Poiseuille flow (**d**) (Ramazanov, 1985)

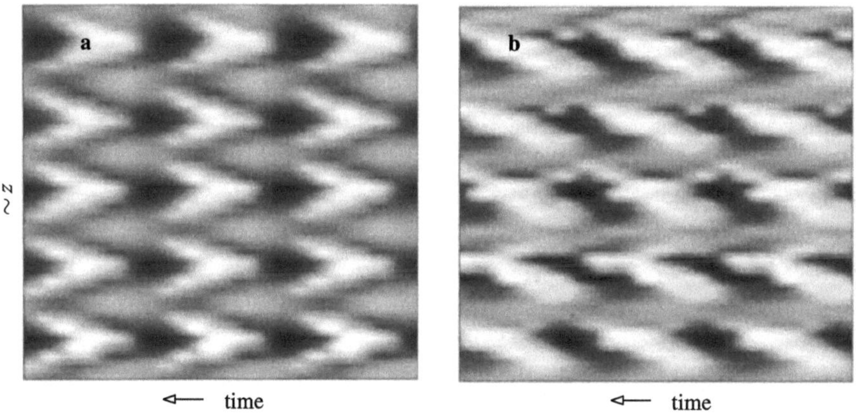

Fig. 12.4 Streamwise velocity disturbances in t–z-planes at the transition on straight (**a**) and swept (**b**) wings (Chernoray et al, 2001)

of them has already developed into a ring vortex and tears off from two 'legs' extending down towards the wall, while the others are at different stages of formation of pre-toroidal Ω-vortices at the tip of the Λ-structure.

The subharmonic route (N-type) of the transition usually dominates at smaller disturbance amplitudes than the K-regime. In this case, the nonlinear processes of generation of multiple harmonics of the primary wave do not have time to operate, while the subharmonic comes from the parametric resonance with the primary wave. Such a scenario is more probable at a low degree of background laminar flow perturbations; e.g., in the conditions of glider flight.

In mechanical systems, the parametric resonance (characterized by subharmonic growth) is observed at an appropriate phase tuning between the oscillations involved in the interaction, with the anti-resonance (characterized by subharmonic attenuation) occurring at a large phase mismatch. The subharmonic amplitude grows su-

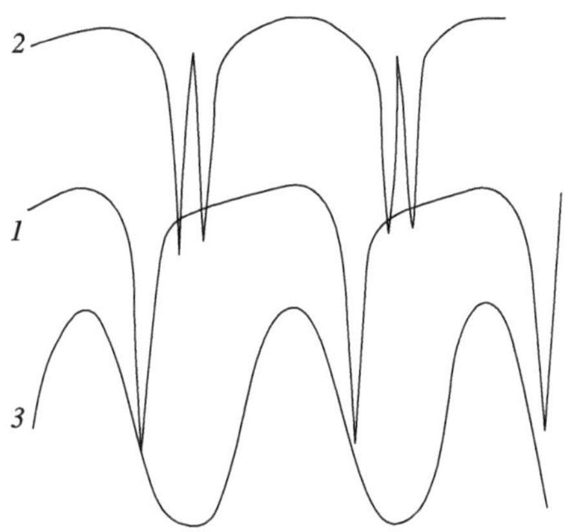

Fig. 12.5 Generation and multiplication of spikes in oscilloscope traces of the boundary-layer velocity oscillations: initial section (*1*), downstream section (*2*) and vibrating ribbon forcing signal (*3*) (Klebanoff et al, 1962)

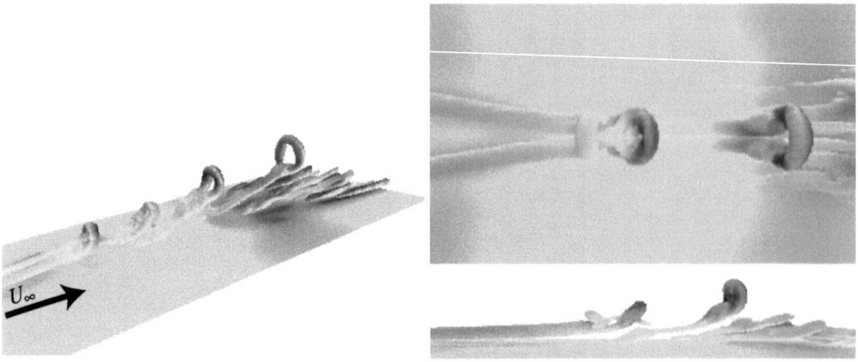

Fig. 12.6 Direct numerical simulation of laboratory experiments of Kachanov (1994) by Rist et al (1998)

perexponentially (as a doubled exponential function) and it can even exceed the amplitude of the forcing wave without an essential backward influence or feedback.

A classical weakly nonlinear resonance model explaining subharmonic growth in boundary layers is the resonant triad of the primary two-dimensional wave and a pair of oblique subharmonic waves (Craik, 1971). According to kinematic considerations, the wave triad can occur if the frequencies of the oscillations and the real parts of the wavevectors satisfy the relation

$$\omega_1 = \omega_2 + \omega_3\,, \quad \kappa_1 = \kappa_2 + \kappa_3\,.$$

For the channel flow and in the boundary layer, Craik's triad is formed if one chooses $\omega_2 = \omega_3 = 1/2\omega_1$, $\alpha_2 = \alpha_3 = 1/2\alpha_1$, and $\beta_2 = -\beta_3$. The dynamical condition of realization of such triads is quite strict, as the wave dispersion quickly destroys the synchronism unless the subharmonic amplification is fast enough to trigger further transition. This restriction as well as some other discrepancies between the model results and experimental observations stimulated further studies related to the wave resonances.

In laboratory experiments, the N-regime appears as a wide-band amplification of low frequencies centered about the subharmonic $\omega_{1/2} = \omega_1/2$ of the dominant wave ω_1 (the fundamental one). The subharmonic becomes periodic along the transverse coordinate, which corresponds to a pair of oblique waves propagating at equal angles but in opposite directions relative to the basic flow. Later stages of the transition are characterized by gradual filling of the frequency spectrum and by growth of overtones, produced by the wave cross-interactions.

12.2.2 Oblique breakdown

The scenario of oblique breakdown applies to the transition that is basically initiated by three-dimensional modes of the laminar shear layer. The model is focused on the interaction of two symmetric oblique waves of the same frequency ω and the transverse wavenumbers $\pm\beta$. At the initial stage, a set of discrete modes is generated, including stationary streamwise perturbations at 2β, i.e., spaced with the transverse period twice smaller than that of the interacting waves. Through the lift-up mechanism, the streamwise disturbances evolve into streaky structures (see Chap. 9) subjected to secondary instabilities which, finally, result in turbulent motion.

The oblique breakdown with the perturbed flow pattern illustrated in Fig. 12.7 is found in a series of experimental and numerical studies. When comparing this scenario with other routes of laminar flow turbulization, it turns out that the oblique breakdown proceeds rapidly and operates at fairly small initial amplitudes of the laminar flow disturbances; moreover, it can be observed at subcritical Reynolds numbers.

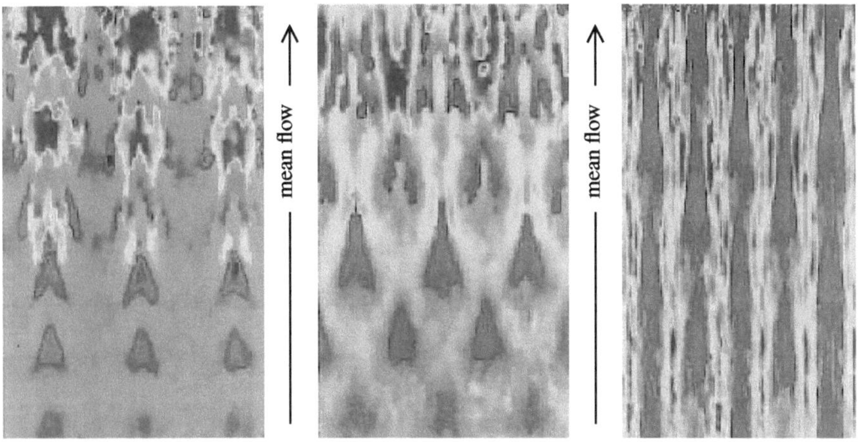

Fig. 12.7 Numerical simulation of the instantaneous streamwise boundary-layer perturbations during the oblique breakdown (*right*) as compared to the K- (*left*) and N- (*middle*) regimes of breakdown (Berlin et al, 1999)

12.2.3 Nonlinearity of locally separating boundary layers

Similarly to near-wall and free shear layers, the unstable flow at boundary-layer separation is subject to nonlinear effects coming after the exponential growth of small-amplitude disturbances. At this stage, a broad spectrum of velocity oscillations is generated through wave interactions including the fundamental and subharmonic (parametric) resonances, wave combinations, oblique breakdown, etc. Their contribution to the separated flow pattern depends on the initial conditions; that is, the frequency and wavenumber spectra originating from preceding growth in the region of linear instability. As a result, quite different unsteady phenomena are found in separation bubbles with variation of the initial spectrum of perturbations.

A key mechanism of 'randomization' in shear layers is excitation of subharmonic disturbances of the fundamental wave amplified at the stage of linear instability. Figure 12.8 illustrates this route of laminar flow breakdown in the separation region over an airfoil by the spectra of velocity fluctuations. The data are obtained under a weak monochromatic forcing of the separated boundary layer modeling the natural flow occurring with a fairly smooth initial spectrum. It is seen that streamwise growth of the controlled instability wave is accompanied by rapid amplification of perturbations centered around the subharmonic $f_0/2$ of the fundamental disturbance f_0. As a result, a broad band of low-frequency oscillations is generated. Furthermore, the resonant coupling of the fundamental wave with background fluctuations produces a wide wave-number spectrum of subharmonic disturbances. That is, in separation bubbles with small dispersion of propagating velocities of the separated layer oscillations, both two-dimensional and oblique subharmonic waves may have comparable resonant growth rates.

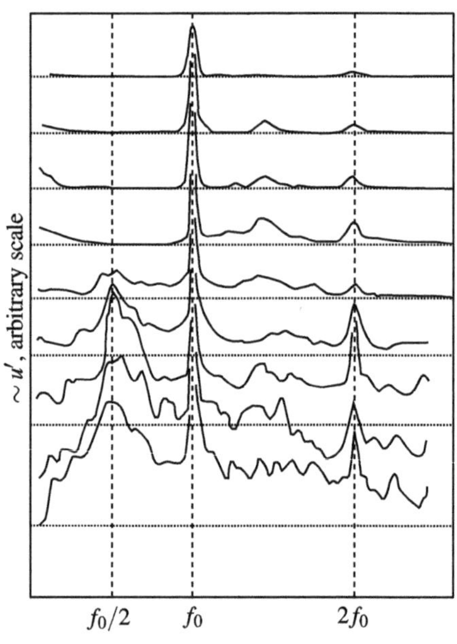

Fig. 12.8 Subharmonic excitation in a separation bubble on an airfoil (Boiko et al, 1989). The boundary layer separates at $x/l = 0.57$ and reattaches at about $x/l = 0.75$, where l is the wing chord length. From *top* to *bottom* $x/l = 0.657$, 0.664, 0.671, 0.679, 0.686, 0.693, 0.700, 0.707, 0.714

The disturbances excited with larger amplitudes in the separation region may evolve in a completely different manner, Fig. 12.9. In this case, growth of the generated perturbations also starts from linear instability waves in the upstream part of the separation bubble. However, comparing these observations with similar data for the subharmonic excitation (see Fig. 12.8), one can find that, upon an increase of the initial amplitude of oscillations, the disturbances become more regular with a delay of the onset of chaotic motion. In terms of vortex dynamics, this corresponds to formation of coherent vortices rolling-up in the separated shear layer and dominating the flow pattern, instead of the laminar–turbulent transition.

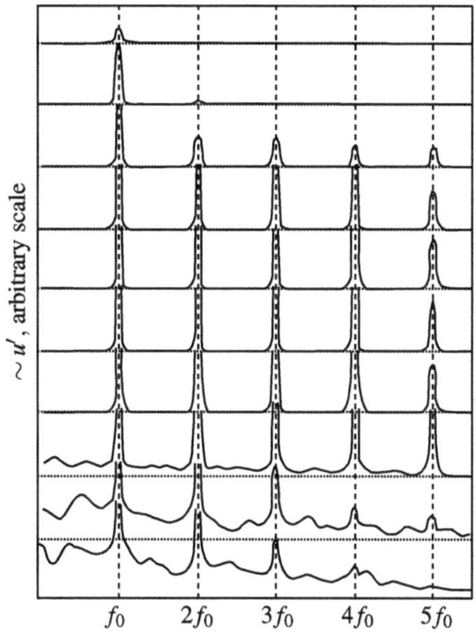

Fig. 12.9 Regular disturbance in the flow configuration of Fig. 12.8 under high-amplitude excitation at $x/l =$ 0.586, 0.614, 0.643, 0.671, 0.700, 0.729, 0.757, 0.786, 0.814, 0.843 (from *top* to *bottom*)

An explanation of the above observation is that the initial spectrum of fluctuations near to separation includes its response to environmental perturbations as well as to unsteadiness of the reattaching flow, while the latter is affected by excitation. As a result, generation of an amplifying periodic disturbance in the separation region may have a pronounced effect on background components as is illustrated in Fig. 12.10. In this example, controlled excitation of an instability wave behind a two-dimensional hump on a flat surface is followed by a decrease in low-frequency separated flow oscillations matching the frequency band of the subharmonic resonance. Such an interplay between the forced perturbation and the initial spectrum of random fluctuations might be responsible for the delay in subharmonic generation and removal of turbulent flow downstream, as is observed in Fig. 12.9.

12.3 Transition to turbulence in boundary layers at a high free-stream disturbance level

At the free-stream turbulence level (Tu) above approximately 5%, the transition in a flat-plate boundary layer occurs at the minimum Reynolds number at which the self-sustained boundary-layer turbulence can exist, i.e., at $\mathrm{Re}_{\delta^*} \approx 190$ (Arnal, 1992). At lower levels of Tu, different experiments disagree widely concerning the position

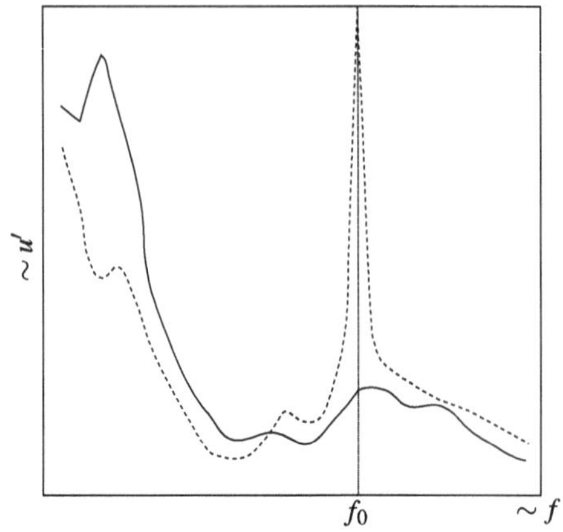

Fig. 12.10 Suppression of background oscillations under controlled excitation of boundary-layer separation behind a two-dimensional hump on a flat plate (Boiko et al, 1991). Natural (*solid line*) and excited (*dotted line*) spectra

and the extent of the transition region, and there seems to be no general correlation between the free-stream turbulence level and Re_T.

For a long time it was conjectured that the transition at high free-stream turbulence is closely related to local boundary-layer separations induced by free-stream fluctuations, so that turbulent spots are generated and their spatial evolution results in a completely turbulent flow (Taylor, 1936). More recent data indicate that the boundary-layer streaks, provided that their initial amplitudes are large enough, are also capable of transforming into turbulent spots, as shown in Fig. 12.11. The transformation occurs through either the stage of secondary instability considered in Chap. 11 or some other mechanisms among which, e.g., interactions of the streaks with the Tollmien–Schlichting waves are probably most important.

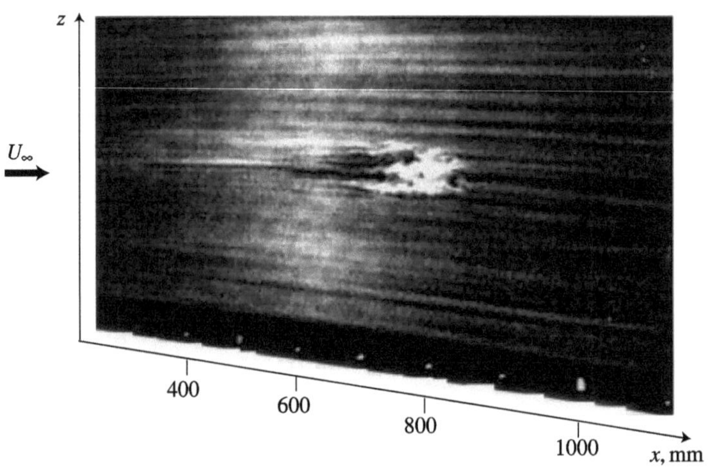

Fig. 12.11 Incipient turbulent spot smoke visualization at the natural transition over a flat plate at Tu = 1.5% (Alfredsson et al, 1996)

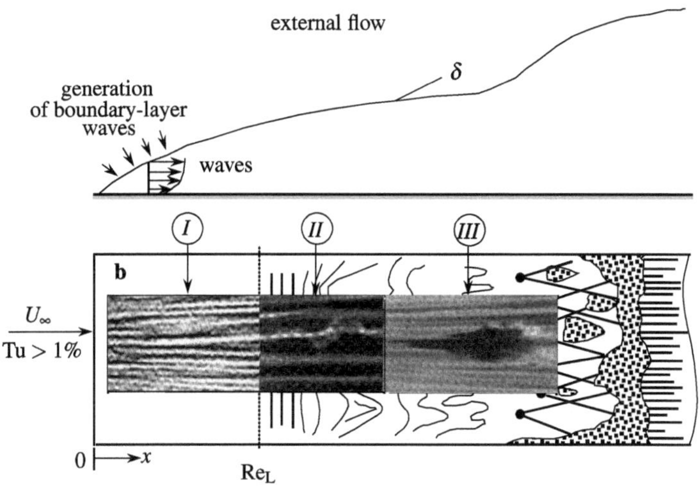

Fig. 12.12 Transition scenarios at high free-stream turbulence (Alfredsson et al, 1996): development of streaks (*I*); origination of incipient turbulent spots (*II*); formation, interaction, and merging of turbulent spots (*III*)

Based on modern detailed experimental investigations, a scenario of the transition at high free-stream turbulence may be proposed as a sequence of stages (Fig. 12.12), as is done for the transition at low free-stream turbulence. The first of them is the formation of streaks by free-stream-localized vortical disturbances. The streaks modulate the boundary layer in the transverse direction. The second stage consists in the following streak development accompanied by generation of high-frequency wave packets and incipient turbulent spots due to various mechanisms including interaction with the Tollmien–Schlichting waves and secondary instability. The third stage includes development and interaction of turbulent spots, completing the laminar–turbulent transition.

12.3.1 Linear Tollmien–Schlichting waves in the presence of streaks

'Natural' Tollmien–Schlichting waves were detected at moderate levels of free-stream turbulence by many investigators. These studies showed that the transition in a flat-plate boundary layer at least up to Tu $\leq$ 3–5% can occur through the development of Tollmien–Schlichting wave packets, which were observed together with the streaks discussed earlier.

An example of amplitude and phase distributions of the waves, extracted by the phase filtering technique, is shown in Fig. 12.13. At Tu = 1.5%, the overall value of u'/U_e in the middle of the boundary layer is 5% of U_e at the most downstream position. In spite of this, the extracted wave exhibits the same characteristics as the

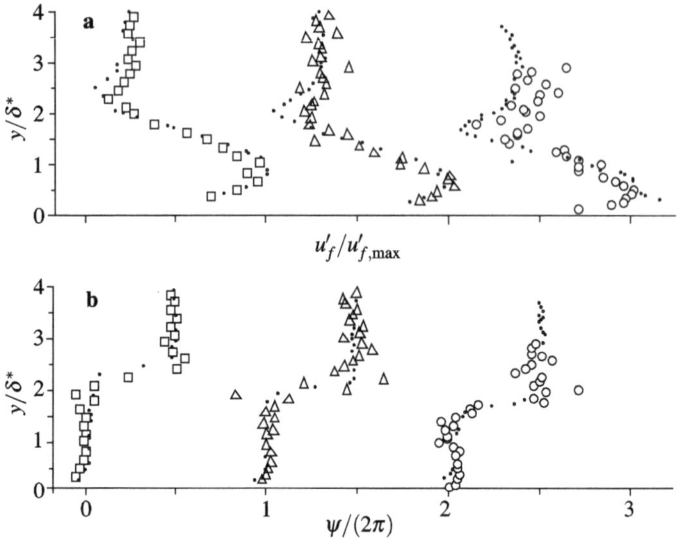

Fig. 12.13 Amplitude (**a**) and phase (**b**) profiles of the Tollmien–Schlichting waves at a high free-stream turbulence level (Tu = 1.5%): $F = 200 \times 10^{-6}$, $\mathrm{Re}_{\delta^*} = 345$ (□), 525 (△), 715 (○); profiles measured in low-turbulent flow (.) (Boiko et al, 1994)

Tollmien–Schlichting wave in an undisturbed boundary layer. The contribution of random fluctuations increases downstream, resulting in a certain scatter in the data points. This is particularly obvious close to the amplitude minimum and the corresponding phase shift. It was found that the dependence of the disturbance phase on the streamwise coordinate in these and similar measurements could be approximated by straight lines, and the wave phase velocity was the same as at low Tu.

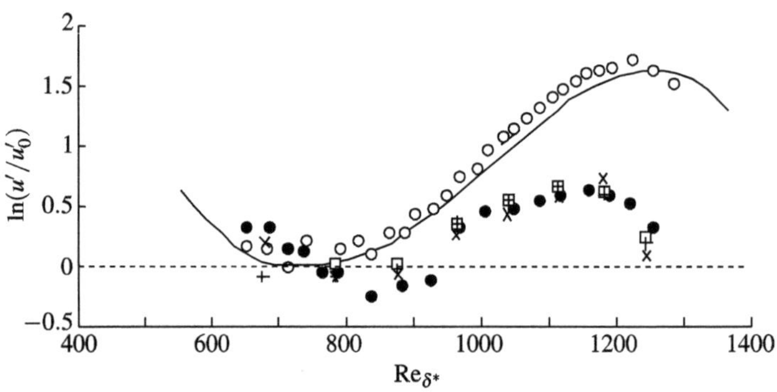

Fig. 12.14 Amplitude evolution of the Tollmien–Schlichting waves with $F = 100 \times 10^{-6}$ normalized to the wave amplitude at branch I of the neutral stability curve: measurement in low-turbulence (○) and high-turbulence conditions at different amplitudes of excitation (•, □, ×, +); calculation by PSEs (*solid line*) (Boiko et al, 1994)

Amplification curves for the wave at low and high free-stream turbulence are shown in Fig. 12.14. The decrease of the amplification rate is associated with a delicate mechanism caused by the presence of the streaks. Moreover, the amplification rates of the Tollmien–Schlichting wave appear to be independent of the excitation amplitude. Hence, the waves seem linear perturbations. Although the vibrating ribbon initially sets up a plane wavefront, randomly moving large-scale structures may lead to three-dimensional deformations of it, which cause local changes in both the frequency and phase of oscillations.

12.3.2 Interaction of the streaks with the Tollmien–Schlichting waves

The above-described observations demonstrate that different kinds of interactions may be present, which could transfer energy between the waves or between large-scale flow structures and the waves. Some indication of the nature of such interactions may be obtained by analyzing power spectra of disturbances obtained in the boundary layer. Particularly, it was found that the energy is increased over a broad frequency range as the forcing wave amplitude grows; that is, the energy increase can be attributed to generation of the Tollmien–Schlichting waves at frequencies other than the forcing frequency.

In experiments of Boiko et al (1994), the number of turbulent spots and transitional flow structures at $Re_{\delta^*} = 1260$ increased drastically with growth of the excitation amplitude. For the highest excitation amplitude, 90% of the collected records were considered to be affected by intermittent turbulence. The observed energy growth, therefore, represents an increase of random fluctuations in the laminar portions between the turbulent spots. Similar interaction processes may also be expected to occur when the Tollmien–Schlichting waves are induced naturally by high free-stream turbulence. If it happens randomly in time and space, turbulence will not appear gradually, but rather in the form of localized turbulent spots.

Interaction of a natural streak with a controlled two-dimensional Tollmien–Schlichting wave is illustrated by the smoke visualization in Fig. 12.15. As a result of interaction, a nonlinear wave packet occurs on the streak. The resulting disturbance transforms downstream into a turbulent spot. Interactions between artificially excited streaks and the Tollmien–Schlichting waves were also studied in a series of investigations. Similarly to the natural case, interaction produced a nonlinear wave packet, which gradually developed into a turbulent spot. This process was also observed as both the controlled streak and the waves decayed when they were generated separately. It was revealed that, depending on the streak intensity and the presence of neighboring streaks, the wave packet can be generated either in the region of the streak trailing edge or inside the streak body.

Figure 12.16 shows the structure of a disturbance that appears (before its final transformation into a turbulent spot) during separate and simultaneous development of the wave and the streak. It can be seen that the streak intensity is increased by

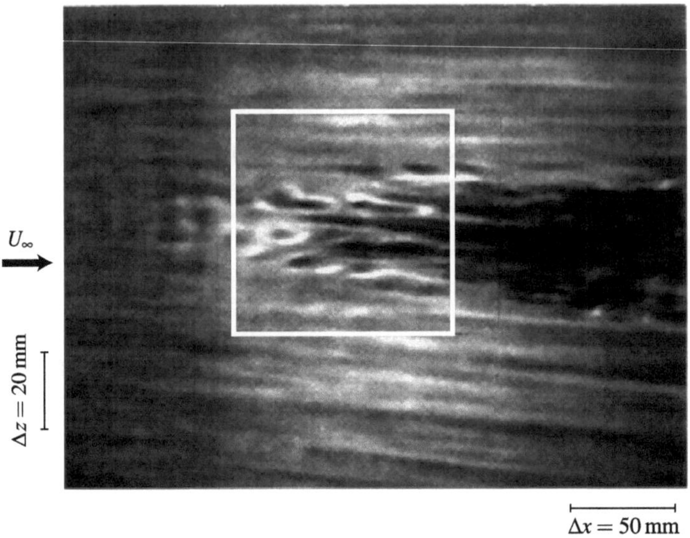

Fig. 12.15 Smoke visualization of interaction between a forced Tollmien–Schlichting wave and a natural streak in a boundary layer at Tu = 1.5%: enlarged image of the generated disturbance is focused in the middle (Matsubara et al, 1996)

more than 10 times due to interaction. It is characteristic that the maximum of the disturbances excited by interaction is located in the region of the maximum transverse velocity shear, that is, in the region subject to secondary instability of stationary vortices considered in Sect. 11.2. Hence, it can be assumed that the breakdown mechanisms of both the streamwise stationary vortices and propagating streaks have much in common and are related to the onset and development of secondary instability.

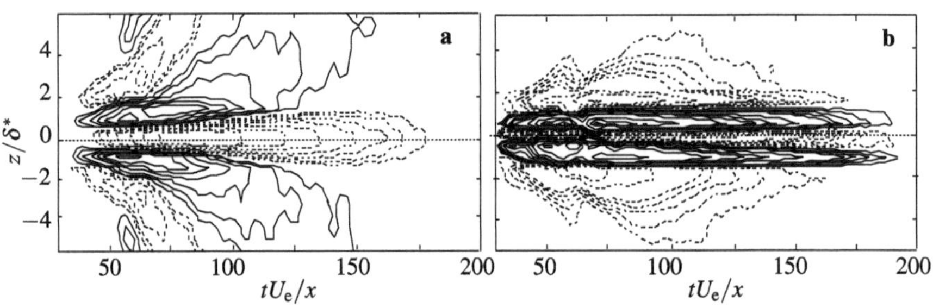

Fig. 12.16 Isolines of velocity fluctuations for a localized disturbance (**a**) and at interaction between the localized disturbance and high-frequency oscillations (**b**) in the z–t-plane: velocity excess (*solid lines*); velocity defect (*dashed lines*) (Bakchinov et al, 1998)

Additional material for this chapter is given in the multimedia files Nos. 11 and 12, which supplement the book, titled 'Visualization of a nonlinear Tollmien–Schlichting wave' and 'Nonlinear wave packets in swept-wing boundary layer.'

References

Alfredsson PH, Bakchinov AA, Kozlov VV, Matsubara M (1996) Laminar–turbulent transition at a high level of a free stream turbulence. In: Duck PW, Hall P (eds) Nonlinear Instability and Transition in Three-Dimensional Boundary Layers, Kluwer, Dordrecht, Fluid Mechanics and its Application, pp 423–436

Arnal D (1992) Transition description and prediction. In: Pironneau O, Rodi W, Ryhming IL, Savill AM, Truong TV (eds) Numerical Simulation of Unsteady Flows and Transition to Turbulence, Cambridge University Press, Cambridge, pp 303–316

Bakchinov AA, Grek GR, Katasonov MM, Kozlov VV (1998) Experimental investigation of the interaction of longitudinal streaky structures with a high-frequency disturbance. Fluid Dyn 33(5):667–675

Berlin S, Wiegel M, Henningson DS (1999) Numerical and experimental investigations of oblique boundary layer transition. J Fluid Mech 393:23–57

Boiko AV, Dovgal AV, Kozlov VV (1989) Nonlinear interactions between perturbations in transition to turbulence in the zone of laminar boundary-layer separation. Soviet J Appl Phys 3(2):46–52

Boiko AV, Dovgal AV, Simonov OA, Scherbakov VA (1991) Effects of laminar–turbulent transition in separation bubbles. In: Kozlov VV, Dovgal AV (eds) Separated Flows and Jets, Springer–Verlag, Berlin, IUTAM Symposium, pp 565–572

Boiko AV, Westin KJA, Klingmann BGB, Kozlov VV, Alfredsson PH (1994) Experiments in a boundary layer subjected to free stream turbulence. Part 2. The role of TS-waves in the transition process. J Fluid Mech 281:219–245

Chernoray VG, Bakchinov AA, Kozlov VV, Löfdahl L (2001) Experimental study of K-regime of breakdown in straight and swept wing boundary layers. Phys Fluids 13(7):2129–2132

Craik ADD (1971) Non-linear resonant instability in boundary layers. J Fluid Mech 50:393–413

Kachanov YS (1994) Physical mechanisms of laminar-boundary-layer transition. Ann Rev Fluid Mech 26:411–482

Klebanoff PS, Tidstrom KD, Sargent LM (1962) The three-dimensional nature of boundary-layer instability. J Fluid Mech 12:1–34

Maslowe SA (1985) Shear flow instabilities and transition. In: Swinney HL, Golub JP (eds) Hydrodynamic Instabilities and the Transition to Turbulence, Topics in Applied Physics, vol 45, 2nd edn, Springer–Verlag, Berlin, chap 7, pp 181–228

Matsubara M, Kozlov VV, Alfredsson PH, Bakchinov AA, Westin KJA (1996) On flat plate boundary layer perturbations at high free stream turbulence level. In: Proc. Eighth Internat. Conf. on Methods of Aerophysical Research, RAS. Sib. Branch Inst. Theoret. Appl. Mech., Novosibirsk, vol 1, pp 174–179

Ramazanov MP (1985) Development of finite-amplitude disturbances in Poiseuille flow. In: Kozlov VV (ed) Laminar–Turbulent Transition, Springer–Verlag, Berlin, IUTAM Symposium, pp 183–190

Rist U, Müller K, Wagner S (1998) Visualisation of late-stage transitional structures in numerical data using vortex identification and feature extraction. In: Carlomango GM, Grant I (eds) Eighth International Symposium on Flow Visualisation, Published and distributed on CD by the editors, Sorrento, Paper No. 109.

Saric WS, Kozlov VV, Levchenko VY (1984) Forced and unforced subharmonic resonance in boundary-layer transition. AIAA Paper 84–0007

Schlichting H, Gersten K (2000) Boundary layer theory, 8th edn. Springer–Verlag, Berlin

Schmid PJ, Henningson DS (2000) Stability and transition in shear flows. Springer–Verlag, Berlin

Taylor GI (1936) Statistical theory of turbulence. V. Effects of turbulence on boundary layer. Theoretical discussion of relationship between scale of turbulence and critical resistance of sphere. Proc R Soc Lond A 156:307–317

Zhigulev VN, Tumin AM (1987) Onset of turbulence. Dynamical theory of excitation and development of instabilities in boundary layers. Nauka. Sib. Otd., Novosibirsk, In Russian.

Further Reading

Arnal D, Juillen JC (1977) Étude expérimentale et théorique de la transition de la couche limite. Rech Aérospat 2:75–88

Arnal D, Juillen JC (1978) Contribution expérimentale à l'étude de la réceptivité d'une couche limite laminaire, à la turbulence de l'écoulement général. CERT RT 1/5018 AYD, ONERA

Arnal D, Michel R (eds) (1990) Laminar–Turbulent Transition, IUTAM Symposium, Springer–Verlag, Berlin

Bake S, Kachanov YS, Fernholz HH (1996) Subharmonic K-regime of boundary-layer breakdown. In: Henkes RAWM, van Ingen JL (eds) Transitional Boundary Layers in Aerodynamics, Proc. Colloquium, Elsevier, Amsterdam, pp 81–88

Bake S, Fernholz HH, Kachanov YS (2000) Resemblance of K- and N-regimes of boundary-layer transition at late stages. Eur J Mech B/Fluids 19(1):1–22

Benney DJ, Bergeron EF (1968) A new class of nonlinear waves in parallel flows. Stud Appl Math 48(3):181–204

Benney DJ, Gustavsson LH (1981) A new mechanism for linear and nonlinear hydrodynamic stability. Stud Appl Math 64:185–209

Berlin S, Lundbladh A, Henningson DS (1994) Spatial simulations of oblique transition in a boundary layer. Phys Fluids A 6(6):1949–1951

Bertolotti FP, Herbert T, Spallart PR (1992) Linear and nonlinear stability of the Blasius boundary layer. J Fluid Mech 242:441–474

Cohen J, Breuer KS, Haritonidis JH (1991) On the evolution of a wave packet in a laminar boundary layer. Part 1. Weak disturbances. J Fluid Mech 225:575–606

Corke TC, Mangano RA (1989) Resonant growth of three-dimensional modes in transitioning boundary layers. J Fluid Mech 209:93–150

Dovgal AV, Boiko AV (2000) Effect of harmonic excitation on instability of laminar separation bubble on an airfoil. In: Fasel and Saric (2000), pp 675–680

Ehrenstein U, Koch W (1991) Three-dimensional wavelike equilibrium states in plane Poiseuille flow. J Fluid Mech 228:111–148

Elofsson PA, Alfredsson PH (1998) An experimental study of oblique transition in plane Poiseuille flow. J Fluid Mech 358:177–202

Elofsson PA, Alfredsson PH (2000) An experimental investigation of oblique transition in a Blasius boundary layer flow. Eur J Mech B/Fluids 19(5):615–636

Fasel HF, Saric WS (eds) (2000) Laminar–Turbulent Transition, IUTAM Symposium, Springer–Verlag, Berlin

Fujimura K (1991) Method of centre manifold and multiple scales in the theory of weakly nonlinear stability of fluid motions. Proc R Soc Lond A 434:719–733

Gaster M (1990) The nonlinear phase of wave growth leading to chaos and breakdown to turbulence in a boundary layer as an example of an open system. Proc R Soc Lond A 430:3–24

Gaster M, Grant T (1975) An experimental investigation of the formation and development of a wave-packet in a laminar boundary layer. Proc R Soc Lond A 347:253–269

Goldshtik MA, Shtern VN (1977) Hydrodynamic stability and turbulence. Nauka. Sib. Otd., Novosibirsk, in Russian.
Goldstein ME, Choi SW (1989) Nonlinear evolution of interacting oblique waves on two-dimensional shear layers. J Fluid Mech 207:97–120, see also Corrigendum, J. Fluid Mech., 216, 659–663.
Goldstein ME, Lee SS (1992) Fully coupled resonant-triad interaction in an adverse-pressure-gradient boundary layer. J Fluid Mech 245:523–551
Grek GR, Kozlov VV (1992) Interaction of Tollmien–Schlichting waves and localized disturbances. Sib Fiz-Techn Zh 34(5):68–76, in Russian.
Grek GR, Kozlov VV, Ramazanov MP (1987) Laminar–turbulent transition at high free stream turbulence. Preprint 8–87, RAS. Sib. Branch Inst. Theoret. Appl. Mech., Novosibirsk, in Russian.
Grek GR, Kozlov VV, Ramazanov MP (1989) Laminar–turbulent transition at high free stream turbulence in gradient flow. Izv Sib Otd Akad Nauk SSSR, Ser Tekhn Nauk 3:66–70
Grek GR, Kozlov VV, Ramazanov MP (1991a) Investigation of boundary layer stability in a gradient flow with a high degree of free-stream turbulence. Fluid Dyn 25(2):207–213
Grek HR, Dey J, Kozlov VV, Ramazanov MP, Tuchto ON (1991b) Experimental analysis of the process of the formation of turbulence in the boundary layer at higher degree of turbulence of windstream. Tech. Rep. 91–FM–2, Indian Institute of Science, Bangalor
Haberman R (1972) Critical layers in parallel flows. Stud Appl Math 51(2):139–161
Henningson DS, Johansson AV, Lunbladh A (1990) On the evolution of localized disturbances in laminar shear flows. In: Arnal and Michel (1990), pp 279–284
Herbert T (1983) On perturbation methods in nonlinear stability theory. J Fluid Mech 126:167–186
Herbert T (1983) Secondary instability of plane channel flow to subharmonic three-dimensional disturbances. Phys Fluids 26(4):871–874
Herbert T (1984) Analysis of the subharmonic route to transition in boundary layers. AIAA Paper 84–0009
Herbert T (1988) Secondary instability of boundary layers. Ann Rev Fluid Mech 20:487–526
Joseph DD (1976) Stability of Fluid Motion, Springer Tracts on Natural Philosophy, vol 1. Springer–Verlag, Berlin
Kachanov YS (1987) On the resonant nature of the breakdown of a laminar boundary layer. J Fluid Mech 184:43–74
Kachanov YS (1994) Physical mechanisms of laminar-boundary-layer transition. Ann Rev Fluid Mech 26:411–482
Kachanov YS, Levchenko VY (1984) The resonant interaction of disturbances at laminar–turbulent transition in a boundary layer. J Fluid Mech 138:209–247
Kachanov YS, Kozlov VV, Levchenko VY (1978) Nonlinear development of a wave in a boundary layer. Fluid Dyn 12(3):383–390
Kachanov YS, Kozlov VV, Levchenko VY (1982) Origin of Turbulence in Boundary Layer. Nauka. Sib. Otd., Novosibirsk

Kachanov YS, Ryzhov OS, Smith FT (1993) Formation of solitons in transitional boundary layers: Theory and experiment. J Fluid Mech 251:273–297

Kendall JM (1985) Experimental study of disturbances produced in a pre-transitional laminar boundary layer by weak free stream turbulence. AIAA Paper 85–1695

Kendall JM (1990) Boundary layer receptivity to freestream turbulence. AIAA Paper 90–1504

Kendall JM (1991) Studies on laminar boundary layer receptivity to freestream turbulence near a leading edge. In: Reda DC, Reed HL, Kobayashi R (eds) Boundary Layer Stability and Transition to Turbulence, ASME, pp 23–30

Kleiser L, Zang TA (1991) Numerical simulation of transition in wall-bounded shear flows. Ann Rev Fluid Mech 23:495–537

Koch W, Bertolotti FP, Stolte A, Hein S (2000) Nonlinear equilibrium solutions in a three-dimensional boundary layer and their secondary instability. J Fluid Mech 406:131–174

Kosorygin VS, Polyakov NP (1990) Laminar boundary layers in turbulent flows. In: Arnal and Michel (1990), pp 573–578

Kovasznay LS, Komoda H, Vasudeva BR (1962) Detailed flow field in transition. In: Proc. Heat Transfer and Fluid Mech. Institute, Stanford University, Stanford, pp 1–26

Li F, Malik MR (1995) Fundamental and subharmonic secondary instabilities of Görtler vortices. J Fluid Mech 297:77–100

Masad JA, Nayfeh AH (1992) Effect of a bulge on the subharmonic instability of subsonic boundary layers. AIAA J 30(7):1731–1737

Maucher U, Rist U, Wagner S (2000) Secondary disturbance amplification and transition in laminar separation bubbles. In: Fasel and Saric (2000), pp 657–662

Moston J, Stewart PA, Cowley SJ (2000) On the nonlinear growth of two-dimensional Tollmien–Schlichting waves in a flat-plate boundary layer. J Fluid Mech 425:259–300

Nagata M (1990) Three-dimensional finite-amplitude solutions in plane Couette flow: Bifurcation from infinity. J Fluid Mech 217:519–527

Nayfeh AH, Ragab SA (1987) Effect of a bulge on the secondary instability of boundary layers. AIAA Paper 87–0045

Nishioka M, Asai M, Iida S (1980) An experimental investigation of the secondary instability. In: Eppler R, Fasel H (eds) Laminar–Turbulent Transition, Springer–Verlag, Berlin, IUTAM Symposium, pp 37–46

Orszag SA, Patera AT (1983) Secondary instability of wall-bounded shear flows. J Fluid Mech 128:347–385

Rist U (1994) Nonlinear effects of 2D and 3D disturbances on laminar separation bubbles. In: Lin SP, Phillips WRC, Valentine DT (eds) Nonlinear Instability of Nonparallel Flows, Springer–Verlag, Berlin, IUTAM Symposium, pp 324–333

Rist U, Fasel HF (1995) Direct numerical simulation of controlled transition in a flat-plate boundary layer. J Fluid Mech 298:211–248

Rist U, Maucher U (1994) Direct numerical simulation of 2D and 3D instability waves in a laminar separation bubble. In: AGARD-CP–551 Application of Direct

and Large Eddy Simulation to Transition and Turbulence, Chania, Crete, Greece, pp 34.1–34.7
Rist U, Maucher U, Wagner S (1996) Direct numerical simulation of some fundamental problems related to transition in laminar separation bubbles. In: Désidéri JA, Hirsch C, Tallec PL, Pandolfi M, Périaux J (eds) Computational Fluid Dynamics '96, Wiley, New York, pp 319–325
Rotenberry JM (1993) Finite amplitude steady waves in the Blasius boundary layer. Phys Fluids A 5(7):1840–1842
Sandham ND, Kleiser L (1992) The late stages of transition to turbulence in channel flow. J Fluid Mech 245:319–348
Schmid PJ, Henningson DS (1992) A new mechanism for rapid transition involving a pair of oblique waves. Phys Fluids A 4(9):1986–1989
Sen PK, Venkateswarly D (1983) On the stability of plane Poiseuille flow to finite-amplitude disturbances, considering the higher-order Landau coefficients. J Fluid Mech 133:179–206
Smith FT (1987) Non-linear effects and non-parallel flows: The collapse of separated motion. In: Dwoyer DL, Hussaini MY (eds) Stability of Time Dependent and Spatially Varying Flows, Springer–Verlag, Berlin, pp 104–147
Stewartson K, Stewart JT (1971) A nonlinear instability theory for a wave system in plane Poiseuille flow. J Fluid Mech 48:529–545
Stuart JT (1960) On the nonlinear mechanisms of wave disturbances in stable and ubstable parallel flows. Part 1. The basic behaviour in plane Poiseuille flow. J Fluid Mech 9:353–370
Suder KL, O'Brien JE, Reshotko E (1988) Experimental study of bypass transition in a boundary layer. TM 100913, NASA
Watson J (1960) On the nonlinear mechanisms of wave disturbances in stable and ubstable parallel flows. Part 2. The development of a solution for plane Poiseuille flow and Couette flow. J Fluid Mech 9:371–389
Westin KJA, Henkes RAWM (1997) Application of turbulence models to bypass transition. Trans ASME Ser I: J Fluids Eng 119(4):859–866
Wray A, Hussaini MY (1984) Numerical experiments in boundary-layer stability. Proc R Soc Lond A 392:373–389
Zahn JP, Toomre J, Spiegel EA, Gough DO (1974) Nonlinear cellular motions in Poiseuille channel flow. J Fluid Mech 64:319–345
Zelman MB, Maslennikova II (1993) Tollmien–Schlichting-wave resonant mechanism for subharmonic-type transition. J Fluid Mech 252:449–478
Zhigulev VN, Sidorenko NV (1985) On the issue of the nonlinear theory of the instability waves. In: Kozlov VV (ed) Laminar–Turbulent Transition, Springer–Verlag, Berlin, IUTAM Symposium, pp 81–86

Chapter 13
Generation of turbulence

Late stages of the transition to turbulence are characterized by the appearance of turbulent spots, turbulent bursts, and intermittency, the phenomena being universal for the near-wall flows of this book. *Turbulent spots* are spatially localized regions with turbulent fluctuations, propagating downstream in the surrounding laminar fluid. *Turbulent bursts* are intervals in oscilloscope traces of disturbances with turbulent fluctuations superimposed on less intensive low-frequency oscillations of laminar flow. *Intermittency* in time (in space) is an alteration of laminar and turbulent regions in time (in space) at a fixed point in space (in time). These phenomena are considered in more detail in this chapter.

13.1 Wave combinations and intermittency

In many experimental studies of the 'natural' transition in boundary layers, intensive disturbances preceding the final flow turbulization were observed. The disturbances have typical frequencies much lower than that of the primary Tollmien–Schlichting wave or its subharmonic. These low-frequency fluctuations quickly amplify in the region of formation of spikes and turbulent spots. Their increase followed by a decay is normally found in the transition region and is even used as a criterion of the transition 'point' or Re_T at hot-wire measurements based on detection of the downstream maximum of the streamwise velocity perturbations.

The low-frequency large-amplitude perturbations can be a consequence of wave combinations, for example, a result of nonlinear interaction of a pair of the Tollmien–Schlichting waves at frequencies f_1 and f_2, followed by generation of spectral component at $\Delta f = |f_2 - f_1|$. The Tollmien–Schlichting waves observed in experiments on natural transition represent a wave packet with a central frequency f_0 detuned by frequencies less than approximately Δf_0 corresponding to the low-frequency amplitude modulation (Fig. 13.1**a**). Kachanov et al (1982) modeled this situation by introducing two waves at frequencies f_1 and f_2 equal to $f_0 \pm \Delta f_0/2$ of

A.V. Boiko et al., *Physics of Transitional Shear Flows*,
Fluid Mechanics and Its Applications 98, DOI 10.1007/978-94-007-2498-3_13,

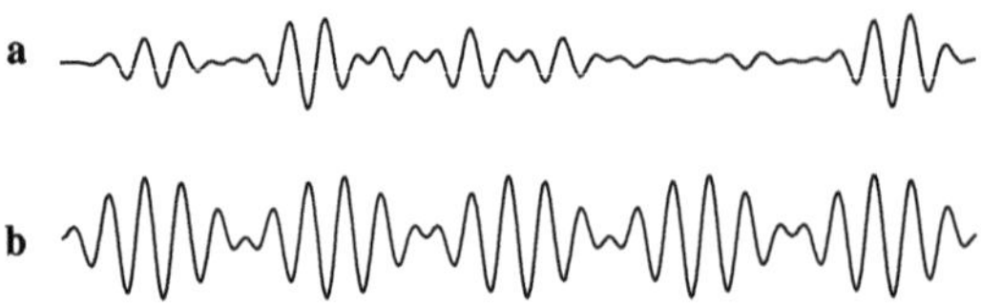

Fig. 13.1 Oscilloscope traces of the hot-wire signal (streamwise disturbance velocity) acquired in the middle of the boundary layer: **a** natural transition (Arnal and Juillen, 1977); **b** simulation of the Tollmien–Schlichting wave packet (Kachanov et al, 1982)

an identical amplitude (Fig. 13.1**b**). Their combinational interaction gradually leads to the onset of turbulence and, in essence, does not differ from the multi-wave case.

Figure 13.2 shows a sequence of frequency spectra in the flat-plate boundary layer measured in different streamwise sections at a constant distance from the wall. Initially, in the two-dimensional flow, all harmonics gradually amplify together with primary peaks f_1 and f_2—the difference mode Δf and its harmonics $2\Delta f$, $3\Delta f$ grow

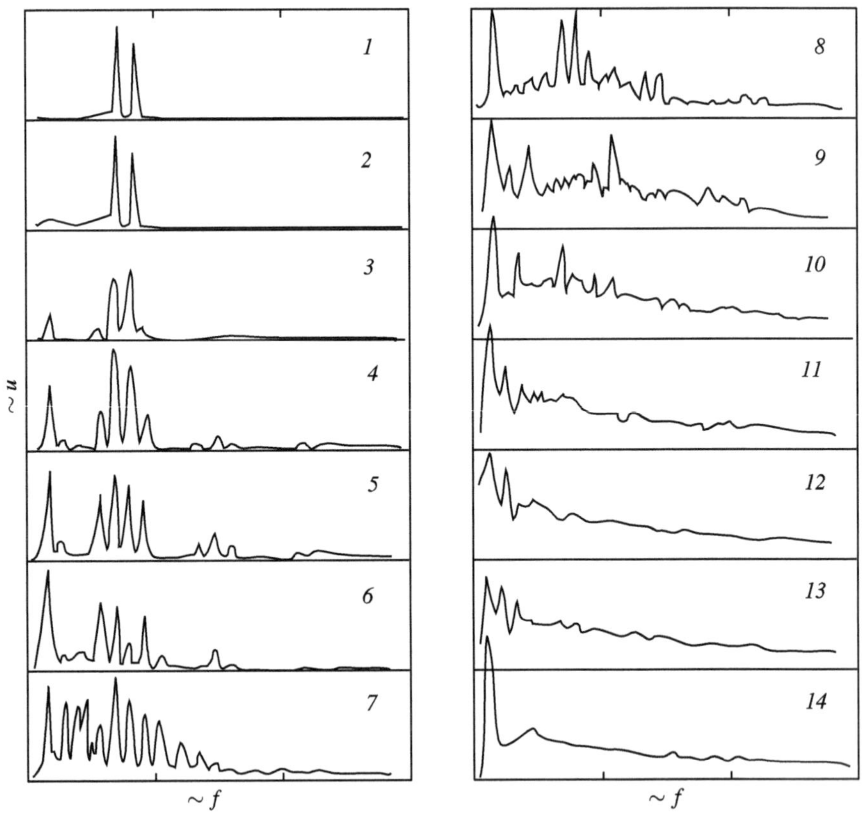

Fig. 13.2 Sequence of spectra of fluctuations for two interacting waves at different downstream positions marked by numbers *1*,...,*14* (Kachanov et al, 1982)

most rapidly. Up to the latest stages of the transition, peaks at $f = mf_2 \pm nf_1$, where $m,n = 0,1,2\ldots$, can be clearly identified, being the result of nonlinear interactions.

The intermittency of bursts of the fluctuations and turbulent spots with the frequency Δf is well explained within the framework of a phenomenological quasi-stationary treatment of boundary-layer unsteadiness. A superposition of two harmonic waves f_1 and f_2 excited in the boundary layer with amplitudes u'_1 and u'_2 can be considered as a single quasi-harmonic wave at $f_0 \sim f_1, f_2$, whose amplitude slowly varies in time from the maximum $u'_1 + u'_2$ to the minimum $|u'_1 - u'_2|$, if $\Delta f = |f_1 - f_2| \ll f_1, f_2$. Thus, the period of the amplitude variations is $T_\Delta = 1/\Delta f \gg T_0 = 1/f_0$. As a result of the slow, quasi-stationary alternations of the amplitude, the transition region moves periodically upstream and downstream, producing oscillatory motion with the period T_Δ. This motion is sensed by a fixed probe located at a certain point in the boundary layer as alternating turbulent and laminar portions in time with the frequency $\Delta f = 1/T_\Delta$; i.e., the intermittency is observed.

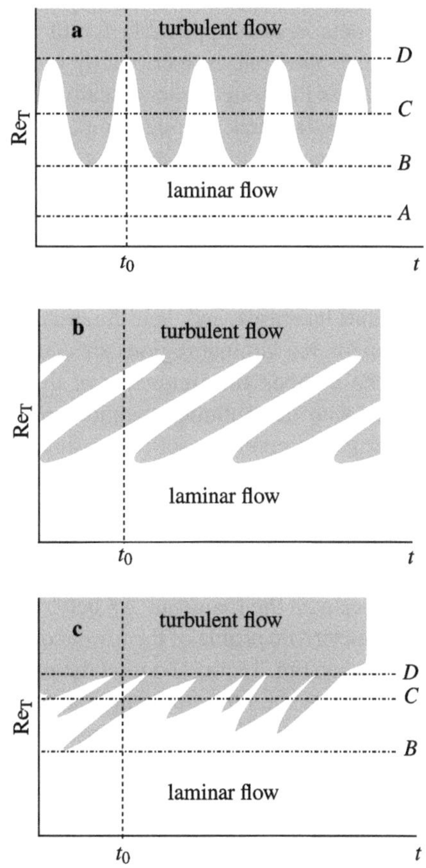

Fig. 13.3 Quasi-stationary oscillations of the transition Reynolds number Re_T with time due to modulations of the Tollmien–Schlichting wave amplitude: without appearance of turbulent spots in space (**a**); with presence of turbulent spots in space (**b**); in the case of natural transition with random modulation of the Tollmien–Schlichting waves (**c**) (Kachanov et al, 1982)

This phenomenon is illustrated in Fig. 13.3**a**. The region of laminar flow breakdown, associated with the transition Reynolds number Re_T, which can be defined as, for example, the end of the region of plane development of the instability wave, periodically moves in space. The horizontal sections *A*, *B*, *C*, and *D* are the positions of an observer who sees, e.g., in section *C*, an alternation of laminar and turbulent states in time (intermittency). The amplitude of the difference harmonics produced by the alternation reaches a maximum at a factor of intermittency of about 0.5 (Fig. 13.3**a**, section *C*) and decays only to the end of the transition region (section *D*). At a time instant t_0, the Reynolds numbers $Re < Re_T$ correspond to laminar regions, $Re > Re_T$ to turbulent ones, and there is no intermittency in space. Hence, the appearance of intermittency in oscilloscope traces as the alternation of laminar and turbulent regions in time yet does not mean formation of turbulent spots in space. However, the dependence of Re_T on t can be similar to that illustrated in Fig. 13.3**b**. Then turbulent spots are observed in space, since the regions with turbulent and laminar flow alternate at fixed time instant t_0.

The quasi-stationary treatment of the mechanism of formation of turbulent bursts and spots is also applicable for the natural transition. Its main difference from the controlled one is aperiodicity of variations of the transition Reynolds number $Re_T(t)$, as illustrated schematically in Fig. 13.3**c**. These oscillations are also related to the low-frequency modulation of the primary wave. In the crests of such composite beatings, the turbulent bursts appear similar to those observed at superposition of two waves. However, in oscilloscope traces the wave amplitudes in various crests are different. In those where the amplitude reaches a maximum, turbulent bursts appear earlier. Downstream they are also generated at other crests. So the turbulent bursts rarely appear at first in section *B*. In section *C*, the frequency of the turbulent portions increases, and their durations are different. In section *D*, the flow is almost turbulent, but laminar regions are occasionally formed.

The number and frequency of the turbulent bursts vary downstream in accordance with the Tollmien–Schlichting wave modulation. Results obtained in other experiments confirm this conclusion. Since the external disturbances determining the modulation of the Tollmien–Schlichting wave frequently have a random character, it is no wonder that the factors of intermittency are described by a normal distribution law of random quantities.

Due to the dependence of Re_T on time, the mean flow velocity profile $U(y)$ changes with the frequency Δf between the turbulent and laminar states. This yields a characteristic profile of the low-frequency modulation u' along the wall-normal coordinate y, Fig. 13.4**a**. The modulation has a maximum near the wall, decreases with y, revealing the other maximum at the boundary-layer edge, and finally decays in the free stream. The quasi-stationary treatment predicts oscillations at the wall and at the boundary-layer edge with opposite phases, i.e., an increase of U at the wall is accompanied by a decrease of U near the boundary-layer edge, and vice versa. This phenomenon is confirmed experimentally, as illustrated in Fig. 13.4**b**.

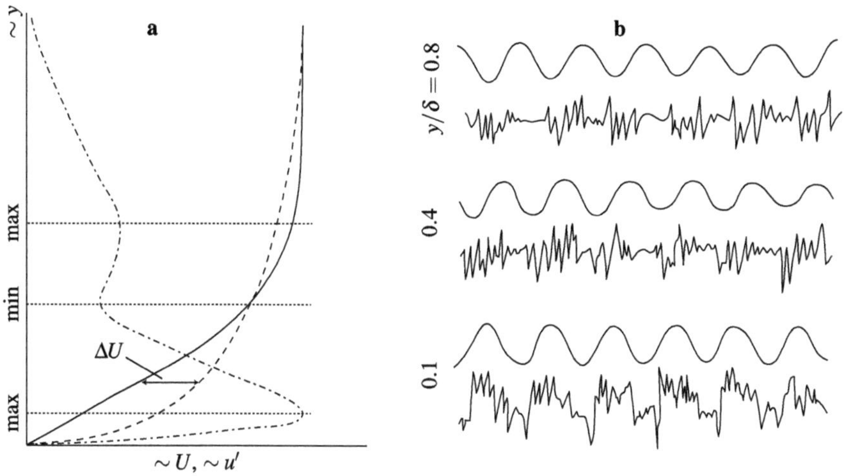

Fig. 13.4 Scheme of alternation of laminar (*solid line*) and turbulent (*dashed line*) states in a quasi-stationary treatment of transition and the corresponding (not in the same amplitude scale) profile of low-frequency fluctuations Δf (*dash-dotted line*) (**a**); phase shift of the difference harmonic (*upper curves*) relative to the turbulent spot on oscilloscope traces, depending on the distance from the wall (**b**) (Kachanov et al, 1982)

13.2 Turbulent spots

In many cases, the transition process in boundary layers and channel flows passes through formation of the so-called turbulent spots (Fig. 13.5). The turbulent spot is easy to generate and visualize, for example, in a laminar flow driven by gravity on an inclined plate by dropping on the free water surface. A strong-enough initial disturbance in the near-wall shear flow can produce a turbulent spot without the stage of instability wave amplification even in the subcritical region. The spot propagates downstream, grows, and eventually covers the entire near-wall region. Continuous development of several turbulent spots in space and time also leads downstream—at their merging—to a completely turbulent flow.

13.2.1 Isolated turbulent spots

The spots were observed for the first time in the boundary layer by Emmons (1951), who came to the following conclusions:

1. The spot formation is 'point-like' (actually, the turbulent spot arose in a region smaller than the local boundary-layer thickness);
2. The boundary between the fluid in the turbulent spot and that in the ambient flow is sharp;
3. The growth of the spot is uniform;

4. The spots do not interact in the bulk of the boundary layer—the merging spots produce a union rather then their superposition or a new formation.

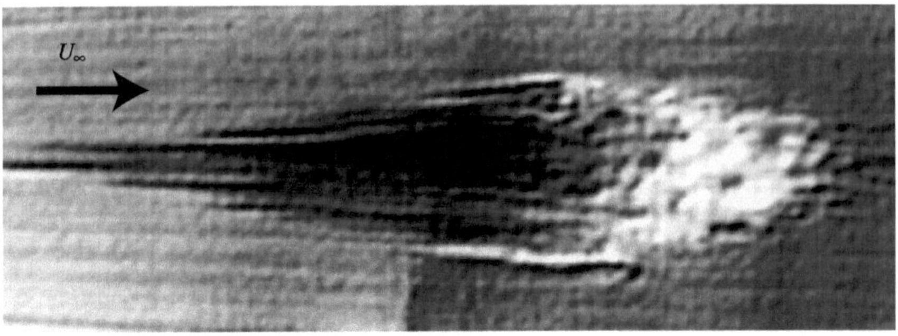

Fig. 13.5 Smoke visualization of a turbulent spot in a flat-plate boundary layer (Matsubara et al, 1996)

The development of turbulent spots was first studied in detail by Schubauer and Klebanoff (1956), and then by Elder (1960). Emmons' main conclusions were confirmed. It appeared that, during its downstream development, the spot has conical similarity irrespective of the method and place of its excitation. Typical transverse semi-angle of expansion of the spot is 9–12°, depending on its precise definition. The spots develop almost independently of one another—their interaction in the majority of the boundary layer is minor. The critical amplitude of the disturbance initiating spot formation across the most part of the boundary layer appeared to be approximately equal to 20% of the free-stream velocity. Later studies were focused on the structure of turbulent spots, the processes of their development and merging, as well as on separation of their mean characteristics from turbulent fluctuations.

Using an ensemble-averaging technique ensures a considerable advance in the knowledge of mechanisms of turbulent spot development, their internal structure, and velocities of propagation. An ensemble-averaged distribution of the streamwise disturbance velocity in the turbulent spot plane of symmetry measured at various distances from the wall is shown in Fig. 13.6. The curves characterize two regions of the spot: velocity excess $\langle U\rangle - U_{\text{lam}} > 0$ and velocity defect $\langle U\rangle - U_{\text{lam}} < 0$, where $\langle U\rangle$ is the ensemble-averaged velocity in the spot and U_{lam} is the mean local velocity in the laminar boundary layer in the absence of the spot (horizontal lines).

By repetition of the measurements at other streamwise positions, it is possible to determine characteristic velocities of spot propagation (Fig. 13.7). It can be seen that the velocity of its trailing front is much lower than that of the leading front, which is the reason for growth of the spot streamwise size. The velocities of the leading and trailing fronts are 0.89 and 0.57 of U_e, respectively. The coincidence of these results obtained in different experimental facilities and flow regimes testifies that the spot does not 'remember' the initial conditions and in this sense it is a turbulent 'eigenmode' of the boundary layer. It is interesting to note that the straight lines in Fig. 13.7 intersect at the point of the 'effective' turbulent spot origin, which does not

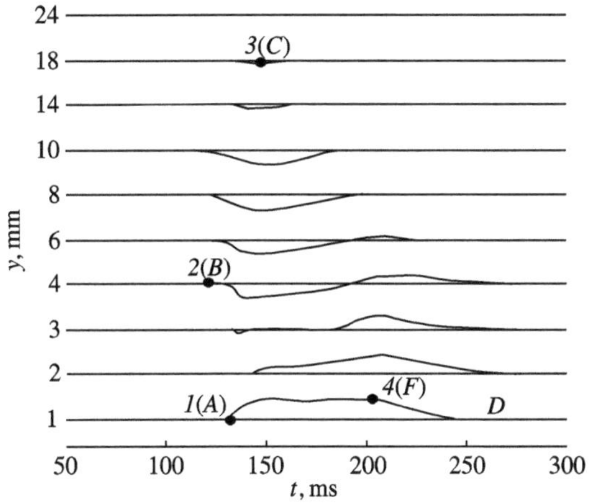

Fig. 13.6 Ensemble-averaged oscilloscope traces across the turbulent spot in the plane of its symmetry in a flat-plate boundary layer (Grek et al, 1987). Characteristic regions and fronts of the spot are marked by numbers, as by Wygnanski et al (1982), and letters as by Cantwell et al (1978): *1* (*A*), spot leading front near the wall; *2* (*B*), leading front forward-most region; *3* (*C*), maximum spot height; *4* (*F*), spot trailing edge; *D*, region of velocity relaxation to the undisturbed state

coincide with the disturbance generation point. This means that, at the short initial stage of the spot formation (the so-called 'incipient' turbulent spot stage), its self-similarity is evidently lost. It is natural that, this effective spot origin has different coordinates in other conditions, which can be found experimentally.

Linear expansion of turbulent spots in the transverse direction in gradientless flow was observed in many investigations. Narasimha et al (1984) carried out special studies concerning the influence of the streamwise pressure gradient on transverse development of the turbulent spots. It was found that it leads to a nonlinear dependence of their size on the streamwise coordinate.

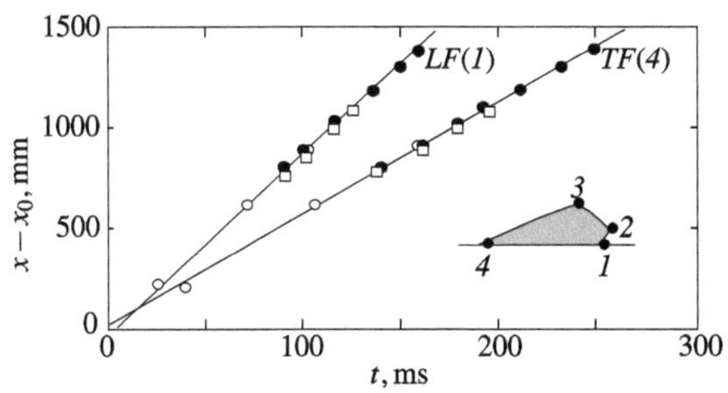

Fig. 13.7 Velocity of propagation of turbulent spot fronts: $x_0 = 300$ mm; $U_e = 10$ m/s. LF(*1*), leading front, TF(*4*), trailing front. Experimental data of Grek et al (1987), Tu $< 0.04\%$ (□); Grek et al (1987), Tu $= 1\%$ (○); Wygnanski et al (1982) (•). For the other notations see the legend to the previous figure

13.2.2 Interaction of turbulent spots

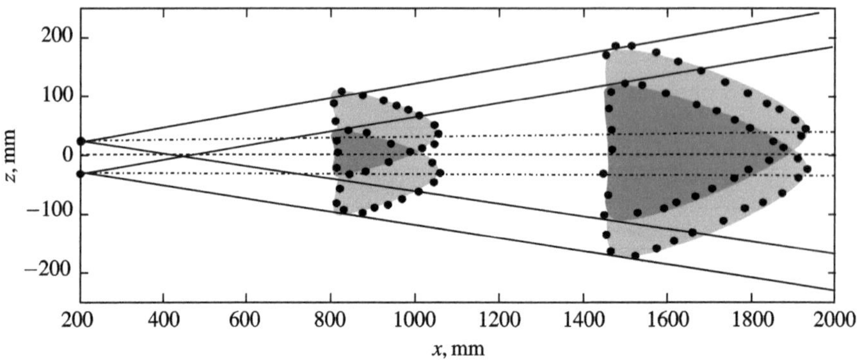

Fig. 13.8 Scheme in the x–z-plane of development and merging of two turbulent spots spaced in the transverse direction (Grek et al, 1987)

An experimental study of interaction of two turbulent spots was carried out for the first time by Elder (1960) in a flat-plate boundary layer at $\mathrm{Re}_{\mathbf{x}} = 4 \times 10^5$. Turbulent spots were simultaneously initiated by an electrical discharge at two points spaced in the transverse coordinate. The disturbances inside the boundary layer was measured by a hot-wire probe. The result of the study confirmed the conjecture of Emmons (1951) that the spots' geometrical characteristics evolve independently of each other inside the boundary layer. A scheme of merging of turbulent spots in the boundary layer is shown in Fig. 13.8. It can be seen that their overlap increases as they propagate, while their transverse sizes still grow linearly, as if each spot develops irrespective of the other, thereby preserving the characteristics of their individual development.

Apparent independence of the front development of the spot overlap is also confirmed when the spots move one behind the other. The scheme of this interaction is shown in Fig. 13.9**a**. Figure 13.9**b** also presents ensemble-averaged oscilloscope traces acquired inside the boundary layer in an experiment when two axially aligned turbulent spots develop simultaneously penetrating in each other and in two independent experiments when the spots develop separately. It demonstrates that the oscillogram obtained at their interaction (solid line) is virtually a simple union of the other two (dashed lines).

The isolines of velocity fluctuations for interacting and isolated turbulent spots are shown in Fig. 13.10. Two characteristic regions can be indicated: the region of the velocity excess close to the wall and the region of the velocity defect at the boundary-layer edge and above it. The maximum deflection of velocity inside the spot is 25–30% of U_{e}. Careful examination of the experimental results shows that the region of the velocity excess is a geometrical union of the regions that would be obtained if the spots were independent, while in the region with the velocity defect

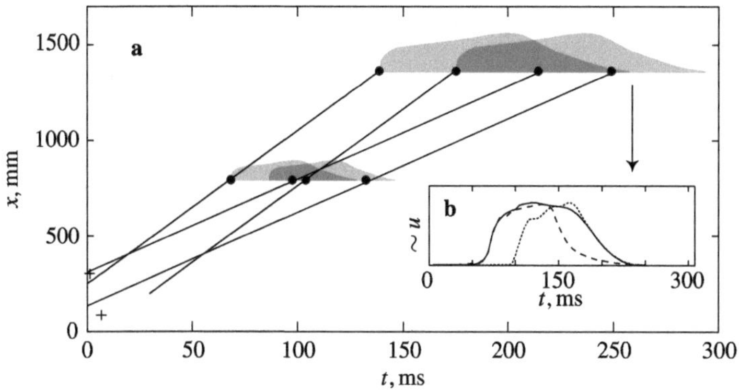

Fig. 13.9 Development and interaction of two turbulent spots following each other; source positions (+) (**a**); ensemble-averaged oscilloscope traces of the turbulent spots excited separately (*dotted line* and *dashed line*) and simultaneously (*solid line*) (**b**) (Grek et al, 1987)

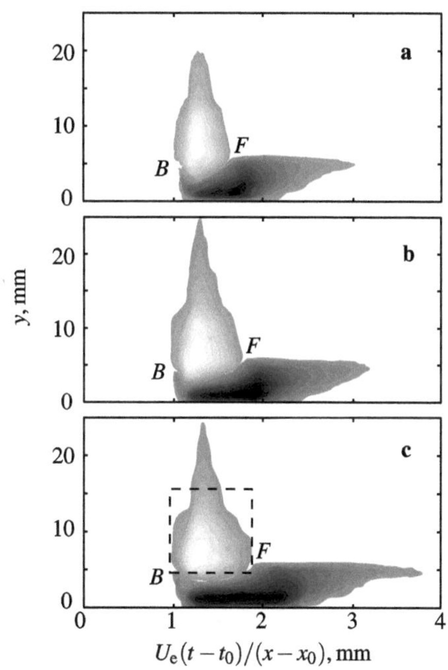

Fig. 13.10 Velocity fluctuations in turbulent spots at $x = 1400$ mm initiated separately at x_0: 90 mm (**a**); 300 mm (**b**); simultaneous initiation (**c**). The 'boundaries' of the turbulent spot are outlined by external contours, which are isolines of velocity differing by 1% from the velocity of laminar fluid. *Dashed box* circumscribes the interaction region (Grek et al, 1987)

at the boundary-layer edge, some minor changes in the back part of the interacting structure can be identified.

13.2.3 Turbulent spots in the presence of other disturbances

Experiments show that both the wavy and the spatially localized boundary-layer perturbations do not have a pronounced effect on the averaged turbulent spot characteristics. In particular, instability waves do not disturb the shape and size of turbulent spots and, vice versa, the growth rates and phase velocities of the Tollmien–Schlichting waves are almost not affected by the spots (Bakchinov et al, 1994). Similarly, turbulent spots are insensitive to streaky structures evolving in the boundary layer at high external-flow turbulence, at least up to Tu = 1.5% (Fig. 13.11).

Figure 13.12 shows the structure and propagation velocity of the ensemble-averaged turbulent spot fronts at low and high free-stream turbulence measured in its

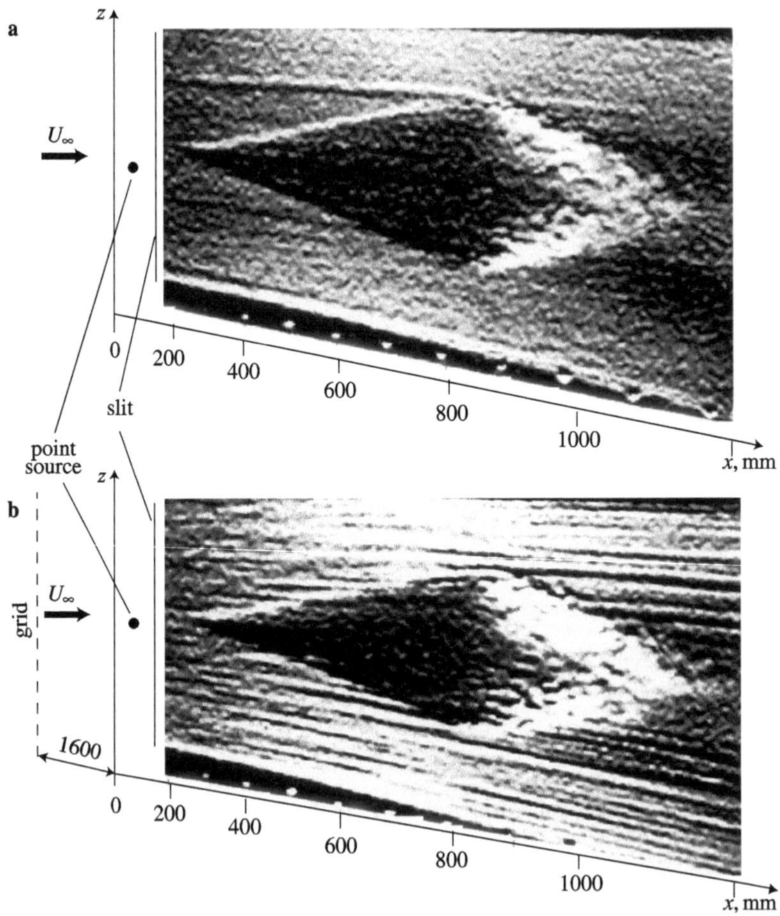

Fig. 13.11 Smoke visualization of the turbulent spot in a flat-plate boundary layer at low (**a**) and high (**b**) free-stream turbulence (Matsubara et al, 1996)

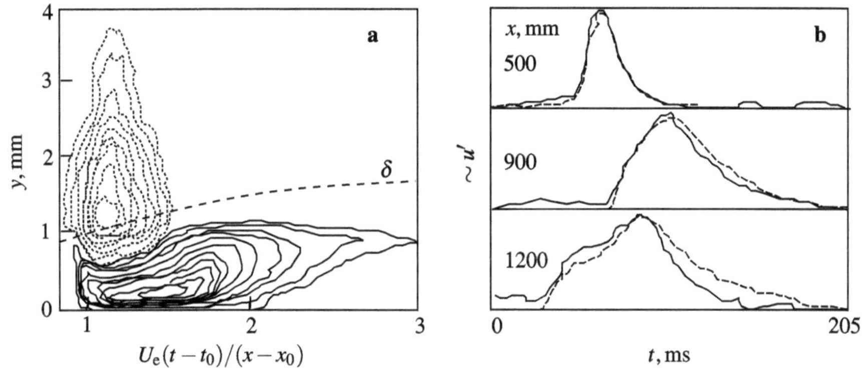

Fig. 13.12 Development of the ensemble-averaged turbulent spot at low and high free-stream turbulence: velocity defect (*dotted lines*) and excess (*solid lines*) along its center plane (**a**); oscilloscope traces obtained inside the boundary layer demonstrating propagation characteristics of the turbulent spot fronts (**b**). Tu $\leq 0.04\%$ (*solid line*), Tu $= 1\%$ (*dashed line*), $x_0 = 300$ mm, $U_e = 10$ m/s (Grek et al, 1987, 1988)

plane of symmetry for different streamwise positions (Grek et al, 1987). The turbulent spot structure includes the velocity excess located near the wall and the velocity defect spreading above the boundary layer. The velocity defect region protrudes far above the laminar boundary-layer edge δ. In both cases, the fronts of the turbulent spot coincide to within experimental accuracy related to the ensemble-averaging technique.

13.3 Super-late stage of the transition: 'Deterministic turbulence'

The idea of deterministic turbulence is based on the following hypotheses:

1. The laws of development of disturbances in boundary layers are deterministic (in particular, molecular chaos plays no significant role);
2. Stochasticity of boundary-layer motion follows exclusively from stochastic external disturbances, which grow downstream due to different instabilities.

The flows in which absolute or global instabilities are realized are excluded from the consideration. Hence, when all instabilities are convective, it is conjectured that the flow can be essentially deterministic even after the transition to turbulence. In experiments of Borodulin et al (2007, 2008), efforts were undertaken to generate and investigate to some extent deterministic turbulence in a flat-plate boundary layer experimentally.

To this end, the experiments were performed in a flat-plate boundary layer in a wind tunnel with an extremely low natural free-stream turbulence level (less than 0.01% of free-stream velocity) to get rid of uncontrolled perturbations as much as

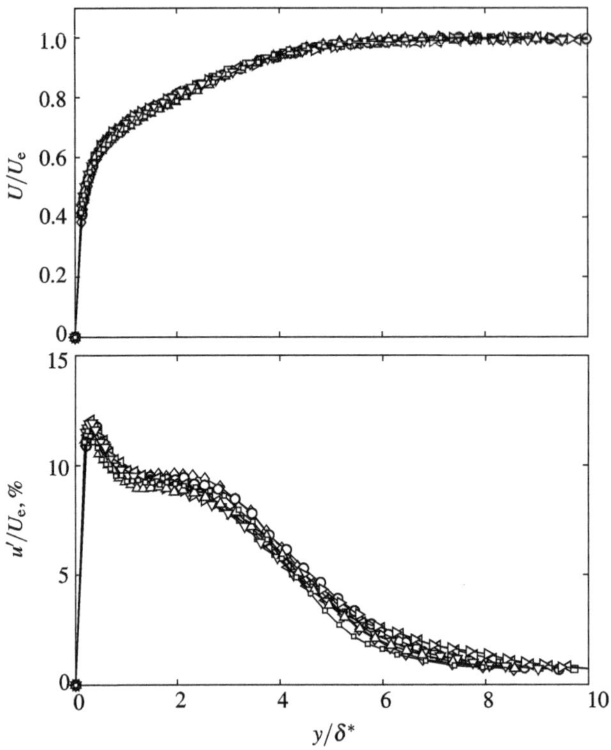

Fig. 13.13 Profiles of mean velocity U (**a**) and streamwise disturbance velocity u' (**b**) in the regime of 'deterministic turbulence' measured at different transverse sections (Borodulin et al, 2007)

possible. Controlled disturbances in the form of plane monochromatic Tollmien–Schlichting waves with an admixture of broad-band noise were introduced into the boundary layer. The noise at small time intervals (about $t = l/U_e$, where l is the flat-plate length and U_e is the free-stream velocity) behaved as the white noise, but was repeated periodically with a period $T \gg t$. As a result of the experiment, after the ensemble averaging of measured oscilloscope traces inside the boundary layer, a field $U(x,y,z) + u'(x,y,z,t)$ of streamwise velocity was acquired.

It appeared that both the mean velocity U and disturbance u' fields are homogeneous along the transverse coordinate z and have classical turbulent profiles, see Fig. 13.13. The frequency spectrum of the disturbances also corresponds to a turbulent flow with Kolmogorov's and Heisenberg's regions of '$-5/3$' and '-7,' Fig. 13.14. However, the coefficient of coherence in the whole region of measurements was not less than 50%, that is, the flow was essentially deterministic. Hence, it was shown that deterministic (reproducible) turbulence exists in nature and can be generated experimentally in some boundary-layer flows. Apart from important philosophical meaning, this result provides investigators with an additional tool of subsequent turbulence research.

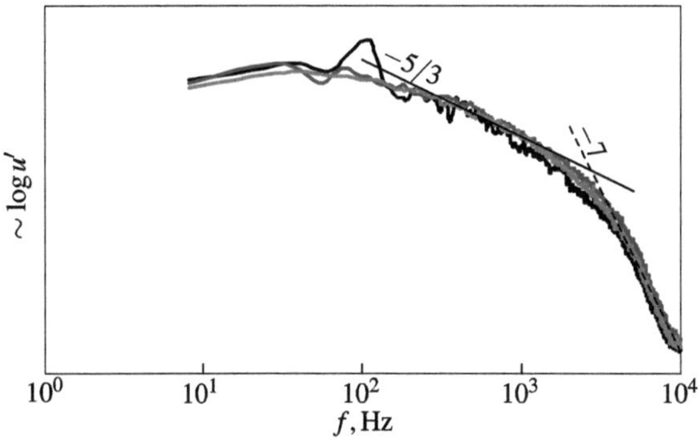

Fig. 13.14 Three frequency spectra in the regime of 'deterministic turbulence' (Borodulin et al, 2008)

> Additional material for this chapter is given in the multimedia file No. 13, which supplements the book, titled 'Origination and evolution of turbulent spots.'

References

Alfredsson PH, Bakchinov AA, Kozlov VV, Matsubara M (1996) Laminar–turbulent transition at a high level of a free stream turbulence. In: Duck PW, Hall P (eds) Nonlinear Instability and Transition in Three-Dimensional Boundary Layers, Kluwer, Dordrecht, Fluid Mechanics and its Application, pp 423–436

Arnal D, Juillen JC (1977) Étude expérimentale et théorique de la transition de la couche limite. Rech Aérospat 2:75–88

Bakchinov AA, Grek GR, Kozlov VV (1994) Development of instability waves between artificial turbulent regions. Thermophys Aeromech 1(1):45–50

Borodulin VI, Kachanov YS, Roschektayev AP (2007) The deterministic wall turbulence is possible. In: Palma JMLM, Lopes AS (eds) Advances in turbulence XI. Springer–Verlag, Dordrecht, pp 176–178

Borodulin VI, Kachanov YS, Roschektayev AP (2008) Investigation of LEBU-device effect on turbulent boundary layer structure by means of 'deterministic turbulence method'. In: Fomin VM (ed) Proc. Ninth Internat. Conf. on Methods of Aerophysical Research, Parallel, Novosibirsk, pp 47–52

Cantwell D, Coles D, Dimotakis P (1978) Structure and entrainment in the plane of symmetry of a turbulent spot. J Fluid Mech 87:641–671

Elder JW (1960) An experimental investigation on turbulent spot and breakdown to turbulence. J Fluid Mech 9:235–246

Emmons HW (1951) The laminar–turbulent transition in a boundary layer – Part 1. J Aeronaut Sci 18(7):490–498

Grek GR, Kozlov VV, Ramazanov MP (1987) Laminar–turbulent transition at high free stream turbulence. Preprint 8–87, RAS. Sib. Branch Inst. Theoret. Appl. Mech., Novosibirsk, in Russian.

Grek GR, Kozlov VV, Ramazanov MP (1988) Laminar–turbulent transition in a presence of a high level of free-stream turbulence. Fluid Dyn 23(6):829–834

Kachanov YS, Kozlov VV, Levchenko VY (1982) Origin of Turbulence in Boundary Layer. Nauka. Sib. Otd., Novosibirsk

Matsubara M, Kozlov VV, Alfredsson PH, Bakchinov AA, Westin KJA (1996) On flat plate boundary layer perturbations at high free stream turbulence level. In: Proc. Eighth Internat. Conf. on Methods of Aerophysical Research, RAS. Sib. Branch Inst. Theoret. Appl. Mech., Novosibirsk, vol 1, pp 174–179

Narasimha R, Subramanian C, Narayanan MA (1984) Turbulent spot growth in favourable pressure gradients. AIAA J 2:837–839

Schubauer GB, Klebanoff PS (1956) Contributions on the mechanics of boundary layer transition. TR 1289, NACA

Wygnanski IJ, Haritonidis JH, Zilberman M (1982) On the spreading of a turbulent spot in the absence of a pressure gradient. J Fluid Mech 123:69–90

Further Reading

Alfredsson PH, Bakchinov AA, Kozlov VV, Matsubara M (1996) Laminar–turbulent transition at a high level of a free stream turbulence. In: Duck PW, Hall P (eds) Nonlinear Instability and Transition in Three-Dimensional Boundary Layers, Kluwer, Dordrecht, Fluid Mechanics and its Application, pp 423–436

Barrow J, Barnes FH, Ross MAS (1984) The structure of a turbulent spot in Blasius flow. J Fluid Mech 149:319–337

Burnel S, Cougat P (1972) Étude du développement des instabilités naturelles dans la couche limite laminaire en écoulement incompressible. Compt Rendus Acad Sci., Sèr A 274(16):1251–1254

Kachanov YS, Kozlov VV, Levchenko VY (1978) Nonlinear development of a wave in a boundary layer. Fluid Dyn 12(3):383–390

Knapp CF, Roache PJ (1968) A combined visual and hot-wire anemometer investigation of boundary layer transition. AIAA J 6(1):29–36

Vasudevan KP, Dey J, Prabhu A (2001) Spots propagation characteristics in laterally strained boundary layers. Exp Fluids 30(5):488–491

Wygnanski IJ, Sokolov M, Friedman D (1975) On transition in a pipe. Part 2. The equilibrium puff. J Fluid Mech 69:283–304

Wygnanski IJ, Haritonidis JH, Kaplan RE (1979) On a Tollmien–Schlichting wave packet produced by a turbulent spot. J Fluid Mech 92:505–528

Appendix A
Basic engineering aspects of the laminar–turbulent transition

In the previous chapters, the laminar–turbulent transition was discussed in detail as a continuous process starting from excitation of small-amplitude shear-layer disturbances up to establishment of a developed turbulent flow. Finally, we briefly comment applications of basic knowledge on instability and transition in shear flows to engineering problems of the laminar–turbulent transition prediction and control.

A.1 Transition prediction

Basically, the goal of transition prediction is to clarify whether it takes place in a flow under consideration and to find the corresponding transition Reynolds number $\mathrm{Re_T}$. In early studies, the laminar–turbulent transition in boundary layers was predicted through empirical correlations between the basic flow parameters and $\mathrm{Re_T}$. One of the correlations was proposed by Michel (1951) and relates the Reynolds number based on the momentum thickness Re_ϑ to the transition Reynolds number as

$$\mathrm{Re}_\vartheta = 1.174\left(1+\frac{22.400}{\mathrm{Re_T}}\right)\mathrm{Re_T^{0.46}}\,.$$

This formula is applicable for attached boundary layers on airfoils with Reynolds numbers based on the wing chord length $\mathrm{Re}_l \gtrsim 2\times 10^6$. There exist also some correlation techniques taking into account the level of free-stream turbulence.

If the transition is modeled without resorting to empirical correlations, the most justified are the methods based on the concept of hydrodynamic stability. In this case, the prediction procedure should include three basic elements:

1. Determination of the structure of initial boundary-layer disturbances excited by external perturbations.
2. Calculation of linear development of small disturbances in the boundary layer.
3. Determination and calculation of the dominant nonlinear processes that characterize the beginning of final laminar flow breakdown.

A.V. Boiko et al., *Physics of Transitional Shear Flows*,
Fluid Mechanics and Its Applications 98, DOI 10.1007/978-94-007-2498-3,

To date certain techniques have been developed, allowing calculation of the initial amplitudes of Tollmien–Schlichting waves in some practical situations. The computations of the linear growth of boundary-layer disturbances are quite well advanced, and the adequacy of the description of Tollmien–Schlichting wave, cross-flow, and Görtler vortex amplification by the linear theory of hydrodynamic stability is confirmed experimentally. The calculation of the nonlinear stage of the transition is a more difficult problem; however, it can be sometimes bypassed in practice. Particularly, at a low free-stream turbulence level, nonlinear processes are usually very fast, and the development of small-amplitude disturbances described by the linear stability theory takes place in the major part of a transitional boundary layer. This enables the use of the linear theory for prediction of the transition 'point,' neglecting the details of nonlinear phenomena.

According to experimental data and estimations, the nonlinear processes usually begin to play a noticeable role at amplitudes of the Tollmien–Schlichting waves of about $u' \approx 1\%$ of external-flow velocity U_e. Since the nonlinear zone is relatively short, it is possible to admit the linear growth of disturbances in calculations a little above their real saturation amplitude and accept a transition 'point' at which the amplitude of a Tollmien–Schlichting wave of a certain frequency (defined during the calculations) reaches the value $u' = 2$–4% of U_e. The exponential growth of disturbances in the linear region makes the calculated Reynolds number insensitive to some arbitrariness in selection of the limit value $\mathrm{Re_T}$. In such a way, knowing the initial amplitude and frequency characteristics of the Tollmien–Schlichting waves (from an experiment or a solution of a relevant receptivity problem), it is possible to calculate the transition position with reasonable accuracy. This idea constitutes the essence of the so-called e^N-method originally suggested by van Ingen (1956), Smith and Gamberoni (1956), which can be applied to two-dimensional boundary layers as well as to three-dimensional and separated flows dominated by convective instabilities.

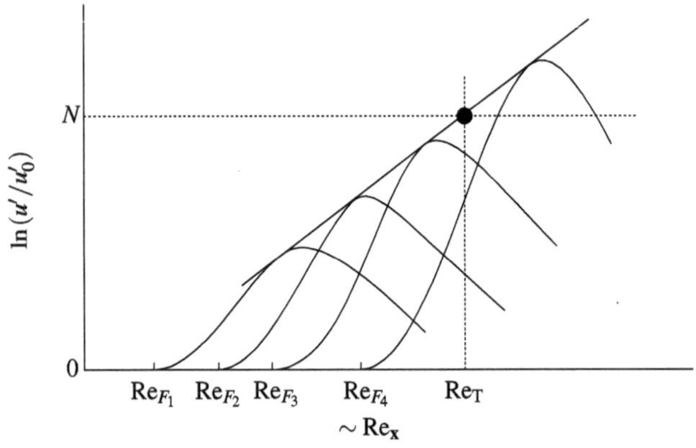

Fig. A.1 To the e^N-method: amplification curves for instability waves with different frequency parameters F

According to the method, $\mathrm{Re_T}$ is determined by the minimum Reynolds number at which an envelope of the growth rate curves for instability waves of different frequencies (see, Fig. A.1)

$$\ln\left(u'/u'_0\right) = -\int_{\mathrm{Re}_F}^{\mathrm{Re}} \alpha_{\mathrm{i}} \mathrm{dRe}_{\mathbf{x}} ,$$

where Re_F corresponds to branch I of the neutral stability curve for the wave with the frequency parameter F, reaches a predefined value N. Usually for plane and axisymmetric flows at a low level of free-stream turbulence, N is about 8–10.

The value of N varies depending on external-flow perturbations. For example, the results of experiments on the transition on gliders are well generalized by the e^N-method, but with the exponent $N = 15$. Such a high value of the exponent, compared with other flight experiments, is stipulated by the absence of disturbing sources, e.g., a propulsion system. The influence of free-stream turbulence, assuming its isotropy $\overline{u'^2} \approx \overline{v'^2} \approx \overline{w'^2}$, so that $\mathrm{Tu} = \sqrt{\overline{u'^2}}/U_{\mathrm{e}}$, on the transition in a flat-plate boundary layer can be expressed by the interpolation formula

$$N = -8.43 - 2.4\ln(\mathrm{Tu}) ,$$

which provides acceptable accuracy in the range $\mathrm{Tu} \approx 0.1$–2%.

For transition prediction at a high free-stream turbulence level, methods based on consideration of spatial growth of optimal disturbances can be used. Let us assume that:

1. Initial kinetic energy of the optimal disturbances in the boundary layer is proportional to kinetic energy of isotropic disturbances of the external flow $E_{\mathrm{e}} = \mathrm{Tu}^2$ that models the receptivity of flow to external vortical perturbations;
2. Kinetic energy of the unstable disturbances is related to E_{e} as $E = \overline{G}\mathrm{Re}E_{\mathrm{e}}$, where $\overline{G}$ is independent of Re, i.e., the growth is in accordance with the growth of the optimal disturbances;
3. Transition to turbulence occurs when kinetic energy of the disturbances reaches certain value $E = E_{\mathrm{T}}$.

Combining these requirements yields

$$\mathrm{Tu}\sqrt{\mathrm{Re_T}} = K ,$$

where K is a constant. Experiments testify that K is in the range of 1131–1506 at $\mathrm{Tu} = 0.9$–6.0.

The experimental observations of the transition in swept-wing boundary layers show the presence of quite extended regions of linear-waves development, which enables extension of the e^N-method to three-dimensional flows. The transition to turbulence in a swept-wing boundary layer can occur in the region of the negative pressure gradient as a result of crossflow instability, or downstream where there is an amplification of the Tollmien–Schlichting waves. Then it is possible to apply the e^N-method to both kinds of instability and to consider that the transition occurs if its criterion is satisfied for at least one of them. The main difficulty of application of the e^N-method for three-dimensional flows is the necessity of determining the direction of the disturbance growth, as outlined in Sect. 6.4.2. Nevertheless, Cebeci (1999), Cebeci et al (1991) applied the procedure and showed the efficiency of the e^N-method for the ONERA–D airfoil with the angles of sweep 49–60° and angles of attack 0–6° at $\mathrm{Re}_l = 1.2\text{–}1.6 \times 10^6$. Crouch and Ng (2000) proposed a further modification of the e^N-method to account for both the crossflow instability and the flow receptivity, which demonstrated promising results at experimental verification.

We also notice that streamwise surface curvature in the absence of centrifugal instabilities can be taken into account through local profiles of the basic flow. However, it is known that the Görtler vortices together with the Tollmien–Schlichting and crossflow waves promote an earlier transition to turbulence. Experiments on concave surfaces with domination of the Görtler vortices indicate that the transition occurs at the Görtler number based on boundary-layer momentum thickness depending on the free-stream turbulence level as

$$\mathrm{Gö}_\vartheta = 9\mathrm{e}^{-17.3\mathrm{Tu}} .$$

Further advances in transition prediction are related to more accurate account of the nonparallel flow effects and curvature with the help of parabolized stability equations. This approach is is attractive because of the low cost of the calculations comparable with the standard e^N-method. The approach also admits account of the flow receptivity in a natural manner through modification of initial and boundary conditions.

A.2 Outline of the linear control theory

The primary goal of the laminar–turbulent transition control is neutralizing disruptive shear-layer perturbations, thereby prolonging the laminar flow state. Research data on this subject are widely reported in original studies and reviewing papers while here we notice some fundamentals only.

The control strategies can be categorized based on the type of flow actuation as passive or active, depending on the absence or presence of energy consumption by the controlling device from external sources. Another classification scheme is based on the means by which the actuation changes in response to modifications in the flow. In such a way, the control strategy can be open-loop (without feedback) or closed-loop (with feedback). In open-loop control, actuator characteristics are designed in advance and remain fixed during its operation. Note that passive control is always the open-loop one.

In the control theory, the governing equations (the Navier–Stokes momentum equations and the continuity equation in our case) are called the descriptor system. In generalized form it can be written as

$$\mathscr{E}\frac{\partial \mathbf{q}}{\partial t} = \mathscr{A}\mathbf{q} + \mathscr{B}\mathbf{g} + \mathscr{D}\mathbf{w} ,$$

where $\mathbf{q}$ is called the state vector consisting of disturbance velocities and pressure, $\mathbf{w}$ is the vector of initial conditions, and $\mathbf{g}$ is the control vector of forcing by which the system is manipulated with the actuator described by $\mathscr{B}$. The sensor gives information (usually limited) about the flow state as

$$\mathbf{y} = \mathscr{C}\mathbf{q}\,,$$

where $\mathbf{y}$ is the vector of measured values and $\mathscr{C}$ describes the sensor characteristics. To conduct the control in an open-loop manner, it is usually assumed that $\mathbf{y} = \mathbf{q}$ and forcing is changed through variations of actuator parameters, while the resulting measurements $\mathbf{y}$ are monitored to reach an objective.

A closed-loop control system, in addition to the actuator and the sensor, requires a controller. To close the loop, the control system must estimate the input given by $\mathbf{y}$ and try to tune the forcing in real time according to the measurements:

$$\mathbf{g} = \mathscr{K}\mathbf{y}$$

to reach an objective that is used to construct the feedback kernel $\mathscr{K}$.

It is frequently required to solve an optimization problem to design the actuator with account of given constraints, for example, to find a control technique that minimizes efforts to suppress the worst initial condition. The problem of optimization of a specified objective functional is called the optimal control problem. However, the presence of the control can make the worst initial condition different. This implies that both must be simultaneously optimized, to ensure control robustness. This approach is called the robust control.

References

Cebeci T (1999) An engineering approach to the calculation of aerodynmic flows. Springer–Verlag, Berlin

Cebeci T, Chen HH, Arnal D, Huang TT (1991) Three-dimensional linear stability approach to transition on wings and bodies of revolution at incidence. AIAA J 29(12):2077–2085

Crouch JD, Ng LL (2000) Variable N-factor method for transition prediction in three-dimensional boundary layers. AIAA J 38(2):211–216

van Ingen JL (1956) A suggested semi-empirical method for the calculation of the boundary layer transition region. VTH 74, Department of Aerospace Engineering, Delft University of Technology

Michel R (1951) Étude de la transition sur les profiles d'aile; établissement d'un critère de determination de point de transition et calcul de la traînée de profilé incompressible. Rep. 1/1578A, Onera

Smith AMO, Gamberoni N (1956) Transition, pressure gradient and stability theory. Rep. ES 26388, Douglas Aircraft Co., El Segundo, Cal.

Further Reading

Abu-Ghanamm BJ, Shaw R (1980) Natural transition of boundary layers — the effects of turbulence, pressure gradient, and flow history. J Mech Eng Sci 22:213–228

Airiau C, Bottaro A, Walther S, Legendre D (2003) A methodology for optimal laminar flow control: Application to the damping of Tollmien–Schlichting waves in a boundary layer. Phys Fluids 15(5):1131–1145

Arnal D, Casalis G (2000) Laminar-turbulent transition prediction in three-dimensional flows. Progr Aerosp Sci 36:173–191

Bewley MHTR, Henningson DS (2003) Linear feedback control and estimation of transition in plane channel flow. J Fluid Mech 481:149–175

Bewley TR (2001) Flow control: New challenges for a new renaissance. Progr Aerosp Sci 37:21–58

Bewley TR, Liu S (1998) Optimal and robust control and estimation of linear paths to transition. J Fluid Mech 365:305–349

Bewley TR, Temam R, Ziane M (2000) A general framework for robust control in fluid mechanics. Physica D 138:360–392

Cathalifaud P, Luchini P (2001) Algebraic growth in boundary layers: Optimal control by blowing and suction at the wall. Eur J Mech B/Fluids 19:469–490

Collis SS, Joslin RD, Seifert A, Theofilis V (2004) Issues in active flow control: Theory, control, simulation, and experiment. Progr Aerosp Sci 40:237–289

Corbett P, Bottaro A (2001) Optimal control of nonmodal disturbances in boundary layers. Theoret Comput Fluid Dyn 15:65–81

van Driest ER, Blumer CB (1963) Boundary layer transition: Freestream turbulence and pressure gradient effects. AIAA J 1:1303–1306

Dryden HL (1948) Recent advances in mechanics of boundary layer flow. Adv Appl Mech 1:1–40

Forest AE (1977) Engineering predictions of transitional boundary-layers. CP 224, AGARD

Henningson DS, Åkervik E (2008) The use of global modes to understand transition and perform flow control. Phys Fluids 20(7):031302, 1–15

Jaffe NA, Okamura TT, Smith AMO (1970) Determination of spatial amplification factors and their application to prediction transition. AIAA J 8(2):301–308

Joshi SS, Speyer JL, Kim J (1997) A systems theory approach to the feedback stabilization of infinitesimal and finite-amplitude disturbances in plane Poiseuille flow. J Fluid Mech 332:157–184

Joslin RD, Gunzburger MD, Nicolaides RA, Erlebacher G, Hussaini MY (1997) Self-contained automated methodology for optimal flow control. AIAA J 35(5):816–824

Liepman HW (1945) Investigation of boundary layer transition on concave walls. Wartime Rep. W-87, NACA

Mack LM (1975) A numerical method for the prediction of high speed boundary-layer transition using linear theory. Tech. Rep. SP-347, NASA

Michel R, Arnal D, Coustols E (1985) Stability calculations and transition criteria on two- and three-dimensional flows. In: Kozlov VV (ed) Laminar–Turbulent Transition, Springer—Verlag, Berlin, IUTAM Symposium, pp 455–461

Runiyan LJ, Gerge-Falvy D (1979) Amplification factors corresponding to transition on an unswept wing in free flight and on a swept wing in wind tunnel. AIAA Paper 79–0267

Schmid PJ, Henningson DS (1994) Optimal energy density growth in Hagen–Poiseuille flow. J Fluid Mech 277:197–225

Walther S, Airiau C, Bottaro A (2001) Optimal control of Tollmien–Schlichting waves in a developing boundary layer. Phys Fluids 13(7):2087–2096

White FM (1974) Viscous fluid flow. McGraw-Hill, New York

Index

A.V. Boiko et al., *Physics of Transitional Shear Flows*,
Fluid Mechanics and Its Applications 98, DOI 10.1007/978-94-007-2498-3,

MIX
Papier aus verantwortungsvollen Quellen
Paper from responsible sources
FSC® C105338
FSC
www.fsc.org

If you have any concerns about our products,
you can contact us on
ProductSafety@springernature.com

In case Publisher is established outside the EU,
the EU authorized representative is:
Springer Nature Customer Service Center GmbH
Europaplatz 3, 69115 Heidelberg, Germany

Printed by Libri Plureos GmbH
in Hamburg, Germany